T0092181

Artificial Intelligence and Data Science in Environmental Sensing

Cognitive Data Science in Sustainable Computing

Artificial Intelligence and Data Science in Environmental Sensing

Edited by

Mohsen Asadnia
School of Engineering, Macquarie University, Sydney, NSW, Australia

Amir Razmjou
UNESCO Centre for Membrane Science and Technology, School of Chemical Engineering, University of New South Wales, Sydney, NSW, Australia

Amin Beheshti
School of Computing, Macquarie University, Sydney, NSW, Australia

Series editor

Arun Kumar Sangaiah
School of Computing Science and Engineering, Vellore Institute of Technology (VIT), Vellore, India

ACADEMIC PRESS

An imprint of Elsevier

Academic Press is an imprint of Elsevier
125 London Wall, London EC2Y 5AS, United Kingdom
525 B Street, Suite 1650, San Diego, CA 92101, United States
50 Hampshire Street, 5th Floor, Cambridge, MA 02139, United States
The Boulevard, Langford Lane, Kidlington, Oxford OX5 1GB, United Kingdom

Notices
Knowledge and best practice in this field are constantly changing. As new research and experience broaden our understanding, changes in research methods, professional practices, or medical treatment may become necessary.

Practitioners and researchers must always rely on their own experience and knowledge in evaluating and using any information, methods, compounds, or experiments described herein. In using such information or methods they should be mindful of their own safety and the safety of others, including parties for whom they have a professional responsibility.

To the fullest extent of the law, neither the Publisher nor the authors, contributors, or editors, assume any liability for any injury and/or damage to persons or property as a matter of products liability, negligence or otherwise, or from any use or operation of any methods, products, instructions, or ideas contained in the material herein.

Library of Congress Cataloging-in-Publication Data
A catalog record for this book is available from the Library of Congress

British Library Cataloguing-in-Publication Data
A catalogue record for this book is available from the British Library

ISBN: 978-0-323-90508-4

For information on all Academic Press publications visit our website at https://www.elsevier.com/books-and-journals

Publisher: Mara Conner
Editorial Project Manager: Joshua Mearns
Production Project Manager: Omer Mukthar
Cover Designer: Matthew Limbert

Typeset by TNQ Technologies

Contents

3. Intelligent geo-sensing for moving toward smart, resilient, low emission, and less carbon transport

Omid Ghaffarpasand, Ahmad Miri Jahromi, Reza Maleki, Elika Karbassiyazdi and Rhiannon Blake

4. Language of response surface methodology as an experimental strategy for electrochemical wastewater treatment process optimization

A. Yagmur Goren, Yaşar K. Recepoğlu and Alireza Khataee

5. Artificial intelligence and sustainability: solutions to social and environmental challenges

Firouzeh Taghikhah, Eila Erfani, Ivan Bakhshayeshi, Sara Tayari, Alexandros Karatopouzis and Bavly Hanna

8. Tuning swarm behavior for environmental sensing tasks represented as coverage problems

Shadi Abpeikar, Kathryn Kasmarik, Phi Vu Tran, Matthew Garratt, Sreenatha Anavatti and Md Mohiuddin Khan

9. Machine learning applications for developing sustainable construction materials

Hossein Adel, Majid Ilchi Ghazaan and Asghar Habibnejad Korayem

10. The AI-assisted removal and sensor-based detection of contaminants in the aquatic environment

Sweta Modak, Hadi Mokarizadeh, Elika Karbassiyazdi, Ahmad Hosseinzadeh and Milad Rabbabni Esfahani

11. Recent progress in biosensors for wastewater monitoring and surveillance

Pratiksha Srivastava, Yamini Mittal, Supriya Gupta, Rouzbeh Abbassi and Vikram Garaniya

12. Machine learning in surface plasmon resonance for environmental monitoring

Masoud Mohseni-Dargah, Zahra Falahati, Bahareh Dabirmanesh, Parisa Nasrollahi and Khosro Khajeh

Contributors

Rouzbeh Abbassi, School of Engineering, Faculty of Science and Engineering, Macquarie University, Sydney, NSW, Australia

Shadi Abpeikar, School of Engineering and Information Technology, University of New South Wales, Canberra, ACT, Australia

Hossein Adel, School of Civil Engineering, Iran University of Science and Technology, Tehran, Iran

Sreenatha Anavatti, School of Engineering and Information Technology, University of New South Wales, Canberra, ACT, Australia

Ivan Bakhshayeshi, School of Computing, Macquarie University, Sydney, NSW, Australia; Faculty of Science and Engineering, Macquarie University, Sydeny, NSW, Australia

Nima Bayat-Makou, The Edward S. Rogers Sr., Department of Electrical and Computer Engineering, University of Toronto, Toronto, ON, Canada

Amin Beheshti, School of Computing, Macquarie University, Sydney, NSW, Australia

Rhiannon Blake, School of Geography, Earth, and Environmental Sciences, University of Birmingham, Birmingham, United Kingdom

Bahareh Dabirmanesh, Department of Biochemistry, Faculty of Biological Sciences, Tarbiat Modares University, Tehran, Iran

Eila Erfani, Faculty of Engineering and Information Technology, University of Technology Sydney, Sydney, NSW, Australia

Milad Rabbabni Esfahani, Department of Chemical and Biological Engineering, The University of Alabama, Tuscaloosa, AL, United States

Karu P. Esselle, School of Electrical and Data Engineering, University of Technology Sydney (UTS), Sydney, NSW, Australia

Zahra Falahati, Department of Biological Sciences, Institute for Advanced Studies in Basic Sciences (IASBS), Zanjan, Iran

Helia Farhood, School of Computing, Macquarie University, Sydney, NSW, Australia

Vikram Garaniya, Australian Maritime College, College of Sciences and Engineering, University of Tasmania, Launceston, TAS, Australia

Matthew Garratt, School of Engineering and Information Technology, University of New South Wales, Canberra, ACT, Australia

Omid Ghaffarpasand, School of Geography, Earth, and Environmental Sciences, University of Birmingham, Birmingham, United Kingdom

Ebrahim Ghasemy, Centre Énergie Matériaux Télécommunications, Institut National De La Recherché, Varennes, QC, Canada

A. Yagmur Goren, Department of Environmental Engineering, Izmir Institute of Technology, Urla, Izmir, Turkey

Supriya Gupta, Environment and Sustainability Department, CSIR-Institute of Minerals and Materials Technology, Bhubaneswar, Odisha, India; Academy of Scientific and Innovative Research (AcSIR), CSIR-Human Resource Development Centre, CSIR-HRDC Campus, Ghaziabad, India

Asghar Habibnejad Korayem, School of Civil Engineering, Iran University of Science and Technology, Tehran, Iran

Bavly Hanna, Faculty of Engineering and Information Technology, University of Technology Sydney, Sydney, NSW, Australia

Ahmad Hosseinzadeh, Centre for Technology in Water and Wastewater, University of Technology Sydney, Sydney, NSW, Australia

Majid Ilchi Ghazaan, School of Civil Engineering, Iran University of Science and Technology, Tehran, Iran

Ahmad Miri Jahromi, Computational Biology and Chemistry Group (CBCG), Universal Scientific Education and Research Network (USERN), Tehran, Iran; Department of Petroleum Engineering, Amirkabir University of Technology (Tehran Polytechnic), Tehran, Iran

Alexandros Karatopouzis, Faculty of Engineering and Information Technology, University of Technology Sydney, Sydney, NSW, Australia

Elika Karbassiyazdi, Centre for Technology in Water and Wastewater, University of Technology Sydney, Sydney, NSW, Australia

Kathryn Kasmarik, School of Engineering and Information Technology, University of New South Wales, Canberra, ACT, Australia

Khosro Khajeh, Department of Biochemistry, Faculty of Biological Sciences, Tarbiat Modares University, Tehran, Iran

Md Mohiuddin Khan, School of Engineering and Information Technology, University of New South Wales, Canberra, ACT, Australia

Alireza Khataee, Department of Environmental Engineering, Gebze Technical University, Gebze, Turkey; Research Laboratory of Advanced Water and Wastewater Treatment Processes, Department of Applied Chemistry, Faculty of Chemistry, University of Tabriz, Tabriz, Iran

Mohammad Khedri, Computational Biology and Chemistry Group (CBCG), Universal Scientific Education and Research Network (USERN), Tehran, Iran

Ahmed A. Kishk, The Department of Electrical and Computer Engineering, Concordia University, Montreal, QC, Canada

Ali Lalbakhsh, The School of Engineering, Macquarie University, Sydney, NSW, Australia; School of Electrical and Data Engineering, University of Technology Sydney (UTS), Sydney, NSW, Australia

Reza Maleki, Computational Biology and Chemistry Group (CBCG), Universal Scientific Education and Research Network (USERN), Tehran, Iran; Department of Chemical Engineering, Shiraz University, Shiraz, Iran

Yamini Mittal, Environment and Sustainability Department, CSIR-Institute of Minerals and Materials Technology, Bhubaneswar, Odisha, India; Academy of Scientific and Innovative Research (AcSIR), CSIR-Human Resource Development Centre, CSIR-HRDC Campus, Ghaziabad, India

Sweta Modak, Department of Chemical and Biological Engineering, The University of Alabama, Tuscaloosa, AL, United States

Masoud Mohseni-Dargah, Department of Biochemistry, Faculty of Biological Sciences, Tarbiat Modares University, Tehran, Iran; School of Engineering, Macquarie University, Sydney, NSW, Australia

Hadi Mokarizadeh, Department of Chemical Engineering, Amirkabir University of Technology, Tehran, Iran

Parisa Nasrollahi, Department of Nanobiotechnology, Faculty of Biological Sciences, Tarbiat Modares University, Tehran, Iran

Arman Nedjati, Industrial Engineering Department, Quchan University of Technology, Quchan, Iran

Matineh Pooshideh, School of Computing, Macquarie University, Sydney, NSW, Australia

Yaşar K. Recepoğlu, Department of Chemical Engineering, Izmir Institute of Technology, Urla, Izmir, Turkey

Nabi Rezvani, School of Computing, Macquarie University, Sydney, NSW, Australia

Roy B.V.B. Simorangkir, Tyndall National Institute, Cork, Ireland

Pratiksha Srivastava, Australian Maritime College, College of Sciences and Engineering, University of Tasmania, Launceston, TAS, Australia

Firouzeh Taghikhah, College of Asia and the Pacific, Australian National University, Canberra, ACT, Australia; Faculty of Engineering and Information Technology, University of Technology Sydney, Sydney, NSW, Australia

Sara Tayari, Faculty of Engineering and Information Technology, University of Technology Sydney, Sydney, NSW, Australia

Phi Vu Tran, School of Engineering and Information Technology, University of New South Wales, Canberra, ACT, Australia

Mohammad Yazdi, Centre for Risk, Integrity, and Safety Engineering (C-RISE), Faculty of Engineering and Applied Science, Memorial University of Newfoundland, St. John's, NL, Canada; School of Engineering, Faculty of Science and Engineering, Macquarie University, Sydney, NSW, Australia

Esmaeil Zarei, Centre for Risk, Integrity, and Safety Engineering (C-RISE), Faculty of Engineering and Applied Science, Memorial University of Newfoundland, St. John's, NL, Canada

Editor Bio

Mohsen Asadnia is an Associate Professor and group leader in Mechatronics-biomechanics at Macquarie University, Australia. He received his PhD degree in Mechanical Engineering from Nanyang Technological University, Singapore. Prior to joining Macquarie University, Mohsen had several teaching and research roles with the University of Western Australia, Massachusetts Institute of Technology, and Nanyang Technological University. His work has resulted in over 150 peer-reviewed journal articles published in prestigious journals, including *Nature Communication*, *Advanced Materials*, and *Nano-Micro Letters*. His research interest lies in environmental/biomedical sensors, artificial hearing implant devices, microfluidics, artificial intelligence, and bio-inspired sensing.

Amir Razmjou received his PhD in Chemical Engineering from the University of New South Wales, Australia, in 2012, and since then he has accrued multidisciplinary skills to develop innovative technologies for biomedical and environmental applications. His surface architecturing skills using functional nanostructured materials alongside biofunctionalization have helped him to develop innovative membranes for desalination and water treatment and nanobiosensors.

Amin Beheshti is a Full Professor of Data Science and the Director of AI-enabled Processes Research Centre, School of Computing, Macquarie University. Amin is also the Head of the Data Analytics Research Lab and Adjunct Academic in Computer Science at UNSW Sydney. Amin completed his PhD and postdoctoral degrees in Computer Science and Engineering at UNSW Sydney and holds master's and bachelor's degrees in Computer Science both with First Class Honours. He is the leading author of several authored books in data, social, and process analytics and has co-authored with other high-profile researchers.

Preface

Industrialization and population growth have resulted in significant environmental implications such as global warming, waste disposal, ocean acidification, deforestation, etc. A sustainable clean energy future requires systems with zero carbon and water footprints, which requires advanced materials and autonomous processes. The Digital revolution has fast-forwarded the transition from fossil fuel to a renewable civilization, which can reverse the damage caused by human activities to the environment. Sensors are the pillars of the digital revolution which generates data for developing advanced mathematical models and autonomous processes. Novel sensing technologies and advancements in data processing are our greatest tools to fight against various ways that humans have affected the environment such as overpopulation, pollution, burning fossil fuels, deforestation, etc. Using these technologies, we can find ways to reduce climate change, soil erosion, poor air and water quality, and keep our planet "green" for the next generations. In the last decade, considerable research works have been carried out on developing sensitive, low-powered, and durable sensors for environmental sensing. Artificial intelligence (AI) and big data processing techniques and algorithms made it possible to create continuous monitoring systems to minimize the effect of human activities on the environment. We devoted this book to exploring new opportunities and possibilities in using advanced devices and AI for various environment sensing applications which will be published in the name of "*Artificial Intelligence and Data Science in Environmental Sensing.*"

This book is divided into 12 chapters which are 1. Smart sensing technologies for wastewater treatment plants; 2. Recent advancement in antennas for environmental sensing; 3. Intelligent geo-sensing for moving toward smart, resilient, low emission, and less carbon transport; 4. Language of Response Surface Methodology (RSM) as an experimental strategy for electrochemical wastewater treatment process optimization; 5. Artificial intelligence and sustainability: solutions to social and environmental challenges; 6. Application of multiattribute decision making tools for site analysis of offshore wind turbines; 7. Recent Advances of Image Processing Techniques in Agriculture; 8. Applications of Swarm Intelligence in Environmental Sensing; 9. Machine learning applications for developing sustainable construction materials; 10. The AI-assisted removal process of contaminants in the aquatic environment; 11. Recent progress in biosensors and data processing systems for wastewater monitoring and surveillance; 12. Machine learning in surface plasmon resonance for environmental monitoring. The authors sincerely thank various researchers who contributed to the chapters by sharing their findings and knowledge.

Mohsen Asadnia
Amir Razmjou
Amin Beheshti

Chapter 1

Smart sensing technologies for wastewater treatment plants

Reza Maleki[1], Ahmad Miri Jahromi[1], Ebrahim Ghasemy[2], Mohammad Khedri[1]

[1]*Computational Biology and Chemistry Group (CBCG), Universal Scientific Education and Research Network (USERN), Tehran, Iran;* [2]*Centre Énergie Matériaux Télécommunications, Institut National De La Recherché, Varennes, QC, Canada*

1. Introduction

The most important goals of constructing wastewater treatment systems include the protection of homogeneity, environmental protection, preventing the pollution of water sources, and the reuse of treated wastewater in sectors such as agriculture and industry [1,2]. So, reducing water pollutants and improving water quality in the wastewater treatment plant (WWTP) is essential. The establishment of WWTPs alone does not solve environmental concerns, but in order to reach the desired environmental standards, the performance of treatment plants must be constantly monitored and evaluated. Typical parameters that are considered to evaluate the performance of WWTPs are biological oxygen demand (BOD), suspended solids, soluble solids, and pH of wastewater [3–6]. If these parameters meet the standards, effluents could be used in sectors such as agriculture and industry. This can help solve the water shortage crisis to some extent [7].

The complex composition of wastewater has different diffusion properties and concentrations of pollutants and effluents in WWTPs [8–11]. Wastewater is rich in toxic substances such as lead, copper, nickel, silver, mercury, chromium, zinc, cadmium or tin, nutrients, and organic matter and can also have a wide range of pH [12,13]. Complex natural phenomena, human activities, and the process of wastewater treatment have led to great uncertainty in wastewater treatment systems. These uncertainties fluctuate randomly due to the amount, quality, and efficiency of wastewater disposal [14]. Currently, with stricter regulations on effluent quality, the operation of a WWTP has become more difficult and complex. Improper use of a WWTP can lead to general health and environmental problems. The entry of effluent from these treatment plants into water sources can spread various human diseases [15].

Artificial Intelligence and Data Science in Environmental Sensing
https://doi.org/10.1016/B978-0-323-90508-4.00003-4

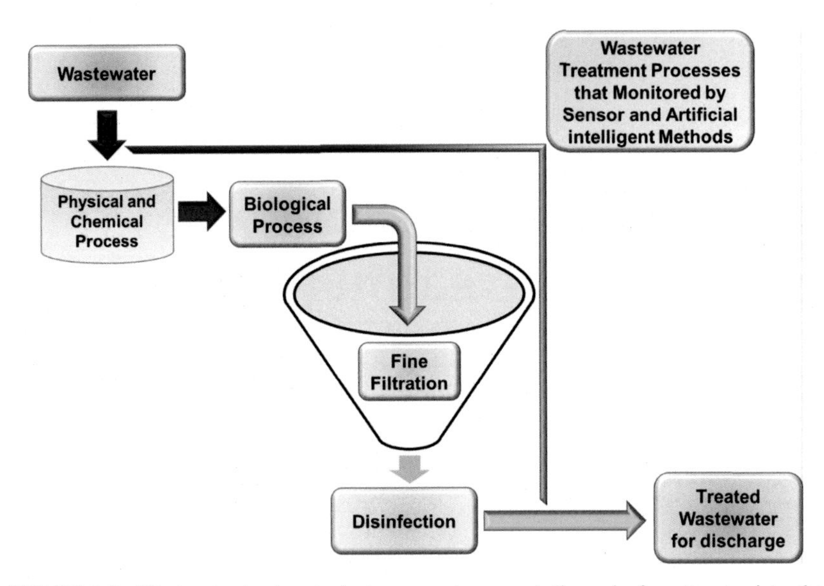

FIGURE 1.1 Wastewater treatment plant process to prevent disposal of wastewater into the environment.

Wastewater treatment operations include a set of complex processes and their dynamics are nonlinear and change over time and can directly overshadow the operation of the treatment plant. Fig. 1.1 shows the simple process of wastewater treatment. Moreover, the input characteristics of each treatment plant vary depending on the area covered. Therefore, the performance of any treatment plant strongly depends on recognizing the main factors affecting the treatment plant. In addition, random disturbances and effective variability will force operators to perform suitable operational controls on the system [12,13]. Also, modern WWTPs face stricter emission restrictions as well as new regulations on energy efficiency and resource recycling [16].

Given the above, today, in addition to the operation of the treatment plant, it is important to pay attention to mathematical models to predict the performance of the treatment plant. The response of the process to any change can be examined by mathematical simulation and can ultimately be achieved with an output stream of optimal quality and low operating costs. As mentioned, due to the variable nature of wastewater, to maintain the stability of treatment processes in optimal conditions, the proper operation of the WWTP is of great importance. However, the WWTP modeling is very difficult due to the nonlinear relationships of effective parameters, but the use of conceptual models to prevent growing concerns about environmental impacts and to help engineers to predict treatment plant behavior as well as complex treatment processes has received tremendous attentions [17–19]. In this regard, artificial intelligence models can be used as an effective tool to simulate the behavior of the treatment system. Therefore, researchers have tried to use artificial intelligence technology in WWTP networks to overcome these problems [20,21]. In this regard, the required information such as pollutant concentration or pH of the wastewater is collected using sensors.

A sensor is a gadget that reacts to the characteristics of the environment and reports in real time as a readable analog or digital signal (for example, an ASCII voltage or data stream) [22]. It is a tool for analysis that, after performing some chemical changes, for example, mixing with a reagent, measures characteristics of the sample [22–24]. Therefore, the sensors are employed for measuring the mentioned parameters in the WWTP [25]. In contrast to traditional sensors, which (ideally) respond to only one environmental property and respond directly to it, the soft sensors are capable of responding to several environmental properties [26]. In other words, the goal analyst has more adjustment parameters as a result of more relevant and complex algorithms than a standard calibration curve. Finally, by reviewing the studies in this work, the applications of artificial intelligence in wastewater treatment are investigated. Also, the role of sensors in intelligent wastewater treatment is investigated.

2. Online estimation

Due to the fact that the WWTPs methods and their importance are increasing day by day, it is necessary to first predict and then analyze the pollutant parameters based on new methods. Many works have been done to closely monitor the level of pollutants in the wastewater treatment processes [27,28]. In addition, continuous monitoring of these contaminants requires the use of sensors with fast response, sufficient sensitivity, and long life. Achieving a quick response has made the use of field measurements inevitable. Since the use of online systems in wastewater treatment has been considered, issues such as robustness, the cost-effectiveness of sensors, which are one of the main components required to use these methods, should be considered. Using this method reduces the cost, delay time, and issues related to the accuracy of measurements [29,30]. Examining the results of the mentioned methods to measure the necessary parameters in wastewater treatment, it can be seen that to obtain the desired result, a method should be used that has the fastest response time and the most accuracy; therefore, the use of systems intelligent has been considered by many researchers and the use of these systems in various fields of medicine and engineering has been expanded [31]. However, WWTP modeling is difficult due to its complexity. Complex physical, chemical, and biological processes involved in the wastewater treatment process caused nonlinear behaviors that are the barrier for describing their behavior with linear mathematical models [14]. It is difficult to describe these processes with mathematical models because of their nonlinear behavior. For this reason, in the last decade, many studies have been conducted on modeling the wastewater treatment process using intelligent methods such as artificial neural network (ANN) modeling in modeling wastewater treatment processes. Alfonso and Redondo [32] proposed an intelligent wastewater treatment method using a neural network to control the treatment plant. The researchers concluded that the use of neural networks in the management of WWTPs can be very beneficial. Cote et al. [33] used a neural network to increase the accuracy of mechanical models used for the activated sludge process (ASP). Pai et al. [34], from a three-layer ANN with six neurons in the middle layer and four information neurons at the entrance, succeeded in predicting the quality of the effluent from a hospital treatment plant in Taiwan. They used characteristics including the temperature, pH, solid solids (SS), and wastewater chemical oxygen demand (COD) at the ANN inlet to predict SS and effluent COD. Their results indicated the proper performance of the designed neural network. Prediction results of soluble oxygen in water dissolved oxygen (DO) (in Serbia by Ranković et al.) suggested that the use of the ANN is appropriate [35]. Fernandez de Canete et al. [26] used artificial intelligence methods at a WWTP to control and estimate an ASP. They concluded that this method can improve effluent quality and reduce the operation costs. In this work, they showed that artificial intelligence methods are suitable for the recognition of operational states and prediction of total nitrogen content, COD, and total suspended solids (TSS). Jagielska et al. [36], in this manner, used neural network primarily based software sensors to discover low-price working modes and efficiently expected overall nitrogen and overall suspended solids, for operators. These models have a distinct ability to learn the relationships of nonlinear functions and also do not require prior structural knowledge of the relationships between important variables and processes for modeling.

Operation of a WWTP can be done by developing a modeling tool to predict the performance of the treatment plant based on past observations of quality-specific parameters. One of the most common mathematical models is the activated sludge model which uses differential equations in the form of a matrix. Although activated sludge models have been developed since 1987 [37], these models still do not have the required performance. Therefore, artificial intelligence techniques have been developed as an alternative to these mathematical models based on ANN, which are more commonly used and provided successful results. Gontarski et al. [38], by creating an ANN, predicted the quality characteristics of industrial wastewater effluents. They concluded that the inflow to the treatment plant and the pH of the incoming wastewater are the most effective parameters in controlling the treatment plant. Hong et al. [39] worked on a type of neural network, called KSOFM2, which was used as an efficient tool to determine the dependencies of process variables and also to predict the behavior of the municipal wastewater treatment system, making it an effective analytical tool and a useful method for diagnosing and understanding the behavior of activated sludge house

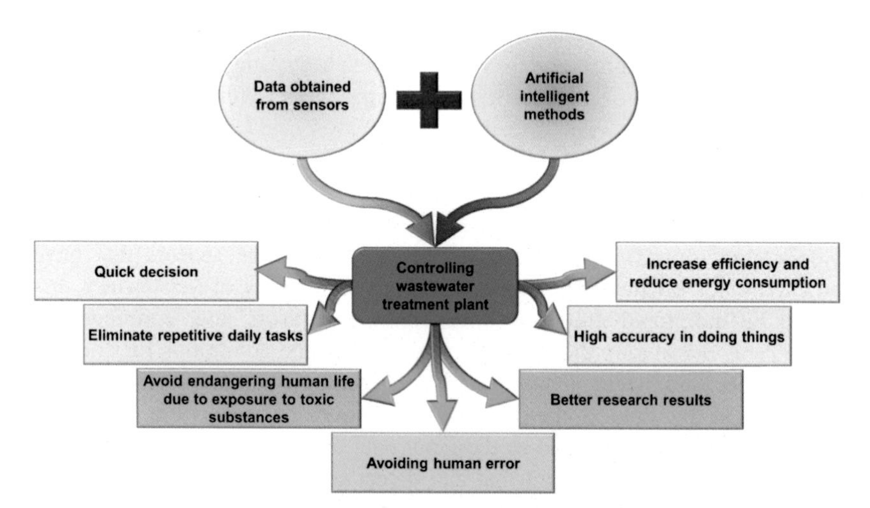

FIGURE 1.2 Advantages of artificial intelligent usage in wastewater treatment plants.

drainage system. Hamed et al. [40] developed two models based on an ANN to predict the output concentration of BOD and SS from a large WWTP (with activated sludge system) in Cairo. Research on ANN introduced a valuable tool for predicting the performance of WWTPs.

Artificial intelligence networks can be used to model WWTP processes. Several key factors have been combined to accelerate the evolution of artificial intelligence in recent years. With the exception of large-scale investment, factors such as obtaining large amounts of data using sensors have led to the development of artificial intelligence in wastewater treatment. Having a lot of data is essential for activating artificial intelligence devices for learning. The abundance of data with the help of sensors in WWTPs provides the necessary basis for the use of artificial intelligence in the control of WWTPs. Artificial intelligence typically relies on past process data. In each WWTP, there are specific key parameters that can be used to evaluate the performance of the treatment plant. This parameter can be achieved using intelligent sensors. Artificial intelligence is an effective approach to deal with the complexities of the wastewater treatment process and its use has many benefits, some of which are mentioned in Fig. 1.2 [41].

3. Fault detection and diagnostics

Water is the most important and fundamental factor in the life of living organisms and in this regard, preventing water pollution is equally important. Water pollution is increase in the concentration of chemical, physical, or biological species that change the properties of water. Water pollutants are very diverse and can contaminate both groundwater and surface water sources

[42,43]. The most important causes of water pollution are organic matter and species, microbes and bacteria, some metal ions, heavy metals, anions such as nitrate and phosphate, industrial and municipal wastewater, insecticides and pesticides, etc. [44,45]. The investigation, detection, and monitoring of any of these pollutants and constant monitoring of water resources and their health are of great importance and this would be almost impossible without proper tools and based on the color and smell of water. The most important and widely used tools for this purpose are sensors and nanosensors that have attracted a lot of attention. Some important parameters of wastewater that should be monitored are showed in Fig. 1.3. Sensors and, consequently, nanosensors are among the tools that can monitor the conditions for controlling various parameters in wastewater treatment, so extensive research has been done in this field. Some different applications of sensors are measuring the COD and TSS to be used as the input data of artificial intelligence [46].

A sensor is actually a tool that can detect some properties related to its environment. The sensors detect events or changes in various quantities and display the result as an output signal corresponding to the resulting changes, which is usually an electrical or optical signal. There are many types of sensors and they have found many applications in different fields. Sensors can detect, collect, and transmit changes to the basic parameters of wastewater treatment with great sensitivity and accuracy to the macroscopic world. Since the accuracy of input data in artificial intelligence has a great impact on the accuracy of predictions, one of the most important characteristics required by sensors is that they must have high sensitivity and detection power to be able to trust their data. This issue is considered as one of the most basic requirements for the use of artificial intelligence in wastewater treatment. The amount of DO, minerals in the water, pH, and water temperature are some of the parameters that are usually important in the wastewater treatment

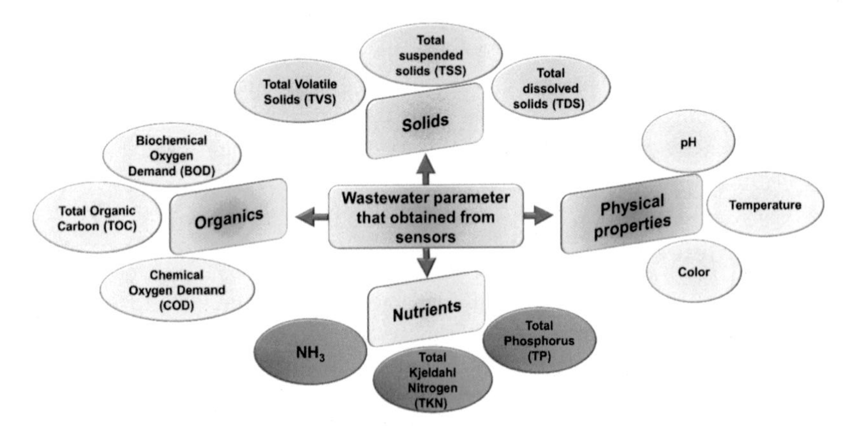

FIGURE 1.3 Some important parameters of wastewater which can be measured by sensors.

process and can be controlled using artificial intelligence and sensors. The amount of DO in physiological and environmental systems indicates the amount of chemical, physical, and biochemical activity [47]. Therefore, it is essential to evaluate the amount of DO in aqueous solutions. The most important and widely used method of measuring the amount of DO in water is the use of sensors. Meanwhile, the design and use of nanosensors to measure the amount of DO in water is growing and developing, and in this regard, every year, various nanosensors to measure the amount of DO in water are introduced and some of them are commercially available [48].

The total dissolved solids (TDS) standard is an indicator for measuring the amount of minerals (chlorides, bicarbonates, calcium, magnesium, and sodium) and small amounts of water-soluble organic compounds. The standard TDS value for drinking water is 500 ppm, and exceeding the TDS value causes an unpleasant odor in the water. Since TDS is also a measure of water hardness, TDS control is also of particular importance in various industries, and in this regard, various sensors have been developed and marketed to measure and monitor TDS in the WWTPs to be controlled by artificial intelligent methods. Today, nanotechnology is also used in the manufacture of TDS sensors, but since almost all commercial TDS sensors are electrochemical sensors and are manufactured to a specific standard, the application of nanotechnology is limited to creating nanocoating on the sensors. So, this sensor could be a suitable detector of the wastewater TDS to be used as the data collector for ANN.

There are different types of sensors, each of which uses a special mechanism to measure the various parameters required for wastewater treatment and they are used as input data in artificial intelligence algorithms. In the following, a number of sensors will be investigated.

3.1 Electrochemical sensors

Electrochemical sensors have various measurement methods such as potentiometric, voltametric, and conductivity methods that are used to measure water pollution. These sensors change their properties due to the interactions performed in contact with the measurable component. The redox process performed on the electrode produces measurable signals [49,50].

3.2 Fiber optic sensors for direct monitoring of water quality

Some are based on fiber technology, sending a beam of light from a laser source (often a single-frequency laser light) into an optical fiber. The properties of light transmitted along the fiber change according to environmental factors, and eventually on the other side of the fiber reaches a detector [51,52]. Fiber optic sensors change the light wavelength due to environmental changes such as changes in various water parameters.

3.3 Sensors based on microwave technology

The use of electromagnetic waves for measurement is one of the successful commercial methods being developed that are used for various industrial applications such as determining the concentration of ions in water [53]. The results obtained from direct monitoring of the amount of nitrate in the effluent show the potential ability of microwave sensors to measure water quality [54].

Although there are different sensors for detecting contaminants and pollutants, nanotechnology enables the creation of new generations of sensors with a high ability to measure one or more pollutants portably that are able to detect pollutants in very small amounts and concentrations [55,56]. In this regard, biological nanosensors and chemical nanosensors are used to detect contamination and monitor water quality [57,58]. These nanosensors can detect a variety of pathogens, including very small amounts of microbes and their toxins, chemical compounds, and ions in water at an amount of one part per billion (ppb) continuously and rapidly. Atar et al. [56] have used nanosensors to detect triclosan in wastewater. These sensors prepare suitable data for the use of artificial intelligence methods in wastewater treatment.

4. Multivariate analysis models

Human beings were able to calculate with logical and mathematical intelligence. In this regard, humans tried to build a machine that could perform mathematical calculations and problem-solving with greater speed and accuracy. But the efforts did not stop there and it was tried to complete this machine to make it smart. To that end, humans identified other aspects of their intelligence to simulate it. They were able to gradually upgrade intelligent systems from systems that were proficient in only one aspect of intelligence to systems that exhibit different aspects of intelligence and use sensors to understand their surroundings and take special action in that environment. Intelligent systems perform their tasks with the help of a concept called Agent, which can be a person, a machine, or even software. Intelligent agents have different levels of intelligence and characteristics. Artificial intelligence is a branch of computer science that studies and develops software and intelligent devices by simulating human abilities in the machine and mimicking intelligent human behaviors. Artificial intelligence is the ability of a system to perform tasks that normally require human intelligence. In an intelligence system, after data collection by some device like sensor, the process of information analysis and decision-making begins. This part is actually the main part and software of intelligent systems, which is usually done using different methods of artificial intelligence. Using intelligent sensors and thus receiving various parameters of the effluent, including data such as the amount of oxygen, the amount of solids in the effluent, etc., as well as their analysis using various methods of artificial intelligence such as artificial intelligence network, artificial intelligence network based on fuzzy logic, etc., can control wastewater treatment.

The ANN has been developed as one of the main techniques of the artificial intelligence system based on the human nervous system and brain [59]. The high performance of biological systems is due to the parallel programming nature of their neurons. An ANN performs this structure by distributing the simulation into small, simple, and interconnected processor units called neurons. In ANN, by processing observational data, the structure of relations and the dependence of the ruler on the phenomenon under consideration are determined. Then, according to the obtained relationships, predictions have been done. Using ANN models, Mjalli et al. [39] predicted the values of COD, BOD, and TSS parameters of Doha treatment plant effluent and found that the ANN model has very high accuracy in predicting and estimating the utilization parameters of wastewater. Shi and Qiao [60] controlled and optimized the concentration of DO and heterotrophic microorganisms in the wastewater treatment process by the neural network model.

The ANN model is a type of computational model that is able to relate the inputs and outputs of a system; though complex and nonlinear, it is determined by a network of nodes that are all connected [61–63]. One of the most common and practical structures in ANN is a multiple-layer perceptron neural network method [64], which usually consists of three layers: an input layer, one or more hidden layers, and an output layer. The input layer receives sets of input data (parameters obtained by sensors from the wastewater), features are processed in hidden layers, and the output layer is used to represent predicted results (Fig. 1.4) [65]. In this method, the number of hidden layers and neurons considered in each layer is determined according to the amount of input and output data.

One of the most important factors in defining ANN is its architectural structure. ANN architecture is a form in which neurons are connected in groups, called layers. A significant advantage of the ANN approach to system modeling is that there is no need for a physical connection to systematically convert the input to output. Instead, in artificial networks, only a set of samples (effluent parameters obtained from sensors) is required. After obtaining the

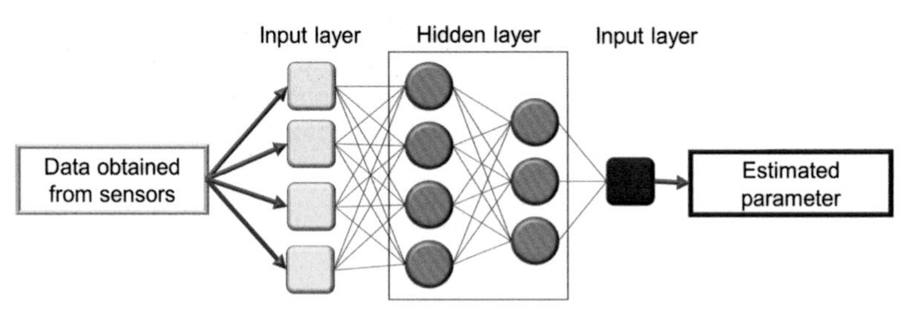

FIGURE 1.4 The simple structure of artificial neural network structure.

values of the effluent parameters by the sensors, using the ANN, the parameters can be predicted according to the relationships between the data. But creating a proper neural network for wastewater treatment is very important. To create this network, it is necessary to count the number of layers, the number of units (neurons) in each layer, and the relationship between the units, select the functions of converting the middle units to the output units, design the training algorithm, and select the initial weights in a special way [66].

The advent of neural network technology has provided promising results in the fields of wastewater treatment and water resources simulation. Using hybrid models or combining different models is a common way to improve the accuracy of predictions. In this regard, many researchers are trying to provide more practical and accurate models by developing this method. In recent years, fuzzy logic has emerged as an important topic in the development of ANN to use fuzzy logic instead of binary logic. Using fuzzy logic, artificial intelligence can more closely resemble human intelligence or the collective intelligence of creatures. In binary logic, everything is considered 0 and 1; but the real world is different from this point of view, and as a result, the use of binary logic in the ANN causes errors in predictions.

Neural-fuzzy networks are a special class of hybrid models that consist of a combination of ANN and fuzzy logic systems. The combination of fuzzy logic and the ANN has created the neural-fuzzy network method. Although ANN and fuzzy systems are structurally different, due to the unique characteristics of each, these two systems are complementary to each other. In this regard, by combining fuzzy systems and neural networks, the learning capabilities of neural networks will enter fuzzy systems. The fuzzy neural network has features such as the ability to learn, classify, and compile data. The structure of a fuzzy neuron is like a definite space neuron, except that some or all of its parameters and components are expressed in fuzzy logic. In fact, the theory and logic of fuzzy sets have been used to describe human thinking and reasoning in a mathematical framework. Due to the use of neural-fuzzy networks of fuzzy perceptual system knowledge and its ability to learn, the neural-fuzzy system is able to accurately model the uncertainty and internal inaccuracy of data in addition to the ability to learn neural-fuzzy networks. Another advantage of neural-fuzzy networks compared to neural networks is that they can be reasoned using logical rules in particular situations. By introducing fuzzy logic in the ANN and using it in the WWTP, more accurate and reliable results can be obtained to control the wastewater parameters. Nadiri et al. [5] used a fuzzy logic model to obtain the effluent parameters. Huang et al. [16] estimated the nutrient concentrations using the genetic algorithm−based neural-fuzzy system (GA-NFS) in a process of biological wastewater treatment. The method of nutrient removal from wastewater treatment was GA-NFS and results indicated that the GA-NFS functioned effectively. The adaptive neuro-fuzzy inference system (ANFIS) model includes two models of neural networks and the fuzzy model. The fuzzy part establishes the relationship between input and output, and the parameters related to the fuzzy part membership functions are determined by neural networks. Therefore, the characteristics of both fuzzy and neural models lie in ANFIS [67,68]. Mingzhi et al. [69] showed that the ANFIS method for treating biological wastewater is better than the ANN method. Actually, the R^2 value of ANFIS and ANN was 0.99 and 0.95. Therefore, ANFIS had less error in predicting the parameters.

As mentioned, in the ANN method, obtaining the structure, the number of hidden layers, and the number of neurons in each layer has a great impact on the accuracy of the predictions. In this regard, it is necessary to calculate the number of optimal layers and neurons. Therefore, it is essential to combine the ANN method and optimization methods such as genetic algorithm or particle swarm optimization (PSO) algorithm to obtain the optimal number of layers and neurons to predict the parameters required for wastewater treatment.

Genetic algorithm and PSO method are in fact a way to solve optimization problems that are inspired by nature [70–73]. Genetic algorithms frequently change a population of individual solutions to a problem, called evolution (see Fig. 1.5). At each stage of this evolution, two members of the population are randomly selected as parents and their offspring are considered as the next generation. In this way, the population evolves toward an optimal solution [74]. Furthermore, the PSO algorithm is one of the most important intelligent optimization algorithms in the field of swarm intelligence. The algorithm was introduced by James Kennedy and Russell C. Eberhart in 1995, inspired by the social behavior of animals such as fish and birds that live together in small

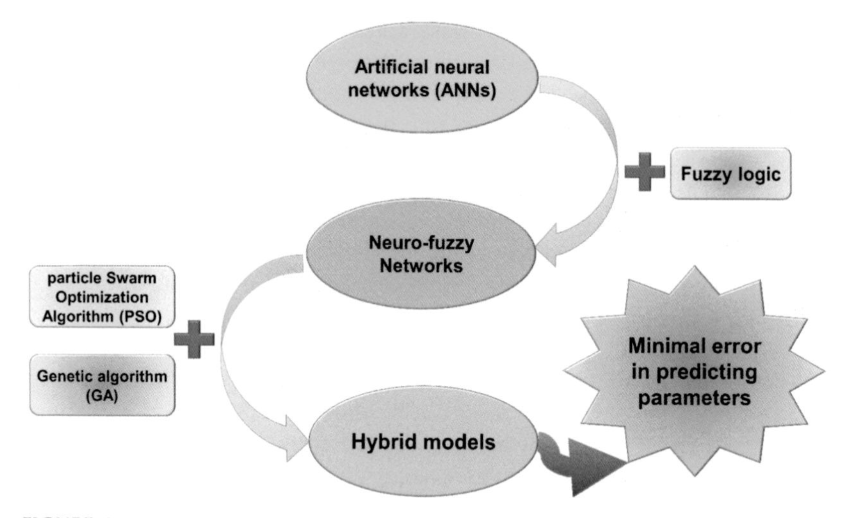

FIGURE 1.5 Progression of artificial methods by using fuzzy logic and particle swarm optimization and genetic algorithms.

and large groups. In the PSO algorithm, the members of the population of the answers are directly related to each other, and by exchanging information with each other and recalling good memories of the past, the problem is solved [75]. Finally, genetic algorithms and PSO algorithms optimize the number of hidden layers and the number of nodes in each layer in the neural network. In fact, the purpose of this work is to minimize the amount of computational error in the wastewater treatment process [61]. Zhu et al. [76] showed that by optimizing the ANN, the output characteristics of WWTPs with advanced treatment systems can be predicted instantly. Piuleac et al. [77] developed a neural network optimization method and genetic algorithm, thereby optimizing a true electrocoagulation process. Validation of optimization results using experimental data in this work showed an error of less than 11%. Kana et al. [78] modeled and optimized the biogas produced from wastewater using the ANN with a genetic algorithm. The results of this study showed the effectiveness of the neural network model genetic algorithm in the nonlinear behavior of the system and optimization of biogas production. Nag et al. [79] used GA-ANN for the efficient remotion of Cd(II) from a medium of scale wastewater industries operating. This method is appreciably well, in terms of the errors minimization. So, they showed the effectiveness of GA-ANN for controlling the WWTP.

5. Conclusion and future direction

Since environmental issues are very important for humans and other living organisms and also due to lack of water resources, proper operation and control of WWTPs have been considered. WWTPs include a variety of equipment and processes for wastewater treatment that generally convert human activities and industrial wastewater into the natural environment without risk to human health or adverse damage. Wastewater generally contains organic matter, minerals, and living organisms. Sewage disposal with acceptable quality characteristics is one of the environmental problems that today's societies face.

One of the methods used to control refineries is to use mathematical models. But in mathematical models, there are limitations in terms of calibration of coefficients and also unsuitability of these models for the nonlinear process of treatment plant effluent. However, using the high ability of ANN in modeling, estimating complex processes, and optimizing various engineering problems, it is hypothesized that the ANN can accurately estimate the values of the quality factors of the treatment plant effluent using the parameters obtained from the sensors. Among artificial intelligence methods (such as fuzzy logic and neural networks and hybrid methods), the use of hybrid methods to treat effluents that have complex and nonlinear behavior is of particular importance. This method provides the basis for improving the productivity of refineries. Fan et al. [80] used PSO-ANN and GA-ANN models to estimate the removal efficiencies of pollutants. They obtained suitable results from these methods as the small average error and the high R^2 value. But in neural networks, the input data need to be very accurate to get acceptable results. In this regard, it is necessary to use various sensors to obtain the parameters needed to control wastewater treatment. Also, Nag et al. [79] showed the effectiveness of GA-ANN for the removal of Cd(II) from wastewater. Finally, it seems that the development of the capability of sensors to extract wastewater data and also the development of artificial intelligence models will pave the way to use artificial intelligence models in wastewater treatment.

References

[1] A. Feigin, I. Ravina, J. Shalhevet, Irrigation with Treated Sewage Effluent: Management for Environmental Protection, Springer Science & Business Media, 2012.

[2] H. Bozkurt, M.C. van Loosdrecht, K.V. Gernaey, G. Sin, Optimal WWTP process selection for treatment of domestic wastewater—A realistic full-scale retrofitting study, Chem. Eng. J. 286 (2016) 447—458.

[3] C. Wei, et al., Residual chemical oxygen demand (COD) fractionation in bio-treated coking wastewater integrating solution property characterization, J. Environ. Manag. 246 (2019) 324—333.

[4] D. Yordanov, Preliminary study of the efficiency of ultrafiltration treatment of poultry slaughterhouse wastewater, Bulg. J. Agric. Sci. 16 (6) (2010) 700—704.

[5] A.A. Nadiri, S. Shokri, F.T.-C. Tsai, A.A. Moghaddam, Prediction of effluent quality parameters of a wastewater treatment plant using a supervised committee fuzzy logic model, J. Clean. Prod. 180 (2018) 539—549.

[6] C.W. Chan, G.H. Huang, Artificial intelligence for management and control of pollution minimization and mitigation processes, Eng. Appl. Artif. Intell. 16 (2) (2003) 75—90.

[7] H. Hong, P. Tsangaratos, I. Ilia, J. Liu, A.-X. Zhu, W. Chen, Application of fuzzy weight of evidence and data mining techniques in construction of flood susceptibility map of Poyang County, China, Sci. Total Environ. 625 (2018) 575—588.

[8] Z. Long, et al., Microplastic abundance, characteristics, and removal in wastewater treatment plants in a coastal city of China, Water Res. 155 (2019) 255—265.

[9] F. Ejeian, et al., Biosensors for wastewater monitoring: a review, Biosens. Bioelectron. 118 (2018) 66—79.

[10] R. Sharma, N. Verma, Y. Lugani, S. Kumar, M. Asadnia, Conventional and advanced techniques of wastewater monitoring and treatment, in: Green Sustainable Process for Chemical and Environmental Engineering and Science, Elsevier, 2021, pp. 1—48.

[11] P. Srivastava, R. Abbassi, A.K. Yadav, V. Garaniya, M.A. Jahromi, A review on the contribution of an electron in electroactive wetlands: electricity generation and enhanced wastewater treatment, Chemosphere (2020) 126926.

[12] R. Loos, B.M. Gawlik, G. Locoro, E. Rimaviciute, S. Contini, G. Bidoglio, EU-wide survey of polar organic persistent pollutants in European river waters, Environ. Pollut. 157 (2) (2009) 561—568.

[13] R. Loos, et al., EU-wide monitoring survey on emerging polar organic contaminants in wastewater treatment plant effluents, Water Res. 47 (17) (2013) 6475—6487.

[14] S. Long, et al., A Monte Carlo-based integrated model to optimize the cost and pollution reduction in wastewater treatment processes in a typical comprehensive industrial park in China, Sci. Total Environ. 647 (2019) 1–10.

[15] L. Rizzo, et al., Urban wastewater treatment plants as hotspots for antibiotic resistant bacteria and genes spread into the environment: a review, Sci. Total Environ. 447 (2013) 345–360.

[16] M. Huang, Y. Ma, J. Wan, X. Chen, A sensor-software based on a genetic algorithm-based neural fuzzy system for modeling and simulating a wastewater treatment process, Appl. Soft Comput. 27 (2015) 1–10.

[17] O.I. Abiodun, A. Jantan, A.E. Omolara, K.V. Dada, N.A. Mohamed, H. Arshad, State-of-the-art in artificial neural network applications: a survey, Heliyon 4 (11) (2018) e00938.

[18] K. Dieguez-Santana, H. Pham-The, P.J. Villegas-Aguilar, H. Le-Thi-Thu, J.A. Castillo-Garit, G.M. Casañola-Martin, Prediction of acute toxicity of phenol derivatives using multiple linear regression approach for Tetrahymena pyriformis contaminant identification in a median-size database, Chemosphere 165 (2016) 434–441.

[19] L. Liu, et al., Tracing the potential pollution sources of the coastal water in Hong Kong with statistical models combining APCS-MLR, J. Environ. Manag. 245 (2019) 143–150.

[20] Y. Liu, H. Xiao, Y. Pan, D. Huang, Q. Wang, Development of multiple-step soft-sensors using a Gaussian process model with application for fault prognosis, Chemometr. Intell. Lab. Syst. 157 (2016) 85–95.

[21] E. Fijani, R. Barzegar, R. Deo, E. Tziritis, K. Skordas, Design and implementation of a hybrid model based on two-layer decomposition method coupled with extreme learning machines to support real-time environmental monitoring of water quality parameters, Sci. Total Environ. 648 (2019) 839–853.

[22] C.M. Thürlimann, D.J. Dürrenmatt, K. Villez, Soft-sensing with qualitative trend analysis for wastewater treatment plant control, Control Eng. Pract. 70 (2018) 121–133.

[23] M. Asadnia, et al., Ca2+ detection utilising AlGaN/GaN transistors with ion-selective polymer membranes, Anal. Chim. Acta 987 (2017) 105–110.

[24] M. Mahmud, et al., Recent progress in sensing nitrate, nitrite, phosphate, and ammonium in aquatic environment, Chemosphere (2020) 127492.

[25] H. Xiao, B. Bai, X. Li, J. Liu, Y. Liu, D. Huang, Interval multiple-output soft sensors development with capacity control for wastewater treatment applications: a comparative study, Chemometr. Intell. Lab. Syst. 184 (2019) 82–93.

[26] J.F. de Canete, P. del Saz-Orozco, R. Baratti, M. Mulas, A. Ruano, A. Garcia-Cerezo, Soft-sensing estimation of plant effluent concentrations in a biological wastewater treatment plant using an optimal neural network, Expert Syst. Appl. 63 (2016) 8–19.

[27] H. Ratnaweera, J. Fettig, State of the art of online monitoring and control of the coagulation process, Water 7 (11) (2015) 6574–6597.

[28] Y. Chen, L. Song, Y. Liu, L. Yang, D. Li, A review of the artificial neural network models for water quality prediction, Appl. Sci. 10 (17) (2020) 5776.

[29] Y. Jiang, L.-L. Xiao, L. Zhao, X. Chen, X. Wang, K.-Y. Wong, Optical biosensor for the determination of BOD in seawater, Talanta 70 (1) (2006) 97–103.

[30] J.R. Stetter, W.R. Penrose, S. Yao, Sensors, chemical sensors, electrochemical sensors, and ECS, J. Electrochem. Soc. 150 (2) (2003) S11.

[31] Z. Ye, J. Yang, N. Zhong, X. Tu, J. Jia, J. Wang, Tackling environmental challenges in pollution controls using artificial intelligence: a review, Sci. Total Environ. 699 (2020) 134279.

[32] N. Moreno-Alfonso, C. Redondo, Intelligent waste-water treatment with neural-networks, Water Policy 3 (3) (2001) 267–271.
[33] M. Côte, B.P. Grandjean, P. Lessard, J. Thibault, Dynamic modelling of the activated sludge process: improving prediction using neural networks, Water Res. 29 (4) (1995) 995–1004.
[34] T. Pai, Y. Tsai, H. Lo, C. Tsai, C. Lin, Grey and neural network prediction of suspended solids and chemical oxygen demand in hospital wastewater treatment plant effluent, Comput. Chem. Eng. 31 (10) (2007) 1272–1281.
[35] V. Ranković, J. Radulović, I. Radojević, A. Ostojić, L. Čomić, Neural network modeling of dissolved oxygen in the Gruža reservoir, Serbia, Ecol. Model. 221 (8) (2010) 1239–1244.
[36] I. Jagielska, C. Matthews, T. Whitfort, An investigation into the application of neural networks, fuzzy logic, genetic algorithms, and rough sets to automated knowledge acquisition for classification problems, Neurocomputing 24 (1–3) (1999) 37–54.
[37] M.T. Sorour, L.M. Bahgat, Application of activated sludge models in traditionally operated treatment plants—a software environment overview, Water Qual. Res. J. 39 (3) (2004) 294–302.
[38] C. Gontarski, P. Rodrigues, M. Mori, L. Prenem, Simulation of an industrial wastewater treatment plant using artificial neural networks, Comput. Chem. Eng. 24 (2–7) (2000) 1719–1723.
[39] Y.-S.T. Hong, M.R. Rosen, R. Bhamidimarri, Analysis of a municipal wastewater treatment plant using a neural network-based pattern analysis, Water Res. 37 (7) (2003) 1608–1618.
[40] M.M. Hamed, M.G. Khalafallah, E.A. Hassanien, Prediction of wastewater treatment plant performance using artificial neural networks, Environ. Model. Softw. 19 (10) (2004) 919–928.
[41] L. Zhao, T. Dai, Z. Qiao, P. Sun, J. Hao, Y. Yang, Application of artificial intelligence to wastewater treatment: a bibliometric analysis and systematic review of technology, economy, management, and wastewater reuse, Process Saf. Environ. Protect. 133 (2020) 169–182.
[42] M. Sophocleous, Interactions between groundwater and surface water: the state of the science, Hydrogeol. J. 10 (1) (2002) 52–67.
[43] C. Cooper, Biological effects of agriculturally derived surface water pollutants on aquatic systems—a review, J. Environ. Qual. 22 (3) (1993) 402–408.
[44] S. Madhav, et al., Water pollutants: sources and impact on the environment and human health, in: Sensors in Water Pollutants Monitoring: Role of Material, Springer, 2020, pp. 43–62.
[45] P. Goel, Water Pollution: Causes, Effects and Control, New Age International, 2006.
[46] H. Haimi, M. Mulas, F. Corona, R. Vahala, Data-derived soft-sensors for biological wastewater treatment plants: an overview, Environ. Model. Softw. 47 (2013) 88–107.
[47] J. Huang, H. Yin, S.C. Chapra, Q. Zhou, Modelling dissolved oxygen depression in an urban river in China, Water 9 (7) (2017) 520.
[48] X.-d. Wang, O.S. Wolfbeis, Optical methods for sensing and imaging oxygen: materials, spectroscopies and applications, Chem. Soc. Rev. 43 (10) (2014) 3666–3761.
[49] V. Velusamy, K. Arshak, O. Korostynska, K. Oliwa, C. Adley, An overview of foodborne pathogen detection: in the perspective of biosensors, Biotechnol. Adv. 28 (2) (2010) 232–254.
[50] V. Velusamy, et al., Label-free detection of Bacillus cereus DNA hybridization using electrochemical impedance spectroscopy for food quality monitoring application, in: 2010 IEEE Sensors Applications Symposium (SAS), IEEE, 2010, pp. 135–138.

[51] O. Korostynska, A. Mason, A. Al-Shamma'a, Monitoring OF nitrates and phosphates IN wastewater: current technologies and further challenges, Int. J. Smart Sens. Intell. Syst. 5 (1) (2012).

[52] K. Paek, H. Yang, J. Lee, J. Park, B.J. Kim, Efficient colorimetric pH sensor based on responsive polymer–quantum dot integrated graphene oxide, ACS Nano 8 (3) (2014) 2848–2856.

[53] O. Korostynska, A. Mason, A. Al-Shamma'a, Microwave sensors for the non-invasive monitoring of industrial and medical applications, Sens. Rev. 34 (2) (2014) 182–191.

[54] A. Mason, O. Korostynska, A. Al-Shamma'a, Microwave sensors for real-time nutrients detection in water, in: Smart Sensors for Real-Time Water Quality Monitoring, Springer, 2013, pp. 197–216.

[55] D.E. Reisner, T. Pradeep, Aquananotechnology: Global Prospects, CRC Press, 2014.

[56] N. Atar, T. Eren, M.L. Yola, S. Wang, A sensitive molecular imprinted surface plasmon resonance nanosensor for selective determination of trace triclosan in wastewater, Sensor. Actuator. B Chem. 216 (2015) 638–644.

[57] P.J. Vikesland, Nanosensors for water quality monitoring, Nat. Nanotechnol. 13 (8) (2018) 651–660.

[58] X. Su, L. Sutarlie, X.J. Loh, Sensors, biosensors, and analytical technologies for aquaculture water quality, Research 2020 (2020).

[59] A. T. C. o. A. o. A. N. N. i. Hydrology, Artificial neural networks in hydrology. I: preliminary concepts, J. Hydrol. Eng. 5 (2) (2000) 115–123.

[60] X. Shi, J. Qiao, Neural network predictive optimal control for wastewater treatment, in: 2010 International Conference on Intelligent Control and Information Processing, IEEE, 2010, pp. 248–252.

[61] F.S. Hoseinian, B. Rezai, E. Kowsari, The nickel ion removal prediction model from aqueous solutions using a hybrid neural genetic algorithm, J. Environ. Manag. 204 (2017) 311–317.

[62] F. Li, J. Qiao, H. Han, C. Yang, A self-organizing cascade neural network with random weights for nonlinear system modeling, Appl. Soft Comput. 42 (2016) 184–193.

[63] M. Zeinolabedini, M. Najafzadeh, Comparative study of different wavelet-based neural network models to predict sewage sludge quantity in wastewater treatment plant, Environ. Monit. Assess. 191 (3) (2019) 1–25.

[64] P.R. Zonouz, A. Niaei, A. Tarjomannejad, Modeling and optimization of toluene oxidation over perovskite-type nanocatalysts using a hybrid artificial neural network-genetic algorithm method, J. Taiwan Inst. Chem. Eng. 65 (2016) 276–285.

[65] B. Raheli, M.T. Aalami, A. El-Shafie, M.A. Ghorbani, R.C. Deo, Uncertainty assessment of the multilayer perceptron (MLP) neural network model with implementation of the novel hybrid MLP-FFA method for prediction of biochemical oxygen demand and dissolved oxygen: a case study of Langat River, Environ. Earth Sci. 76 (14) (2017) 1–16.

[66] H.R. Maier, G.C. Dandy, Neural networks for the prediction and forecasting of water resources variables: a review of modelling issues and applications, Environ. Model. Softw. 15 (1) (2000) 101–124.

[67] J.-S. Jang, ANFIS: adaptive-network-based fuzzy inference system, IEEE Trans. Syst. Man Cybern. 23 (3) (1993) 665–685.

[68] J.-S.R. Jang, C.-T. Sun, E. Mizutani, Neuro-fuzzy and soft computing-a computational approach to learning and machine intelligence [Book Review], IEEE Trans. Automat. Control 42 (10) (1997) 1482–1484.

[69] H. Mingzhi, W. Jinquan, M. Yongwen, W. Yan, L. Weijiang, S. Xiaofei, Control rules of aeration in a submerged biofilm wastewater treatment process using fuzzy neural networks, Expert Syst. Appl. 36 (7) (2009) 10428–10437.

[70] M. Asadnia, L.H. Chua, X. Qin, A. Talei, Improved particle swarm optimization–based artificial neural network for rainfall-runoff modeling, J. Hydrol. Eng. 19 (7) (2014) 1320–1329.

[71] M. Asadnia, A.M. Khorasani, M.E. Warkiani, An accurate PSO-GA based neural network to model growth of carbon nanotubes, J. Nanomater. 2017 (2017).

[72] M. Asadnia, M.S. Yazdi, A. Khorasani, An improved particle swarm optimization based on neural network for surface roughness optimization in face milling of 6061-T6 aluminum, Int. J. Appl. Eng. Res. 5 (19) (2010) 3191–3201.

[73] M. Farahnakian, M.R. Razfar, M. Moghri, M. Asadnia, The selection of milling parameters by the PSO-based neural network modeling method, Int. J. Adv. Manuf. Technol. 57 (1) (2011) 49–60.

[74] M. Al-Obaidi, J.-P. Li, C. Kara-Zaitri, I.M. Mujtaba, Optimisation of reverse osmosis based wastewater treatment system for the removal of chlorophenol using genetic algorithms, Chem. Eng. J. 316 (2017) 91–100.

[75] C. Zhou, H. Gao, L. Gao, W. Zhang, Particle swarm optimization (PSO) algorithm [J], Appl. Res. Comput. 12 (2003) 7–11.

[76] J. Zhu, J. Zurcher, M. Rao, M.Q. Meng, An on-line wastewater quality predication system based on a time-delay neural network, Eng. Appl. Artif. Intell. 11 (6) (1998) 747–758.

[77] C.G. Piuleac, S. Curteanu, M.A. Rodrigo, C. Sáez, F.J. Fernández, Optimization methodology based on neural networks and genetic algorithms applied to electro-coagulation processes, Cent. Eur. J. Chem. 11 (7) (2013) 1213–1224.

[78] E.G. Kana, J. Oloke, A. Lateef, M. Adesiyan, Modeling and optimization of biogas production on saw dust and other co-substrates using artificial neural network and genetic algorithm, Renew. Energy 46 (2012) 276–281.

[79] S. Nag, N. Bar, S.K. Das, Sustainable bioremadiation of Cd (II) in fixed bed column using green adsorbents: application of Kinetic models and GA-ANN technique, Environ. Technol. Innov. 13 (2019) 130–145.

[80] M. Fan, J. Hu, R. Cao, K. Xiong, X. Wei, Modeling and prediction of copper removal from aqueous solutions by nZVI/rGO magnetic nanocomposites using ANN-GA and ANN-PSO, Sci. Rep. 7 (1) (2017) 1–14.

Chapter 2

Advancements and artificial intelligence approaches in antennas for environmental sensing

Ali Lalbakhsh[1,5], Roy B. V. B. Simorangkir[2], Nima Bayat-Makou[3], Ahmed A. Kishk[4], Karu P. Esselle[5]

[1]*The School of Engineering, Macquarie University, Sydney, NSW, Australia;* [2]*Tyndall National Institute, Cork, Ireland;* [3]*The Edward S. Rogers Sr., Department of Electrical and Computer Engineering, University of Toronto, Toronto, ON, Canada;* [4]*The Department of Electrical and Computer Engineering, Concordia University, Montreal, QC, Canada;* [5]*School of Electrical and Data Engineering, University of Technology Sydney (UTS), Sydney, NSW, Australia*

1. Printed antennas for wireless sensor networks

Wireless Sensor Networks (WSNs) are a group of spatially distributed sensors to monitor and record the physical conditions of the environment and organize the data at a central location for measuring environmental conditions, such as humidity, sound, temperature, wind, etc. [1]. The number of sensor nodes in WSNs can be significantly greater than the number of nodes in a wireless ad hoc network, and the sensor nodes are densely located in WSNs [2]. Since the first WSN development in the early 90s, WSN represents an ever-evolving research field in various areas with the challenges of creating miniaturized and smart modules having the capability of a long-term application. A wireless module in WSNs is typically composed of a processing unit, a sensor, a power supply, and a telecommunication part. The telecommunication part of the module plays a significant role. It should use minimal power for a high lifespan, and the antenna size needs to be minimum to achieve a reasonably small platform size [3].

The optical transmission was a good candidate for the wireless interface of the sensors, but the necessity of the line of sight transmission has been a significant drawback of this solution. Therefore, radio wave transmission has been chosen as a suitable wireless communication interface for the WSNs. Different radio frequency (RF) bands have been used for wireless sensors

Artificial Intelligence and Data Science in Environmental Sensing
https://doi.org/10.1016/B978-0-323-90508-4.00004-6

communication that must comply with government regulations and wireless standards. Currently, WSNs use frequency bands including 315 MHz, 433 MHz, 868 MHz (Europe), 915 MHz (North America), and 2.45 GHz industrial-scientific-medical (ISM) band. The frequency band of 2.45 GHz provides implementation flexibility because of the availability of the RF devices at this band [1]. In addition, antenna dimensions are reasonably compact in this frequency range, and employing printed antennas offers easy integration with the other RF components, resulting in miniaturized wireless sensors. Here, some recent progress in designing printed antennas and their performance are discussed.

A circularly polarized square microstrip patch antenna was given in Ref. [4] with a pair of crank-shaped slits at each edge of the square patch. Applying the crank-shaped slits at each edge miniaturizes the patch size while producing circular polarized radiated waves. The antenna was made of two stacked substrates with a relative permittivity of 3.8 and thicknesses of $h1 = 1.6$ mm and $h2 = 3.2$ mm. Tuning the crank-slit dimensions improves the axial ratio of the circularly polarized beam, and adjusting the L-probe dimensions provides the required rerun loss at the operating band; for further axial ratio enhancement, two-pair slots can be employed [5]. The small geometry and tunability of this antenna make it suitable for wireless sensors. Apart from the applications mentioned above, microstrip technology has been widely used in other microwave components, such as planar antennas [6–11], microstrip filters [12–18], power dividers [19–23], amplifiers [24,25], and diplexers [26] due to its desired low profile, low cost, lightweight, low volume, easy integration into microwave circuits, making them suitable for modern WSNs.

Miniaturized inverted-F antennas (IFAs) with Hilbert geometry were designed in Ref. [27] for WSN applications at the 2.45 GHz band. The effectiveness of space-filling geometries, for example, fractal Hilbert curve, has been demonstrated in designing miniature antennas, multiband antennas, high-gain antennas, and high-impedance surfaces [28–30]. The employment of Hilbert geometry in the IFA design provided the antenna with a compact geometry on a single-layer substrate. Hilbert geometry makes electrical length longer, which means the same performance can be achieved using a smaller antenna. The geometry of this antenna with its top and bottom printed layers are shown in Fig. 2.1A. The antenna layout was printed on a conventional and low-cost FR4 dielectric substrate with a thickness of 0.4 mm, a dielectric constant of 4.4, and a loss tangent of 0.02. The total size of the antenna is 35 mm $\times$ 6 mm $\times$ 1.6 mm or $3.5 \times 6 \times 1.6$ mm^3, in which the combination of Hilbert structure with IFA resulted in a size reduction of 77% compared to a standard rectangular patch antenna operating at the same frequency band. The simulated and measured return losses of the antenna are plotted in Fig. 2.1B. With the return loss of 27.6 dB at the center frequency of 2.43 GHz, 9.1% bandwidth is achieved for a return loss better than 10 dB. The radiation beam of the antenna at the X-Z plane is given in Fig. 2.1C, with a peak gain of almost 1.4 dBi. The antenna patterns show an omnidirectional behavior confirming that the ground plane determines the dipole-type patterns. The reduced form factor with the aim of space-filling geometry while producing semi-omnidirectional patterns makes this antenna a competitive candidate for wireless sensors.

Medical body area network (MBAN) is another new sensor-based wireless communications technology that is allocated for in and on a human body wireless communications for noninvasive medical care services [31,32]. MBAN sensors are highly demanded to be flexible, low profile, miniaturized, robust, lightweight, and suitable for continuous monitoring for reliable and high-performance communication. Likewise, in other WSNs, the antenna design of an MBAN system is a determining factor as the system requirements for a reliable and energy-efficient wireless link are directly related to the antenna performance. This is important because the capacitive coupling between the antenna and the body in close operation to a human body degrades the antenna performance [33]. Normal antenna polarization is often required for on-body communications where monopole-type or dipole-type antennas could be easily utilized. However, when these omnidirectional antennas are placed parallel to the human body, the radiation efficiency decreases as a result of bulk power absorption by the body. Therefore, the antenna with a unidirectional radiation pattern is appropriate for use. Typically in-body requires omni pattern on-body communication systems to minimize the effects of the human body and reduce body exposure to electromagnetic (EM) field radiation [34]. There are several approaches to reshape and reconfigure the radiation patterns of different antennas, such as FSSs, to achieve the required radiation patterns [35,36].

Cavity antenna structures were used for on-body communication in Ref. [37], where the radiation patterns are determined almost exclusively by the equivalent magnetic current on the aperture so that the antenna can have relatively high efficiency even in the vicinity of the human body. However, the bulky size, rigidness, and complicated manufacturing procedures of most of the planar and slot antenna designs, could limit their practicality in on-body applications. Additionally, cavity-based antennas can also be used for point-to-point communications suitable for large-scale environmental sensing in conjunction with low earth orbit satellite constellations [38–40]. In Ref. [41], a very low-profile patch-type slot antenna was presented for an on-body wireless biosensor in the MBAN band. In Fig. 2.2A, the configuration of an on-body wireless sensor is shown, which is designed to attach directly to the frontal thorax of a human body to continuously measure the heart rate and transmit it toward a mobile device located in close proximity to the sensor. As shown in Fig. 2.2A, the sensor is composed of the laminated structure, including a sensor cover, a wireless module with three control sections on the flexible PCB, a flexible battery, and electrodes. The side view illustration in Fig. 2.2A shows the wireless sensor having an overall thickness of about 2 mm with the antenna designed on the top of the sensor utilized as the sensor cover.

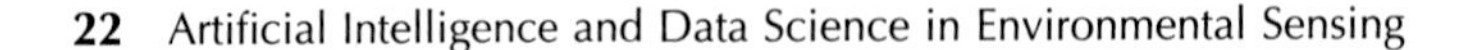

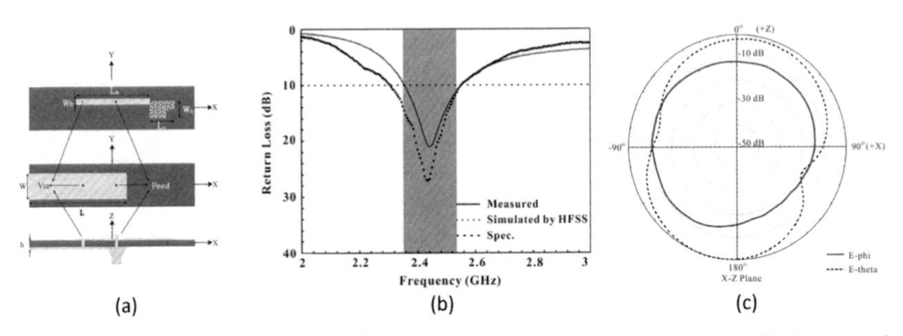

FIGURE 2.1 (A) Configuration of miniaturized Hilbert inverted-F antenna. (B) Measured and simulated return loss. (C) Measured radiation pattern at 2.45 GHz at x-z plane [27].

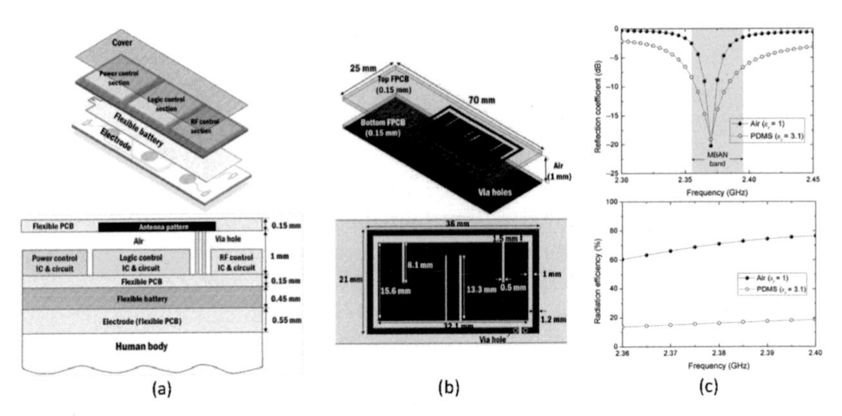

FIGURE 2.2 On-body wireless biosensor in the medical body area network (MBAN) band. (A) Configuration and side view on the human body. (B) Antenna geometry and dimensions, 3D view and top view with dimensions. (C) Reflection coefficient, a radiation efficiency of the antenna with air substrate and PDMS substrate [41].

Using the printed antenna as the sensor cover results in no additional space required and enables flexible chip placement options. In addition, the battery underneath the cover layer can be used as a ground plane for the patch antenna, which is placed on the cover layer.

For the antenna design and optimization of this on-body sensor, the system module, battery, and electrodes of the sensor are removed for simplicity. The FPCB radiating element and the ground plane were made of thin copper layers and polyimide film with electrical characteristics of $\varepsilon r = 2.7$, $\tan\delta = 0.005$, and thickness = 0.15 mm. The antenna consists of a rectangular patch and a loop line surrounding the rectangular patch. The loop provides balanced feeding and induces current to the radiating patch through EM coupling, adjusted by the gap between the loop and the patch for better matching performance. Since a 100-Ω differential feeding mechanism is applied, there is no need to use bulky and lossy balun, which leads to improved receiver noise performance and transmitter power efficiency. The vertical slots inside the patch are employed to reduce the patch size by elongating the induced current path on the patch. This makes the patch operate as a meander line antenna with a ground plane in which the vertical slots are the radiating elements.

As in many biomedical [42–47] and sensing [48–51] applications, polydimethylsiloxane (PDMS)-based elastomer ($\varepsilon r = 3.1$ and $\tan \delta = 0.025$) has been utilized for an encapsulation of the modules. For this purpose, the space between the slotted patch and flexible ground plane is filled with this material that is considered as the antenna substrate. The antenna reflection coefficient and radiation efficiency with and without the PDMS are plotted in Fig. 2.2C. Due to the considerable loss factor of PDMS and its thin substrate thickness (0.008 $\lambda 0$), the antenna reveals low efficiency, around 15% with PDMS, while its efficiency with air substrate is more than 60%. However, the air substrate makes the bandwidth narrower. It should also be considered that a low-profile patch antenna with high input reactance and low input resistance leads to a high-quality factor (Q) that results in low efficiency and a narrow frequency bandwidth [52].

To evaluate the sensor antenna performance in the presence of the human body, a 3D simulation model, with the antenna in close proximity to a multilayered human tissue model was created. The modeled multilayer human tissue consists of equivalent skin, fat, and muscle layers [53]. The presence of body tissue and distance from the body slightly shifts the antenna resonance frequency, but the antenna operating band is still in the MBAN frequency band. Thus, the antenna is not sensitive to the presence of the body, making it suitable for on-body communications deployment.

The absorption loss by the human body model caused a reduction in the body model's maximum gain and total efficiency by around 0.82 dBi and 30%, respectively, in comparison with the free space, where the maximum gain and efficiency are 3.84 dBi and 66.4%, respectively. The conductive medium prevents most of the signal from being radiated toward the body, and the presence of the lossy human body decreases the backward radiation considerably. The antenna behavior with a full sensor structure was further investigated in Ref. [41] that still provides acceptable performance at the MBAN band, making it suitable for a thin and flexible wireless biosensor system.

2. Printed antenna sensors for material characterization

In WSNs, several antenna types are employed to transmit the sensor data through the network. However, the antennas themselves could be used as sensors because their passive sensing materials received considerable attention from researchers. In Ref. [54], an antenna was designed for characterizing both permittivity and loss tangent of a sample under test (SUT). In this antenna sensor, the sample material changes the antenna resonance frequency. Therefore, the sample material's permittivity and loss tangent can be estimated by measuring the reflection coefficient and transmission coefficient at the resonance frequency. The gap in the antenna sensor creates a capacitor, and different permittivity of the sample materials changes its capacitance which can be considered corresponding to the effective electrical length variation of the patch. As a result, the resonance frequency of the antenna will change for different permittivity of the SUTs. If the material is lossy, it is expected to have less power transmitted to the second port, which is used to extract the loss factor of the material. In order to have an effective data transmission at the specified bandwidth with this sensor antenna, a resonance adjuster needs to be utilized to adjust the resonance frequency of the antenna to the frequency of interest after characterizing the materials.

Alzheimer's disease is the most common form of the neurodegenerative disorder that can be early detected by the change of amyloid beta proteins in cerebrospinal fluid (CSF). The changes in protein concentration due to Alzheimer's disease in CSF implicitly affect the CSF dielectric properties. The biological tissues can be considered a dielectric material with EM interaction described by dielectric properties. A sensor comprises an implantable antenna with a large frequency band, and a voltage-controlled oscillator (VCO) was presented in Ref. [55] for detecting loss tangent changes in CSF. In general, for implanted applications where there are stringent regulations on specific absorption rate (SAR) and where maximum efficiency is required (such as wireless power transfer to medical devices), the magnetic-based antennas, such as a loop antenna, are more preferred [56]. This antenna is composed of two meandered transmission lines and a semicircular patch antenna, as shown in Fig. 2.3A. In this work, slits, meandered transmission lines, and resonating structures were applied to the antenna design as miniaturization techniques to reduce the antenna physical form factor. The antenna was made of 35 μm copper cladding and an FR4 substrate (εr = 4.3 and tan δ = 0.025) with a thickness of 0.8 mm. A 65 μm Kapton superstrate with permittivity and tan δ of 3.4 and 0.002, respectively, covered the antenna for preventing direct contact and reducing coupling to near-field tissue. Since the antenna was

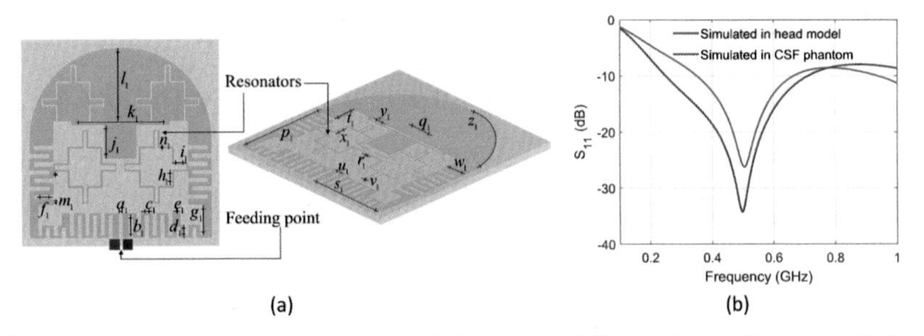

FIGURE 2.3 Implantable antenna sensor. (A) Geometry and (B) simulated reflection coefficient in cerebrospinal fluid cube and head model [55].

embedded in a biological (dielectric) medium, the tissue loading increased the antenna's electrical length, which was insufficient to obtain a good impedance matching.

Fig. 2.3B shows the reflection coefficient of the optimized antenna in the CSF cube and head model. With the aid of a time-domain finite-difference time-domain (FDTD) solver, the antenna was implemented in a 100 mm edge length cubic phantom with dielectric properties of CSF material at sub-1 GHz band. The antenna was embedded in the CSF of a head model with 46 tissues to ensure the robustness of antenna results. Despite having other tissues in the head model, both plots show the wideband performance of the antenna sensor along with the MedRadio band with almost the same center operating frequency. Due to the stable response, it could be used in an implantable sensor along with a stable VCO to detect changes in loss tangent of CSF due to Alzheimer's disease [55].

3. Epidermal antenna for unobtrusive human-centric wireless communications and sensing

Skin-mounted electronics is a new frontier for truly unobtrusive real-time human-centric wireless communications and sensing. Such a system features the skin properties to enable a seamless and nonirritating interface to the body for various applications, including physiological parameters monitoring, human–machine interfaces, and others that traditional wearable electronics cannot address. The concept of an epidermal antenna as one of the enabling components of a skin-mounted wireless system has moved closer to reality because of recent advancements in biologically suitable, flexible materials and devices. In the following, the concept of the epidermal antenna is discussed along with its design considerations.

3.1 Epidermal electronics

Human physiological parameters, such as blood pressure, body temperature, breathing rate, heart rate, blood oxygen saturation, and various electrophysiological signals, represent the operation of a human body and are thus useful as reference values in human health monitoring [57]. Researchers have been interested in sensing human physiological parameters for decades, particularly those that exploit interfaces to human skin. Such a concept originated in 1929 with the discovery of human brain wave recording through the scalp, later known as electroencephalography [58], and has grown ever since, resulting in the ability to monitor various physiological parameters available today. Surprisingly, despite the progress gained in terms of functionality, practically nearly all associated technologies still rely on the concept from the beginning [59]. That is, a variety of interfaces (e.g., electrodes and sensors) are mounted on the skin using mechanical fixtures (e.g., clamps, straps, bands, or glasses), strong adhesive tapes, or microneedles that are connected to separate boxes containing signal conditioning units, data storages, communication modules, and power supplies via wires [60,61]. Despite the significant capabilities demonstrated by these technologies, the current setups are unsuitable for use outside of clinical or laboratory contexts. One problem is that the systems are disintegrated, which complicates the setup and makes the overall shape, size, and weight of the systems incompatible for long-term personal usage. Inconsistent and unreliable adhesion to the skin, discomfort associated with contact pressure, interfacial shear and frictional forces, as well as constraints on wearing sites on the body and some other challenges induced by the mechanical and geometrical mismatch with the body texture have an impact on both the user experience and measurement accuracy.

The concept of epidermal electronics was established in 2011 due to the factors mentioned above [62]. The term refers to a system where all functionalities and associated components, such as electrodes, sensors, electronics, communication modules to power supply, are integrated into an ultrathin membrane with high skin compatibility, allowing for seamless integration with the user's body [62,63]. The notion is basically to develop a device that can mimic as much as possible the characteristics of the human skin (e.g., ultrathin, flexibility, stretchability, self-healing, water/air permeability, etc.) while concurrently performing some additional functionality. Fig. 2.4 illustrates an example of epidermal electronics reported in Ref. [64], which was developed for noninvasive full vital signs monitoring in the neonatal intensive care unit. The developed binodal system captures and transmits electrocardiograms, photoplethysmogram continuously, and skin temperature data, out of which heart rate, heart rate variability, respiratory rate, blood oxygenation (SpO2), and surrogate of systolic blood pressure of an infant can be extracted. The applications of epidermal electronics vary widely from the most popular real-time health care monitoring, diagnostic and therapeutic, to robotics and prosthetics [62,65–67]. In recent years, a great deal of effort has gone into the advancement of epidermal electronics, which includes the transition from single to multimodal functionalities, and most importantly, in the development of materials and fabrication approaches as the key foundations for enabling devices with skin-like form factors and robust integration of all associated components [62,63,65,67,68].

3.2 Epidermal antennas

The epidermal antenna is one of the key enabling technologies of epidermal electronics to replace the wired interconnections of the first generation of human physiological monitoring systems. In the context of epidermal electronics, an antenna may be utilized to enable wireless communication capability with entities inside or outside the body or for the system self-powering [63,69,70]. The term epidermal antenna is often referred to as a single-layer

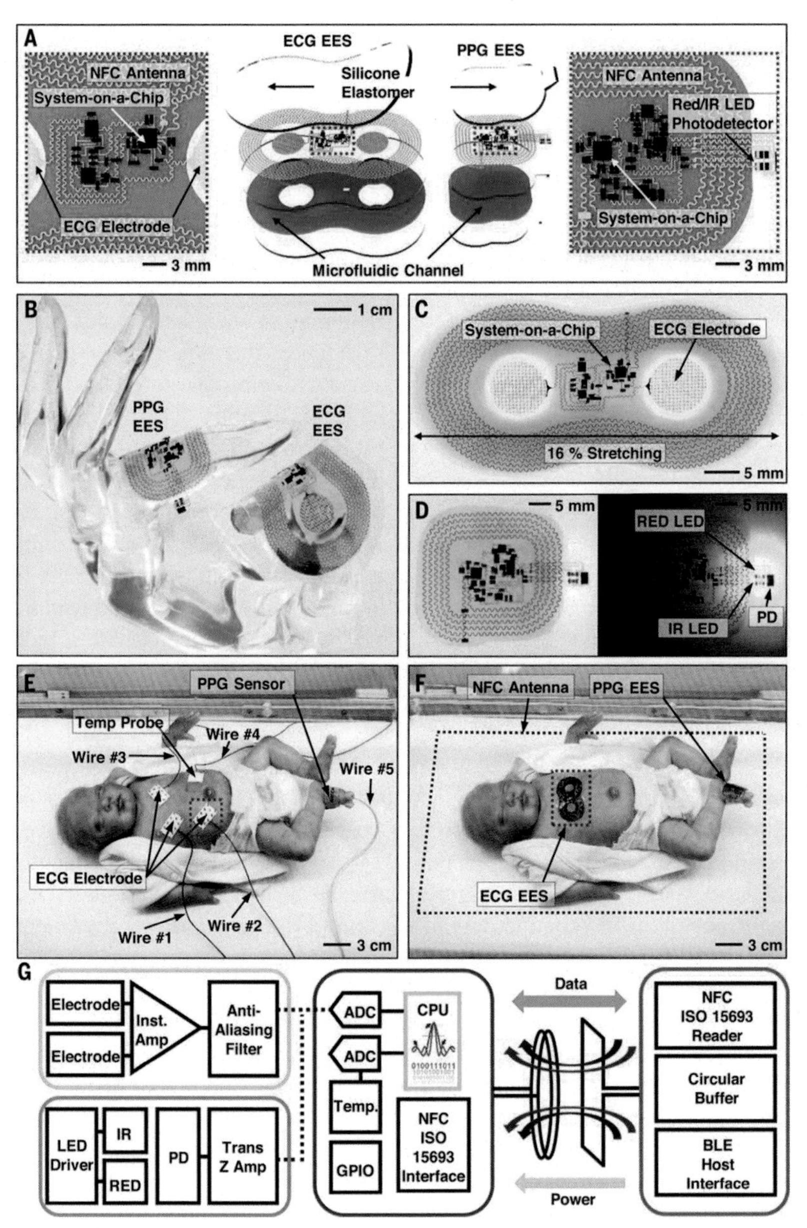

FIGURE 2.4 Ultrathin battery-free binodal (chest and foot) wireless sensing patches for full vital signs monitoring in a neonatal intensive care unit (NICU), developed in Ref. [64]. (A) Schematic illustration. (B) Developed patch wrapped over a finger of a transparent mannequin. (C) ECG EES patch under ~16% horizontal stretching. (D) Photoplethysmogram EES patch a lighted and a dark room. NICU setting using a neonate doll (E) with conventional measurement hardware and (F) with the developed bimodal patch. (G) Functional block diagram.

antenna directly attached over the skin through a thick flexible skin-like membrane [71]. Some of the key challenges in developing the epidermal antenna are listed below, which impose an advancement in terms of antenna and material engineering.

3.2.1 Compensated radio frequency performance

From a generalized EM point of view, a human body is an anisotropic, nonlinear, heterogeneous, and dispersive medium [72], thus not an ideal operating environment for an antenna. Among the issues that face the antenna are the low radiation performance and detuning of impedance (and consequently the resonance frequency). In this instance, the antenna radiation efficiency is hampered by a significant power attenuation due to dielectric and conductive losses of the human body tissue [72]. Detuning occurs because the antenna impedance is greatly influenced by EM properties of the surrounding environment, which, in the case of biological tissues, vary in terms of permittivity and conductivity [73]. In body-centric communications, EM shielding (e.g., utilizing ground plane, electronic band gap, and artificial magnetic conductor structures [74—77]) is frequently employed to isolate the antenna from the human body, resulting in improved wireless communication performance. Unfortunately, this technique leads to an antenna with a bulky profile that is incompatible with epidermal electronics, which aspires for close integration with the human body.

To demonstrate the above issues, EM simulations were performed using ANSYS full-wave high-frequency structure simulator, comparing the performance of a simple square loop antenna designed at ultrahigh frequency 868 MHz band when operated in free space and operated 1 mm above a simplified homogenous phantom (size: $140 \times 80 \times 50$ mm). The phantom mimics very well a human forearm electrically at 868 MHz with an average permittivity of 30 and conductivity of 0.7 S/m [44]. Fig. 2.5 shows the antenna topology and EM simulation setup. A perfect antenna impedance matching was assumed for the sake of generality and to reduce the number of free geometrical parameters to be examined in this study. In the instance of an antenna placed 1 mm above the phantom, as shown in Fig. 2.6, a considerable decrease in radiation efficiency and gain values at 868 MHz can be seen, indicating a large quantity of energy absorbed by the human body. Simulations of varied distances between the loop antenna and the phantom were also carried out, and the effect on the antenna's operating frequency is illustrated in Fig. 2.7, which exhibits the detuning effect discussed earlier.

3.2.2 Induced health and safety issue

Regarding the user's health and safety, there is a limit of human body tissue exposure to EM fields that the antenna must meet, such as the SAR threshold. SAR is a measure of power absorbed per unit mass tissue, which is limited to

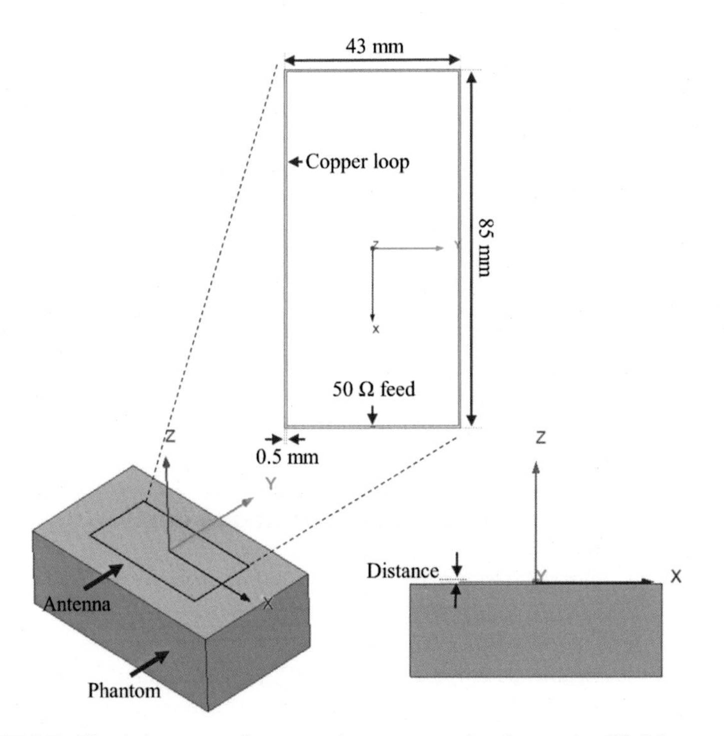

FIGURE 2.5 Simulation setup of a copper loop antenna placed on a simplified homogenous phantom. The inset shows the detailed dimension of the loop used for this study.

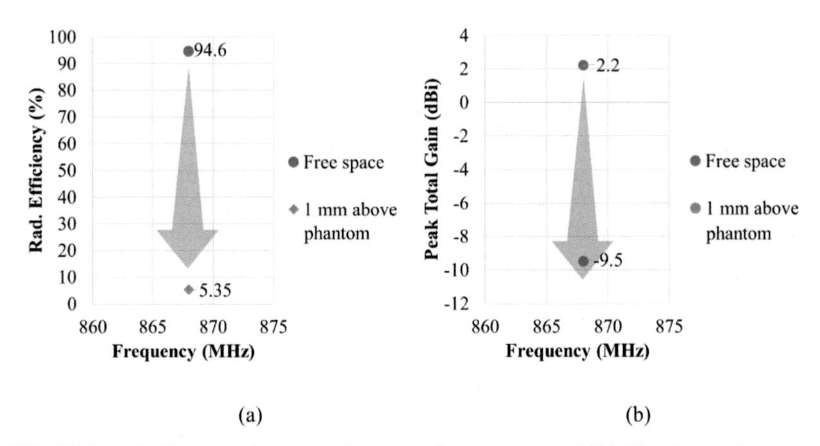

FIGURE 2.6 Radiation performance of a copper loop antenna at 868 MHz when in free space and when placed 1 mm above a homogenous phantom: (A) radiation efficiency, (B) peak total gain.

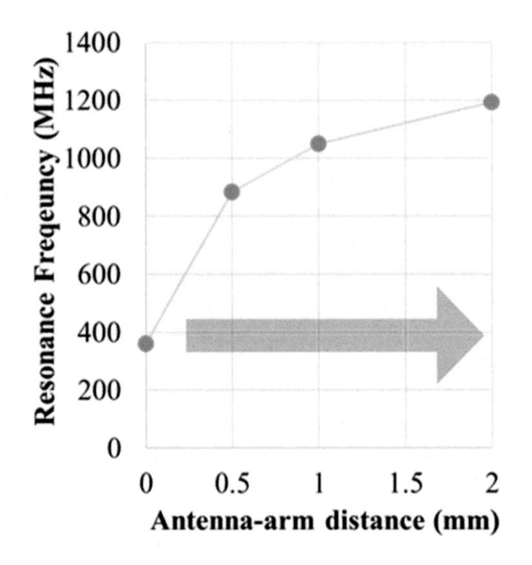

FIGURE 2.7 Effect of varying antenna–phantom distance on the resonance frequency.

2 W/kg in general or 4 W/kg for extremities, for 10-gram averaged human body tissue according to current standard [78,79]. In the case of epidermal antenna, the amount of energy transmitted toward the human body is relatively large due to the lack of shielding. As a result, the amount of power that can be fed into the antenna is limited.

3.2.3 *Imposed additional physical requirements*

The direct placement of the epidermal antenna on the human skin triggers a number of extra physical aspects for the user's comforts and safety, which may or may not be applicable to antennas for other purposes. Human skin can be stretched up to strain values of tens of percent without permanent distortion, allowing the human body to move freely. As a result, the epidermal antenna will be subjected to a range of mechanical stresses, generating strains in many directions. It is hence imperative for the epidermal antenna to be highly flexible, lightweight, and stretchable. In some applications that require prolonged usage, such as wireless health monitoring, the epidermal antenna must also be skin compatible to ensure user's comfort and at the same time water resistant to retain performance. The visually imperceptible epidermal antenna might also be useful for allowing seamless and pervasive health monitoring without interfering with daily activities.

4. Artificial intelligence in antenna design

In the modern antenna design, the need for a highly optimized antenna response with strict requirements of minimum space, cost, and computational cost has dramatically increased. In some cases, achieving all EM requirements along with considering critical nonelectromagnetics considerations may not be possible through the conventional antenna design approaches. Hence, artificial intelligence (AI)—based design methodologies seem a very attractive solution to expedite the design process and meet the strict modern communication requirements.

4.1 Particle swarm optimization in antenna design

AI and particularly evolutionary algorithms (EAs) have played a significant role in developing rigid and flexible antennas used for sensing applications. Swarm intelligence algorithms are a type of EAs based on the collective behavior of self-organized and decentralized systems. Some of the well-publicized swarm intelligence algorithms in the antenna community are particle swarm optimization (PSO) [80], Ant colony optimization [81—84], and invasive weed optimization [85]. The advantage of PSO is its easy implementation, while it is computationally efficient and can be used for both real-valued and discrete-valued problems [86].

The PSO basics and its use in conjunction with the numerical EM simulators have been discussed explicitly [87,88]. In terms of the applications, the PSO has been successfully used to address the undesired side lobe level (SLL) in linear array antennas with a wide application in wireless sensor nodes [89—91]. In Ref. [89], linear array positions were determined by a specialized PSO algorithm for the case of 2—6 elements and different scan angles, beam widths, and antenna elements. In Ref. [91], two fitness functions are proposed: one for positioning nulls in the predetermined direction and the other for the SLL minimization. PSO was used in thinning arrays with 15 and 20 elements, resulting in around 5 dB improvement in the SLL [92]. Other applications of PSO are removing time delay in the aperture antennas [93], eliminating parameters estimates [94], and efficiently dealing with a large number of unknowns based on hybrid periodic boundary conditions in PSO [95], where soft and hard types of boundary conditions were implemented for the PSO.

Moreover, various types of frequency selective surfaces (FSS), such as bandpass FSS filters and quadrature reflection phase surfaces [96], have been designed by an improved simulation-driven PSO algorithm. Furthermore, several multiband and wideband antennas have recently been optimized using PSO. A self-renewing fitness estimation was incorporated with the conventional PSO algorithm in designing an E-type dual-frequency microstrip antenna with less computational cost. In this approach, from the third generation of the algorithm, the PSO fitness of particles is calculated using a prediction model instead of the basic PSO, where the fitness is given by the full-wave EM simulation [97]. Fig. 2.8 shows the flowchart of this efficient PSO algorithm. Concerning the wideband application of PSO, a holistic

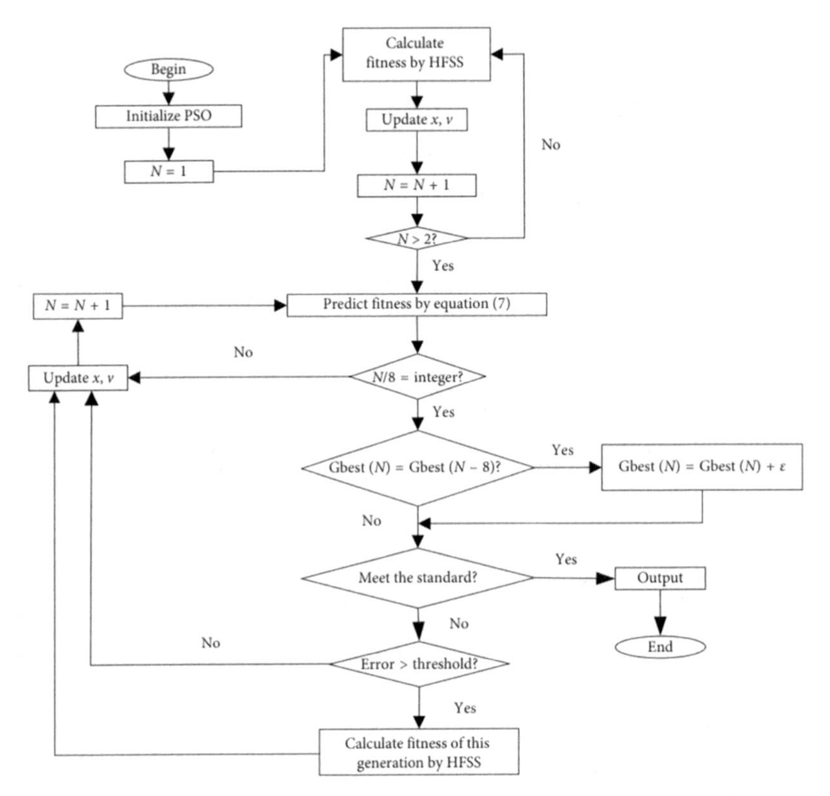

FIGURE 2.8 Flowchart of self-renewing fitness prediction method based on particle swarm optimization (PSO) algorithm [97].

optimization procedure was used to design a lens-like superstrate for a Fabry—Perot resonator antenna. In this technique, the optimization algorithm can retune the antenna parameters in addition to the superstructure to provide a large 3 dB gain bandwidth of 40% [98].

In a different approach, PSO was employed to develop an equivalent circuit for the THz antenna temperature sensor at different operating temperatures [99]. PSO in conjunction with the method of moments was used to determine parameters of a coplanar waveguide-fed monopole antenna for optimum performance over operating bands of 2.51 GHz, 3.98 GHz, and 5.24 GHz, covering the bandwidth requirements of the personal communication systems, 2.4 and 5.2 GHz WLAN systems [100]. A set of Hertzian dipoles for modeling wideband antennas radiation characteristics was obtained using a PSO-based method in Ref. [101]. In this work, the dipole parameters, which are polynomials in the phase constant, are chosen by minimizing differences between the exact or measured near-field data and the dipoles' fields at various frequencies.

4.2 Artificial neural network in antenna design

In conjunction with computer-aided design approaches, artificial neural network (ANN) has been considered a versatile tool in EM and microwave component design [102–109]. In this class of AI-based design approach, the ANN is trained through some initial EM simulations, and after the training, it can replace the computationally costly EM simulations, resulting in a significant reduction in the computational cost of the design. The combination of ANN with FDTD [102], parametric frequency model [107], and Gaussian model [108] have been illustrated in the literature, showing a considerable improvement in the accuracy of antenna modeling.

Another efficient modeling approach combines neural networks with transfer functions, providing accurate parametric modeling of EM responses [103,104,106,109]. One of the advantages of this approach is that it can be used even without accurate equivalent circuits. In conclusion, AI-based optimization/design producers have brought several advantages and novelties in relation to the antennas suitable for sensing applications over the last two decades, improving computational cost, physical features, and the performance of this class of antennas.

References

[1] A.G.P. Kottapalli, et al., Soft polymer membrane micro-sensor arrays inspired by the mechanosensory lateral line on the blind cavefish, J. Intell. Mater. Syst. Struct. 26 (1) (2015) 38–46.

[2] J. Zheng, A. Jamalipour, Wireless Sensor Networks: A Networking Perspective, John Wiley & Sons, 2009.

[3] G. Bacles-Min, P.F. Ndagijimana, Package-antenna for wireless sensor network nodes, in: 2008 Loughborough Antennas and Propagation Conference, IEEE, 2008.

[4] T. Fujimoto, et al., Circularly polarized small microstrip antenna for wireless sensor network, in: 2015 International Symposium on Antennas and Propagation (ISAP), IEEE, 2015.

[5] M.K. Ray, et al., Two-pair slots inserted CP patch antenna for wide axial ratio beamwidth, IEEE Access 8 (2020) 223316–223324.

[6] A. Iqbal, et al., Modified U-shaped resonator as decoupling structure in MIMO antenna, Electronics 9 (8) (2020) 1321.

[7] S. Muhammad, et al., Compact rectifier circuit design for harvesting GSM/900 ambient energy, Electronics 9 (10) (2020) 1614.

[8] M.M. Kamal, et al., Donut-shaped mmWave printed antenna array for 5G technology, Electronics 10 (12) (2021) 1415.

[9] M. Alibakhshikenari, et al., Wideband sub-6 GHz self-grounded bow-tie antenna with new feeding mechanism for 5G communication systems, in: 2019 13th European Conference on Antennas and Propagation (EuCAP), IEEE, 2019.

[10] M.M. Honari, R. Mirzavand, P. Mousavi, A high-gain planar surface plasmon wave antenna based on substrate integrated waveguide technology with size reduction, IEEE Trans. Antenn. Propag. 66 (5) (2018) 2605–2609.

[11] M.M. Honari, A. Abdipour, G. Moradi, Aperture-coupled multi-layer broadband ring-patch antenna array, IEICE Electron. Express 9 (4) (2012) 250–255.
[12] S.K. Bavandpour, et al., A compact lowpass-dual bandpass diplexer with high output ports isolation, AEU-Int. J. Electron. Commun. 135 (2021) 153748.
[13] K. Dehghani, et al., Design of lowpass filter using novel stepped impedance resonator, Electron. Lett. 50 (1) (2014) 37–39.
[14] A. Lalbakhsh, et al., A design of a dual-band bandpass filter based on modal analysis for modern communication systems, Electronics 9 (11) (2020) 1770.
[15] A. Lalbakhsh, et al., A compact lowpass filter for satellite communication systems based on transfer function analysis, AEU-Int. J. Electron. Commun. 124 (2020) 153318.
[16] M.M. Honari, et al., Two-layered substrate integrated waveguide filter for UWB applications, IEEE Microw. Wireless Compon. Lett. 27 (7) (2017) 633–635.
[17] A. Iqbal, et al., Miniaturization trends in substrate integrated waveguide (SIW) filters: a review, IEEE Access 8 (2020) 223287–223305.
[18] A. Iqbal, et al., Multimode HMSIW-based bandpass filter with improved selectivity for fifth-generation (5G) RF front-ends, Sensors 20 (24) (2020) 7320.
[19] S. Lotfi, et al., Wilkinson power divider with band-pass filtering response and harmonics suppression using open and short stubs, Frequenz 74 (5–6) (2020) 169–176.
[20] S. Roshani, S. Roshani, A. Zarinitabar, A modified Wilkinson power divider with ultra harmonic suppression using open stubs and lowpass filters, Analog Integr. Circuits Signal Process. 98 (2) (2019) 395–399.
[21] M.M. Honari, et al., Class of miniaturised/arbitrary power division ratio couplers with improved design flexibility, IET Microw. Antennas Propag. 9 (10) (2015) 1066–1073.
[22] M.M. Honari, et al., Theoretical design of broadband multisection Wilkinson power dividers with arbitrary power split ratio, IEEE Trans. Compon. Packag. Manuf. Technol. 6 (4) (2016) 605–612.
[23] R. Mirzavand, et al., Compact microstrip Wilkinson power dividers with harmonic suppression and arbitrary power division ratios, IEEE Trans. Microw. Theor. Tech. 61 (1) (2012) 61–68.
[24] M. Hookari, S. Roshani, S. Roshani, High-efficiency balanced power amplifier using miniaturized harmonics suppressed coupler, Int. J. RF Microw. Computer-Aided Eng. 30 (8) (2020) e22252.
[25] S. Roshani, S. Roshani, Design of a high efficiency class-F power amplifier with large signal and small signal measurements, Measurement 149 (2020) 106991.
[26] H. Heshmati, S. Roshani, A miniaturized lowpass bandpass diplexer with high isolation, AEU-Int. J. Electron. Commun. 87 (2018) 87–94.
[27] J.-T. Huang, J.-H. Shiao, J.-M. Wu, A miniaturized Hilbert inverted-F antenna for wireless sensor network applications, IEEE Trans. Antenn. Propag. 58 (9) (2010) 3100–3103.
[28] J. Anguera, et al., The fractal Hilbert monopole: a two-dimensional wire, Microw. Opt. Technol. Lett. 36 (2) (2003) 102–104.
[29] N. Bayatmaku, et al., Design of simple multiband patch antenna for mobile communication applications using new E-shape fractal, IEEE Antenn. Wireless Propag. Lett. 10 (2011) 873–875.
[30] B. Mohamadzade, et al., Mutual coupling reduction in microstrip array antenna by employing cut side patches and EBG structures, Prog. Electromag. Res. M 89 (2020) 179–187.

[31] G. Fang, et al., Medical body area networks: opportunities, challenges and practices, in: 2011 11th International Symposium on Communications & Information Technologies (ISCIT), IEEE, 2011.
[32] B. Mohamadzade, et al., Recent developments and state of the art in flexible and conformal reconfigurable antennas, Electronics 9 (9) (2020) 1375.
[33] G.A. Conway, W.G. Scanlon, Antennas for over-body-surface communication at 2.45 GHz, IEEE Trans. Antenn. Propag. 57 (4) (2009) 844–855.
[34] M. Klemm, et al., Novel small-size directional antenna for UWB WBAN/WPAN applications, IEEE Trans. Antenn. Propag. 53 (12) (2005) 3884–3896.
[35] S. Adibi, M.A. Honarvar, A. Lalbakhsh, Gain enhancement of wideband circularly polarized UWB antenna using FSS, Radio Sci. 56 (1) (2021) e2020RS007098.
[36] P. Das, K. Mandal, A. Lalbakhsh, Single-layer polarization-insensitive frequency selective surface for beam reconfigurability of monopole antennas, J. Electromagn. Waves Appl. 34 (1) (2020) 86–102.
[37] N. Haga, et al., Characteristics of cavity slot antenna for body-area networks, IEEE Trans. Antenn. Propag. 57 (4) (2009) 837–843.
[38] M.U. Afzal, K.P. Esselle, A. Lalbakhsh, A methodology to design a low-profile composite-dielectric phase-correcting structure, IEEE Antenn. Wireless Propag. Lett. 17 (7) (2018) 1223–1227.
[39] T. Hayat, et al., Additively manufactured perforated superstrate to improve directive radiation characteristics of electromagnetic source, IEEE Access 7 (2019) 153445–153452.
[40] A. Lalbakhsh, et al., A high-gain wideband ebg resonator antenna for 60 GHz unlicenced frequency band, in: 12th European Conference on Antennas and Propagation (EuCAP 2018), IET, 2018.
[41] T.-W. Koo, et al., Extremely low-profile antenna for attachable bio-sensors, IEEE Trans. Antenn. Propag. 63 (4) (2015) 1537–1545.
[42] B. Mohamadzade, et al., A conformal ultrawideband antenna with monopole-like radiation patterns, IEEE Trans. Antenn. Propag. 68 (8) (2020) 6383–6388.
[43] S.R. Bazaz, et al., A hybrid micromixer with planar mixing units, RSC Adv. 8 (58) (2018) 33103–33120.
[44] R. Hagihghi, et al., A miniaturized piezoresistive flow sensor for real-time monitoring of intravenous infusion, J. Biomed. Mater. Res. B Appl. Biomater. 108 (2) (2020) 568–576.
[45] S.A. Moshizi, et al., Development of an ultra-sensitive and flexible piezoresistive flow sensor using vertical graphene nanosheets, Nano Micro Lett. 12 (2020).
[46] S. Razavi Bazaz, et al., Rapid softlithography using 3D-printed molds, Adv. Meter. Technol. 4 (10) (2019) 1900425.
[47] A. Razmjou, et al., Preparation of iridescent 2D photonic crystals by using a mussel-inspired spatial patterning of ZIF-8 with potential applications in optical switch and chemical sensor, ACS Appl. Mater. Interface 9 (43) (2017) 38076–38080.
[48] S. Azadi, et al., Biocompatible and highly stretchable PVA/AgNWs hydrogel strain sensors for human motion detection, Adv. Mater. Technol. 5 (11) (2020) 2000426.
[49] Z. Changani, et al., Domino P-μMB: a new approach for the sequential immobilization of enzymes using polydopamine/polyethyleneimine chemistry and microfabrication, Adv. Mater. Interfaces 7 (13) (2020) 1901864.
[50] A. Kottapalli, et al., Smart skin of self-powered hair cell flow sensors for sensing hydrodynamic flow phenomena, in: 2015 Transducers-2015 18th International Conference on Solid-State Sensors, Actuators and Microsystems (TRANSDUCERS), IEEE, 2015.

[51] M.A. Raoufi, et al., Development of a biomimetic semicircular canal with MEMS sensors to restore balance, IEEE Sens. J. 19 (23) (2019) 11675–11686.
[52] R.F. Harrington, Effect of antenna size on gain, bandwidth, and efficiency, J. Res. Nat. Bureau Stand. D. Radio Propag. 64 (1) (1960).
[53] A. Christ, et al., The dependence of electromagnetic far-field absorption on body tissue composition in the frequency range from 300 MHz to 6 GHz, IEEE Trans. Micro. Theory Tech. 54 (5) (2006) 2188–2195.
[54] M.M. Honari, et al., A two-port microstrip sensor antenna for permittivity and loss tangent measurements, in: 2019 13th European Conference on Antennas and Propagation (EuCAP), IEEE, 2019.
[55] M. Manoufali, et al., Implantable sensor for detecting changes in the loss tangent of cerebrospinal fluid, IEEE Trans. Biomed. Circuits Syst. 14 (3) (2020) 452–462.
[56] M. Manteghi, A.A.Y. Ibraheem, On the study of the near-fields of electric and magnetic small antennas in lossy media, IEEE Trans. Antenn. Propag. 62 (12) (2014) 6491–6495.
[57] F. Ejeian, et al., Design and applications of MEMS flow sensors: a review, Sens. Actuat. Phys. 295 (2019) 483–502.
[58] T. La Vaque, The history of EEG hans berger: psychophysiologist. A historical vignette, J. Neurother. 3 (2) (1999) 1–9.
[59] M.A. Parvez Mahmud, N. H., S.H. Farjana, M. Asadnia, C. Lang, %J advanced energy materials, Rec. Adv. Nanogen. Driven Self Power. Implant. Biomed. Dev. 8 (2) (2017).
[60] S. Patel, et al., Author archives: madanuuk, J. NeuroEng. Rehabil. 9 (2012) 21.
[61] J.G. Webster, Medical Instrumentation: Application and Design, John Wiley & Sons, 2009.
[62] D.-H. Kim, et al., Epidermal electronics, Science 333 (6044) (2011) 838–843.
[63] X. Huang, et al., Epidermal radio frequency electronics for wireless power transfer, Microsyst. Nanoeng. 2 (1) (2016) 1–9.
[64] H.U. Chung, et al., Binodal, wireless epidermal electronic systems with in-sensor analytics for neonatal intensive care, Science 363 (6430) (2019).
[65] M.L. Hammock, et al., 25th anniversary article: the evolution of electronic skin (e-skin): a brief history, design considerations, and recent progress, Adv. Mater. 25 (42) (2013) 5997–6038.
[66] J.A. Rogers, Electronics for the human body, Jama 313 (6) (2015) 561–562.
[67] J.C. Yang, et al., Electronic skin: recent progress and future prospects for skin-attachable devices for health monitoring, robotics, and prosthetics, Adv. Mater. 31 (48) (2019) 1904765.
[68] W.H. Yeo, et al., Multifunctional epidermal electronics printed directly onto the skin, Adv. Mater. 25 (20) (2013) 2773–2778.
[69] G. León, et al., Wideband epidermal antenna for medical radiometry, Sensors 20 (7) (2020) 1987.
[70] H.A. Damis, et al., Investigation of epidermal loop antennas for biotelemetry IoT applications, IEEE Access 6 (2018) 15806–15815.
[71] S. Amendola, G. Marrocco, Optimal performance of epidermal antennas for UHF radio frequency identification and sensing, IEEE Trans. Antenn. Propag. 65 (2) (2016) 473–481.
[72] J. Peters, G.S.M. Hendriks, Estimation of the electrical conductivity of human tissue, Electromagnetics 21 (7–8) (2001) 545–557.
[73] S. Gabriel, R. Lau, C. Gabriel, The dielectric properties of biological tissues: II. Measurements in the frequency range 10 Hz to 20 GHz, Phys. Med. Biol. 41 (11) (1996) 2251.

[74] A. Lalbakhsh, K. Esselle, S. Smith, Design of a single-slab low-profile frequency selective surface, in: 2017 Progress in Electromagnetics Research Symposium-Fall (PIERS-FALL), IEEE, 2017.
[75] H.R. Raad, et al., Flexible and compact AMC based antenna for telemedicine applications, IEEE Trans. Antenn. Propag. 61 (2) (2012) 524–531.
[76] R.B. Simorangkir, A. Kiourti, K.P. Esselle, UWB wearable antenna with a full ground plane based on PDMS-embedded conductive fabric, IEEE Antenn. Wireless Propag. Lett. 17 (3) (2018) 493–496.
[77] S. Zhu, R. Langley, Dual-band wearable textile antenna on an EBG substrate, IEEE Trans. Antenn. Propag. 57 (4) (2009) 926–935.
[78] 1, I.C. IEEE Standard for Safety Levels with Respect to Human Exposure to Electric, Magnetic, and Electromagnetic Fields, 0 Hz to 300 GHz, The Institute of Electrical and Electronics Engineers, New York, NY, 2019.
[79] Protection, I.C.o.N.-I.R., Guidelines for limiting exposure to electromagnetic fields (100 kHz to 300 GHz), Health Phys. 118 (5) (2020) 483–524.
[80] S.M. Mikki, A.A. Kishk, Particle swarm optimization: a physics-based approach, Synth. Lect. Comput. Electromag. 3 (1) (2008) 1–103.
[81] M. Derigo, T. Stutzle, Ant Colony Optimization, the mit press, 2004.
[82] P. Lalbakhsh, B. Zaeri, A. Lalbakhsh, An improved model of ant colony optimization using a novel pheromone update strategy, IEICE Trans. Info Syst. 96 (11) (2013) 2309–2318.
[83] P. Lalbakhsh, et al., AntNet with reward-penalty reinforcement learning, in: 2010 2nd International Conference on Computational Intelligence, Communication Systems and Networks, IEEE, 2010.
[84] D.Z. Zhu, P.L. Werner, D.H. Werner, Multi-objective lazy ant colony optimization for frequency selective surface design, in: 2018 IEEE International Symposium on Antennas and Propagation & USNC/URSI National Radio Science Meeting, IEEE, 2018.
[85] S. Karimkashi, A.A. Kishk, Invasive weed optimization and its features in electromagnetics, IEEE Trans. Antenn. Propag. 58 (4) (2010) 1269–1278.
[86] J. Kennedy, R.C. Eberhart, A discrete binary version of the particle swarm algorithm, in: 1997 IEEE International Conference on Systems, Man, and Cybernetics. Computational Cybernetics and Simulation, IEEE, 1997.
[87] N. Jin, Y. Rahmat-Samii, Advances in particle swarm optimization for antenna designs: real-number, binary, single-objective and multiobjective implementations, IEEE Trans. Antenn. Propag. 55 (3) (2007) 556–567.
[88] A. Lalbakhsh, K.P. Esselle, Directivity improvement of a Fabry-Perot cavity antenna by enhancing near field characteristic, in: 2016 17th International Symposium on Antenna Technology and Applied Electromagnetics (ANTEM), IEEE, 2016.
[89] P.J. Bevelacqua, C.A. Balanis, Minimum sidelobe levels for linear arrays, IEEE Trans. Antenn. Propag. 55 (12) (2007) 3442–3449.
[90] M.M. Khodier, C.G. Christodoulou, Linear array geometry synthesis with minimum sidelobe level and null control using particle swarm optimization, IEEE Trans. Antenn. Propag. 53 (8) (2005) 2674–2679.
[91] S.U. Rahman, et al., Analysis of linear antenna array for minimum side lobe level, half power beamwidth, and nulls control using PSO, J. Micro. Optoelectron. Electromag. Appl. 16 (2017) 577–591.
[92] S.S. Borah, A. Deb, J. Roy, Design of thinned linear antenna array using particle swarm optimization (PSO) algorithm, Asian J. Converg. Technol 5 (2019).

[93] A. Lalbakhsh, M.U. Afzal, K.P. Esselle, Multiobjective particle swarm optimization to design a time-delay equalizer metasurface for an electromagnetic band-gap resonator antenna, IEEE Antenn. Wireless Propag. Lett. 16 (2016) 912–915.

[94] S.M. Mikki, A.A. Kishk, Quantum particle swarm optimization for electromagnetics, IEEE Trans. Antenn. Propag. 54 (10) (2006) 2764–2775.

[95] S.M. Mikki, A.A. Kishk, Hybrid periodic boundary condition for particle swarm optimization, IEEE Trans. Antenn. Propag. 55 (11) (2007) 3251–3256.

[96] A. Lalbakhsh, et al., A fast design procedure for quadrature reflection phase, in: 2017 Progress in Electromagnetics Research Symposium-Fall (PIERS-FALL), IEEE, 2017.

[97] X. Fan, Y. Tian, Y. Zhao, Optimal design of multiband microstrip antennas by self-renewing fitness estimation of particle swarm optimization algorithm, Int. J. Antenn. Propag. 2019 (2019).

[98] A. Lalbakhsh, et al., Wideband near-field correction of a Fabry–Perot resonator antenna, IEEE Trans. Antenn. Propag. 67 (3) (2019) 1975–1980.

[99] S.H. Zainud-Deen, H.A.E.-A. Malhat, E.A.A. El-Refaay, InSb based microstrip patch antenna temperature sensor for terahertz applications, Wireless Pers. Commun. 115 (1) (2020) 893–908.

[100] H. Pradhan, B.B. Mangaraj, S.K. Behera, Beamforming of linear antenna array using modified PSO algorithm, in: 2018 International Conference on Applied Electromagnetics, Signal Processing and Communication (AESPC), IEEE, 2018.

[101] X.H. Wu, A. Kishk, A. Glisson, Antenna modeling by frequency dependent hertzian dipoles using particle swarm optimization, in: 2006 IEEE Antennas and Propagation Society International Symposium, IEEE, 2006.

[102] H.J. Delgado, M.H. Thursby, A novel neural network combined with FDTD for the synthesis of a printed dipole antenna, IEEE Trans. Antenn. Propag. 53 (7) (2005) 2231–2236.

[103] X. Ding, et al., Neural-network approaches to electromagnetic-based modeling of passive components and their applications to high-frequency and high-speed nonlinear circuit optimization, IEEE Trans. Microw. Theor. Tech. 52 (1) (2004) 436–449.

[104] F. Feng, Q.-J. Zhang, Parametric modeling of millimeter-wave passive components using combined neural networks and transfer functions, in: Global Symposium on Millimeter-Waves (GSMM), IEEE, 2015.

[105] M.B. Jamshidi, et al., An ANFIS approach to modeling a small satellite power source of NASA, in: 2019 IEEE 16th International Conference on Networking, Sensing and Control (ICNSC), IEEE, 2019.

[106] M.B. Jamshidi, et al., A novel neural-based approach for design of microstrip filters, AEU-Int. J. Electron. Commun. 110 (2019) 152847.

[107] Y. Kim, et al., Application of artificial neural networks to broadband antenna design based on a parametric frequency model, IEEE Trans. Antenn. Propag. 55 (3) (2007) 669–674.

[108] K.-C. Lee, Application of neural network and its extension of derivative to scattering from a nonlinearly loaded antenna, IEEE Trans. Antenn. Propag. 55 (3) (2007) 990–993.

[109] S. Roshani, et al., Design and modeling of a compact power divider with squared resonators using artificial intelligence, Wireless Pers. Commun. 117 (3) (2021) 2085–2096.

Chapter 3

Intelligent geo-sensing for moving toward smart, resilient, low emission, and less carbon transport

Omid Ghaffarpasand[1], Ahmad Miri Jahromi[2], Reza Maleki[3], Elika Karbassiyazdi[4], Rhiannon Blake[1]

[1]*School of Geography, Earth, and Environmental Sciences, University of Birmingham, Birmingham, United Kingdom;* [2]*Department of Petroleum Engineering, Amirkabir University of Technology (Tehran Polytechnic), Tehran, Iran;* [3]*Department of Chemical Engineering, Shiraz University, Shiraz, Iran;* [4]*Centre for Technology in Water and Wastewater, University of Technology Sydney, Sydney, NSW, Australia*

1. Introduction

Transport is the backbone of human lives and its history is strongly correlated with the history of civilization. More than 4000 years ago, in the first ancient civilizations, humans utilized cattle to carry goods. In Mesopotamia at that time, rowing was standard in rivers and canals. The invention of the wheel and the discovery of fire revolutionized human life and civilizations in that period. Historical findings evidence the existence of transportation networks and their expansion at different times. For example, in different parts of the vast territory of the Persian Empire (current Iran and neighboring countries) during the Achaemenid period, roads and printing presses were created. The vast and long cobblestone roads in the Roman Empire linked all parts of the center to the capital, the Silk Road, which is connected East Asia to Western Europe.

Furthermore, the advances in navigation and the invention of new vehicles led to the exploration of new areas and caused many changes in those parts of the world. However, before the Industrial Revolution (1870–1800), workforce and livestock were the dominant part of passenger and cargo transport. The invention of the steam engine revolutionized both water and rail transportation. In the early nineteenth century, the first steam locomotives were built in England, whereby rail transportation made it easier to move cargo and carry heavy loads to relatively distant places. Later, with the discovery of oil and its

Artificial Intelligence and Data Science in Environmental Sensing
https://doi.org/10.1016/B978-0-323-90508-4.00011-3

use instead of coal, transportation developments accelerated. In the early twentieth century, the invention of automobiles and mass production changed the face of urban and suburban environments. Main and side roads, junctions, and roundabouts were built in cities, and wide street networks between cities and villages followed by many entertainment/administrative/services places and centers such as gas stations, malls, hotels, etc., were developed. During World War II (1945–39), the first jet aircraft was built in Germany. The jet engine was more powerful and led to the construction of larger passenger aircraft. These would significantly develop aviation transport.

After centuries of advancements, transportation is now one of the major sectors of the new world. Transport is now the primary cause of economic and social dynamicity within human societies. The growth rate of transportation could be introduced as the main proxy of growth patterns and economic activity through any area of the world. Sustainable development is strongly correlated with sustainable transport, for which either has been seeking to establish a reasonable balance between environmental, social, and economic qualities (present and future) [1,2]. Transport has been even recognized as the foundation of sustainable development due to its importance in the economy, industry, politics, and even military purposes; many specialists in the field of sustainable development believe that more efficient transportation is an inclusive development [3–5].

Despite all the economic and social advantages of transport reviewed above, it is one of the leading sources of environmental degradation for the past few decades. For example, along with global economic growth, transport demand and fuel consumption have led to exhaustive emissions instantaneously increasing [6–8]; the transportation sector is responsible for around a quarter of global greenhouse gas emissions. A wide body of research has evidenced the severe environmental drawbacks such as air pollution as well as negative social and economic consequences directly caused by inefficient transportation systems [9].

Due to the dominant role of transportation in the economy, environment, and society, several strategies have been designed and developed to mitigate its environmental drawbacks [10,11]. Those strategies have usually been designed in two approaches of plural and singular mitigations. While the first approach targets the mass of transport and population of vehicles, the latter one is going to provide some instructions for individual drivers to reduce the emission of their vehicles. The implementation of the low emission or clean air zones (LEZs of CAZs) within the urban environments could be considered as examples of plural mitigation strategies. Despite establishing LEZs/CAZs across the urban environments for over a decade, there are many debates around their real efficacy, see for example Ref. [12].

Due to the rapid advancements in positioning, computing, and processing technologies for the past decade, smart technologies/services could introduce themselves as wise solution for mitigating transport emissions within urban environments. Intelligent transportation systems (ITSs) are one of the smart platforms which can considerably assist urban and traffic planners to keep a transport system in a sustainable status. ITS is a general term for the combined application of communication technologies, control, and information processing for transportation systems. Employing ITSs through the urban environments would not just keep the transport system active and alive, but also could save lives, time, money, energy, and environmental benefits. The overall task of ITS is to improve decision-making for traffic planners and other transport users and thus boost not only travel capacity but also the quality of transport services [13–15]. ITS is usually based on three factors: information, communication, and synchronization, which assists users and travelers to make more suitable decisions. The establishment of ITS systems or improving the quality of the existing systems could be itemized through the plural approaches for optimizing transport characteristics.

Eco-driving and eco-routing might be considered as singular mitigation strategies that could reduce the emissions of passenger cars by providing some driving advice to individual drivers [16]. For example, it has been argued that driving at 70 mph has 9% more fuel consumption than driving at 60 mph and up to 15% more fuel consumption than driving at 50 mph. Those services have been constructed on the geo-sensing systems which could determine the location of the moving vehicles in geographical systems. The advancement in geo-sensing systems will be reviewed in the next sections. Geo-sensing technologies have developed not just routing and navigation services but could also boost the quality of existing technologies such as ITSs and introduce new state-of-the-art technologies such as the Internet of Things (IoT). In addition, new smart algorithms and methods such as artificial intelligence (AI) and neural networks could utilize this geodata to address the traditional urban challenges such as congestion and air pollution.

According to the background mentioned above, this script has been designed to review the role of geo-sensing and new smart technologies in the transportation environment. The chapter is organized as such: the contributions of transport to the economy and environment will be discussed in the next section. The history of geo-sensing and the existing geo-sensing technologies and services will be reviewed in Section 3. The role of geo-sensing in green and economic transport will be argued in Section 4. Geographic information system (GIS) and related assets will be introduced in Section 5. New services, technologies, and systems which utilize intelligence geo-sensing will be reviewed in Section 6. The main messages of the chapter will be outlined in Section 7.

2. The role of transport in the economy and environment

Throughout history, one of the main influential factors in countries' economic growth has been transportation process [17]. Transport is determined in three major sectors of infrastructure, operation, and vehicles. Roads, railways, air, and water are part of the infrastructure, while the traffic assets, railway switches, and air traffic control that control the system are part of the operational sector. The vehicles are also disaggregated into public and private vehicles. After the Industrial Revolution and the increase in production and expansion of various industries, access to global markets, especially for developed countries, became the necessary goal. Due to this, the importance of all transportation sectors increased, and with the advancement of science, the emergence of new technologies, and the expansion of road construction, achieving that aim became possible for many countries. As time went on and international trade increased, the two factors of transportation and the economy became intertwined, as the growth of the transportation industry has led to economic growth in many countries. Transportation could build a strong link between agricultural, industrial, commercial, and service activities at the national and international scopes. The production, movement, and distribution of goods depend on the transportation and freight system carried out by land, air, and sea. Currently, the importance of the transportation industry can be seen in various ways, such as reducing travel costs, facilitating the transportation of goods, and increasing the speed of movement and safety, access to global trade, and economic development of countries. Transportation also plays a vital role in income distribution and reducing economic and social inequalities, reducing the effects of poverty and income disparities between villagers and urban dwellers.

Two senior roles are usually determined for the transportation in the economy [18,19]. The first role is the direct impact that transport activity has on economic growth and job creation. The transportation sector as a set of economic activities by providing rail, air, and sea transportation services, like other economic activities, creates extra value and plays a role in economic growth and job creation.

The second function of the transportation sector in the economy is to provide one of the critical economic infrastructures for the economic growth and development of other sectors of the economy. The transportation sector is one of the country's senior and most critical economic infrastructures, which will become an obstacle to economic growth and development if not expanded following the macroeconomic targets. Transport might also have short-term impacts on living costs directly and indirectly through transportation costs of each household and the cost of other goods and services, respectively. Transport has medium-term impacts on living costs that could influence the consumption of alternative services such as communications. The long-term effects of transportation are related to the change in the economic calculation principles of production and development projects. In general, the combination of production activities in each region and, consequently, the composition of the production context, employment composition, production volume, per capita income, and other economic variables can be affected by changes in the price of transportation services.

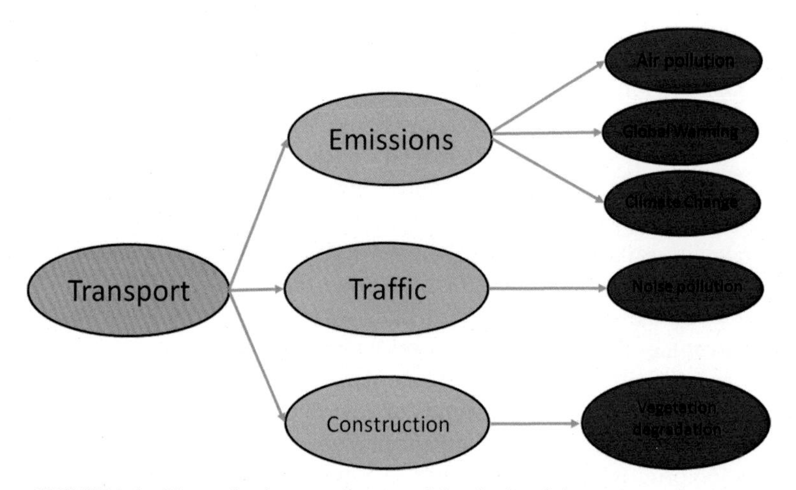

FIGURE 3.1 The predominant environmental drawbacks of the transportation sector.

Despite such outstanding impacts on the economy and social welfare, transport has wide environmental drawbacks which are summarized in Fig. 3.1 and evidenced by many previous investigators such as Refs. [20–22]. Air pollution and subsequent climate change are major drawbacks of transport. According to the World Health Organization (WHO), air pollution is the major cause of around 7 million global deaths per annum [23]. Several diseases and cancers such as stroke, heart diseases, lung cancers, acute and chronic respiratory disease, etc., which are responsible for over 4.2 million death per year, are directly linked to human exposure to ambient air pollution. The combustion of petroleum-derived fuels in the engines of moving vehicles is a leading source of atmospheric air pollutants at low altitudes, for which transport is one of the major fuel and energy consumers of the world. According to the World Economic Forum (WEF) statistics, the annual consumption of energy by the transport sector, measured in 2010, was the equivalent of 2200 million tons of oil [24]. Considering that the total world energy consumption in 2010 was equivalent to about 13,196 million tons of oil, it can be concluded that about 17% of world energy consumption is consumed by the transportation sector. Road transport accounts for the bulk of energy consumption, with passenger cars accounting for about 52% of total consumption and buses and HGVs together accounting for 21%. The share of each air and sea sector is about 10%, and the rail transport sector is about 3% [24]. Undoubtedly, by 2050, the transportation sector will face many considerable challenges, the primary of which is to provide sustainable transportation to over 9 billion people at the lowest possible social cost. Government policy will play a vital role in exploring the future direction. However, by 2050, fuel demand in all modes of transport will enhance by 30%–82% compared to 2010 levels. The bulk of this growth has been in HGVs, buses, trains, ships, and airplanes; therefore, demand for these vehicles is expected to increase by 64%–200% [24].

In addition, the transportation sector will continue to be highly dependent on gasoline, diesel, and jet fuel, with demand for these fuels increasing by 10%–68%. In 2010, CO_2 emissions from the transportation sector were approximately 23% of global CO_2 emissions. Around 41% of CO_2 emissions in the transport sector also belonged to the road transport sector. As transportation demand increases, depending on the fuel composition, CO_2 emissions from the transportation sector are expected to increase by 16%–79%, which is highly dependent on the degree of government intervention and success in developing new, less-polluting fuel systems [24].

Increasing the concentration of greenhouse gases in the atmosphere causes global warming and so climate change, which have been deemed as serious threats to humanity. In recent years, the reduction of carbon and greenhouse gas emissions has been on the agenda of environmental problems; the carbon footprint has been considered a significant and severe problem globally [25]. Carbon footprint is one indicator that determines the emission of greenhouse gases, which is directly and indirectly generated by anthropogenic activity and is expressed in terms of carbon dioxide [26]. The term has been used in the public, private, and media sectors. The dangerous consequences of global warming and the importance of resolving this great crisis can be seen throughout the media and governmental and nongovernmental news [27–29]. Greenhouse gases are evaluated based on the type of gas and the parts that cause the emission. The highest greenhouse gas emissions (76%) are from carbon dioxide; 65% of it is due to the combustion of fossil fuels and industrial processes, and 11% is related to the forestry sector. Methane gas accounts for 16% of emissions, nitrous oxide gas 6% of emissions, and fluorescent gases 2% of emissions [30–32].

According to the Environmental Protection Agency (EPA) statistics, the United States' highest amount of greenhouse gas emissions is related to transportation by 29% [33]. After that, electricity produces 28% of pollutants, industry 22%, commercial and residential 12%, and agriculture 9% (Fig. 3.2), see Ref. [33]. Because transportation plays a major role in emitting greenhouse gases in nature, transportation management is one of the most effective ways to prevent greenhouse gas emissions.

3. Geo-sensing; evolution in the geography

The concepts of "location," "geography," and the basic question of "where am I?" have been formed in the minds of humans thousands of years ago in the Stone Age. Geography is developed and advanced according to the determination of accuracy of the location. Humans firstly tried to determine their locations according to nature coincidences which were periodically repeated. For example, they found that the sun rises and sets every day in the same directions, so they called east and west directions where the sun rises and sets, respectively. However, they had the problem of finding directions at night.

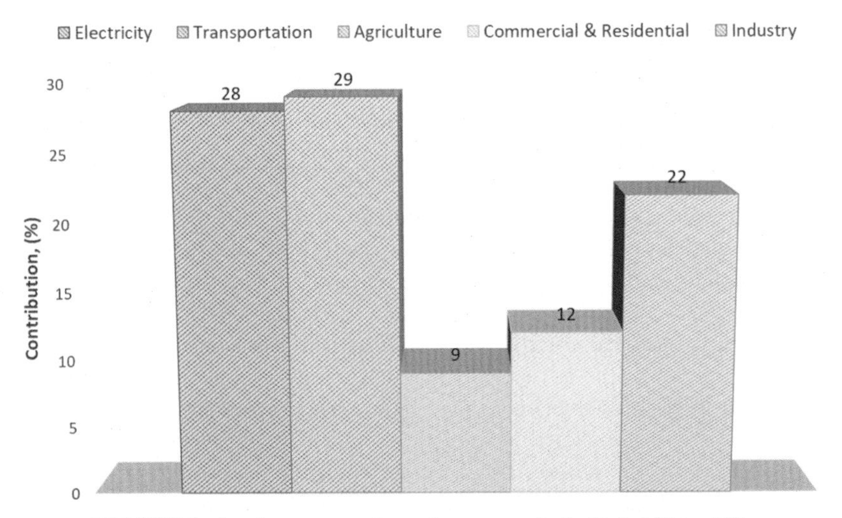

FIGURE 3.2 Leading sources of greenhouse gases in the United States [33].

After the Sun, they were so familiar with a second celestial mass, the moon. Hence, they invented a night orientation using the shape of the moon. At the beginning of the lunar month, the moon appears in the form of a crescent, and in the middle of the lunar month, it turns into a complete disk, and then it becomes a crescent in the opposite direction. In the first half of the lunar month, the outer part of the moon looks like an arrow pointing west. In the second half of the lunar month, the moon is convex to the east [34–36]. With the advancements in astronomy, humans found that the sky and its celestial mass could be an alternative for finding their location and even navigation. However, it had significant failures; they were almost blind in cloudy skies or poor weather. Electromagnetics could provide a revolution in geography. Physicists found that our planet has a permanent magnetic field that could be employed to design new positioning and navigation systems. They invented the compass; accordingly, the compass needle is affected by the earth's magnetic field and indicates the magnetic north/south directions. The compass hand is compassionate, making an error if it is placed near iron or steel objects. It is also sensitive to magnetic fields and can cause gross uncertainty.

The compass is a simple device for displaying the average geographical direction invented in ancient times, but like other ancient inventions, it took a long time for everyone (especially sailors) to realize its value [37]. The first compasses were designed on a square device marked for significant points and constellations. The pointing needle was a spoon-shaped device with a handle pointing south. Later, magnetized needles were used as a direction indicator instead of spoon-shaped limestone. Compasses are used to detect four geographical directions and thus are used in navigation or mapping.

The concept of navigation had been more developed with the help of compasses. Navigation has been defined as finding the way from one place (origin) to another place (destination) at any time. Navigation using the compasses brought a wide revolution inland travel and cruise explorations. For example, in the 11th century, the use of compasses for navigation purposes in cruises became common. Magnetic needle compasses can be used when wet (in water), dry (on a sharp shaft), or suspended (on silk thread), making them a unique tool. They were used mainly by seafarers such as merchants traveling to the Middle East and early navigators seeking to find the magnetic north pole.

Although advanced compasses are still used in ships, airplanes, etc., as a tool for navigation, the advancement in technology introduces new ways for navigation in front of humans in the modern ages. Space exploration proves that space and space-related technologies could be considered promising assets for sustainable, reliable, and accurate navigation. Satellites are a handy tool for positioning on land, sea, and air. They are systems that allow the receiver to determine its position (latitude, longitude, and altitude) with an error of a few meters. This capability is done by transmitting radio waves between the device and the satellite.

During World War II, the Americans felt threatened by an unexpected attack on Pearl Harbor on December 7, 1941 [38]. They began designing a geographic positioning system (GPS) stemming from feeling fear of a sudden start to war and the loss of their colonies. During the Cold War and arms competition period, the ability to target the Soviet Union's missile facilities on Soviet soil and missile launchers, which could be done with great precision, gave the superiority of the American forces in the arms race. This was not challenging for land-based missile sites, but it would be problematic if such a missile strike were to be carried out from the sea or a deepsea submarine. Therefore, determining the exact position of its missile site relative to any point on the planet led to the creation of the GPS. GPS was designed and developed in the United States in 1973. Thus was born the most advanced navigation system globally, which quickly became widespread in determining geographical location and time (https://www.geotab.com/uk/blog/gps-satellites/).

In 1980, the US government declared the use of GPS internationally and universally unimpeded. It is owned by the United States and consists of 24 satellites at an altitude above 20,000 km from Earth. These satellites are continuously orbiting around the Earth in six orbits (each containing four satellites). It takes 12 h for each satellite to complete an orbit [39–41].

The basis of GPS is to send high-frequency radio signals continuously. The receiver needs to receive information simultaneously from at least three satellites to estimate the longitude and latitude of the target and to receive information from at least four satellites to find three-dimensional coordinates. By receiving information from satellites, the receiver calculates the speed, direction, distance traveled, distance remaining to the destination, sunrise and sunset direction, the direction of qibla, and much more helpful information. A GPS receiver, taking this information from three or more satellites, then processes it and shows positions anywhere, 24/7. The receiver compares the time the signal is sent by satellite with its received time. From the difference gap of these two times, the distance of the receiver from the satellite is determined. It repeats this operation with the data received from several other satellites and thus determines the receiver's exact location with a slight difference.

If the volume resulting from the position of the satellites and the ground receiver is larger, the resulting position is more accurate. Satellites and receivers need very accurate clocks to be able to do their job perfectly. The slightest error can increase the measurement by tens or even hundreds of miles. According to the remarkable success of GPS and in the hope of interrupting the US superiority in the positioning services, other countries developed geo-sensing programmes such as GLONASS (Russia), GALILEO (European Union), COMPASS (China), and IRNSS (India).

4. Geographic Information System as a revolution or/and an evolution

As was mentioned earlier, accurate and reliable positioning has become an ordinary and accessible service for not just humans but also any portable or stable objects of the earth for the past few decades. The access to the geographic position of any object caused to translation of several kinds of data into the "geodata." Although many methods and strategies have been introduced to play with geodata, the most promising way to access, compile, and analyze these data is to use the GIS. GIS was developed from 1960 to 1989. It was also commercialized from 1989 to 1999, and from 1999, the software was put into operation and is now widely used. GIS is a platform for storing, maintaining, managing, analyzing, processing, and exporting geographic information and geodata. GIS could well play with geodata which includes two parts of spatial information (indicating the location and shape of features) and descriptive attributes (indicating the features and characteristics of features) [42–44]. It has been considered a revolution in geographical and urban sciences. In the new GIS world, a real and physical space could be considered as several geospatial layers. Each layer has certain attributes, and the spatial (and even temporal) relationship of different layers could be analyzed and discussed. In the GIS space, any feature of the real world could have a certain geographic identity and several corresponding attributes. For example, a layer of trees within a city might include the location of the trees and corresponding information such as the age, pruning time, and even overall conditions.

Today, due to the advancement of GIS technology and computer systems, GIS-based services and technologies are widely used in geology, environmental studies, water resources management, agriculture, forestry, education, urban applications, trade, industry, organizations, etc. [45–51].

According to the wide applications of GIS-based services, different companies have been developing GIS software and platforms. The US company of Environmental Systems Research Institute (ESRI) is the most significant GIS software developer in the world. ESRI (www.esri.com) is an international manufacturer of GIS-related software and spatial data. Jack Dangermond and his wife Laura founded the ESRI in 1969 in California, USA, focusing on land use [52]. The company was originally a small research group whose goal was to organize and analyze geographic information to help planners and decision-makers make environmental decisions. In 1982, ESRI entered the commercial market with ARC INFO software, and in 1986, it introduced PC ARC INFO. Then in 1991, the company introduced ArcView software, which had a user-friendly and straightforward graphical interface. Finally, in 1994, the company launched its website.

In 1996, ESRI introduced the first software component called MapObjects, which was the first platform for developing personalized GIS software. In 1999, the company redesigned its software core with the introduction of ArcInfo 8. In 2004, the company released ArcGIS 8 desktop with development framework extensions and a server platform. The ArcGIS server provides a framework for developers to build applications and web services. In 2007, the company introduced ArcGIS Explorer to provide GIS usage for anyone. In 2012, the company introduced ArcGIS Online, a cloud-based mapping system for organizations. This system offers tools for indexing, visualizing, and sharing spatial information. In the same year, ArcGIS 10.1 was introduced, enabling users to provide GIS resources as a web service and make geographic information available to more users. This company may be considered the most successful company in GIS, which competes with important companies such as Autodesk in the GIS market. The company's products (especially ArcGIS Desktop) account for 40.7% of the global GIS software market. More than one million users in more than 350,000 companies and organizations in 200 countries use the company's products. Among the users of the company's products are the names of federal agencies and organizations responsible for mapping in more than 50 US states. The company also has 10 offices in the United States and 80 offices internationally.

Almost all traffic models of the world are developed in the geo platforms. Meanwhile, it's an outstanding asset for traffic planning and road safety. However, the concept of GIS has been developed in a static world, whereby most of the available GIS datasets are constructed on offline data. The advancement in geo-sensing technologies and data science might be employed to introduce "dynamic GIS frameworks" when both spatial and *temporal* aspects of the real world could be assessed. Then it might be possible to even move from the concept of "geodata" to "geoSTdata" (geospatial and temporal data). GIS frameworks in urban scales have been usually struggling with a huge amount of data, for which many urban features have been linked to several different attributes. Hence, big and even mega datasets would be the further issue in the dynamic GIS platforms. In this regard, big data science and technology on one hand and new and efficient data processing and storing technologies, e.g., cloud computing, should be employed to address those technical challenges.

5. Geo-sensing for moving toward eco-routing and low-emission transport

Transportation deals with the movement of geographical space, and its purpose is to connect places and areas. Thus, transportation is essentially a geographical asset and deals with concepts such as place, location, distance, origin, and destination. Transport networks are affected by both the geographical environment and the geographical space and change it. In other words, transport is a geographic concept and so should be behaved as a geographic asset. In this section, some geographic-based solution which could somehow mitigate the environmental drawbacks of the transport is reviewed.

One of the best ways to reduce fuel and emissions is to optimize routing with the help of a car tracker. A GPS device is used to do this as the device can select the best routes for transit. This will reach the destination from the shortest route and thus reduce fuel consumption and vehicular emissions. High speeds can increase vehicle fuel consumption [53]. Considering this, the use of GPS can control the speed of vehicles, while the GPS device can monitor the speed of vehicles and prevent driving at high speeds [54,55]. This significantly reduces the use of fossil fuels and thus prevents greenhouse gas emissions. Properly maintained engines and equipment are much more efficient than equipment that is not maintained correctly. GPS can play an essential role inadequately maintaining engines and vehicles with vehicle maintenance reports and information about the exact time of engine repair or change of car oil, filters, and spark plugs according to the distance traveled [56]; ordinary repairs of such elements can reduce fuel consumption [57]. As a result, in addition to reducing vehicle fuel consumption and related costs, it increases the durability and efficiency of a vehicle, again reducing the emission of more greenhouse gases into the environment [58].

Geo-sensing technologies could also improve the driving styles of individual drivers to not only improve road safety but also reduce fuel consumption and on-road emissions. The concept of "eco-driving" has been developed in the transport society for the past decade; see Ref. [16]. It has been argued that eco-driving is considerably cost-effective than the other emission mitigation strategies such as fleet retrofit programs. Routing is another important part of driving, and the remarkable roles of that in on-road emission, as well as fuel consumption, have been evidenced by many previous investigators; see for example, Refs. [59,60]. The advancement in geo-sensing technologies could significantly develop both eco-driving and routing technologies, whereby they could be even considered as promising solutions for reducing heavy congestions and vehicle emissions.

6. Intelligent geo-sensing and AI as a new window to the future

ITS is one of the most effective traffic management strategies which has originated from information technology. It could create a new horizon for achieving dynamic and smooth mobility in the communication and information society, providing better services to citizens. The most important benefits of using ITS include increasing the efficiency of mobility and movement for goods and passengers, reducing traffic congestion, increasing the ability to manage transportation structures, increasing safety, reducing operating costs, and reducing environmental problems and fuel costs [61–63]. ITS refers to a set of tools, facilities, and specialties such as traffic engineering concepts, software, hardware, and telecommunication technologies used in a coordinated and integrated manner to improve the efficiency and safety of the transportation system. ITS plays a crucial role in planning land transportation systems today. The use and planning of ITS dates back to the early 1990s, when they began to study the design and development of these systems in developed countries.

ITS operates based on control and information technologies, which are the core of such systems' tasks and functions. From a general perspective, ITS can consist of three main components: intelligent roads, intelligent vehicles, and communication infrastructure. Intelligent geo-sensing technologies, which detect the geospatial position of the moving vehicles, play a vital role in all ITS components. In other words, the vehicles could be intelligent if they could continuously detect their geospatial locations. In further, the roads become intelligent if they could behave as intelligent geo-objects which detect the geospatial presence of the other intelligent moving objects. All new advancements in intelligent transport technologies, such as the IoT, signalized intersections, anonymous vehicles, etc., have been constructed on the communication of the different transportation components in a geospatial space. In IoT technologies, the intelligent vehicles as mobile objects could detect and even communicate with each other according to their geospatial position, and decided with that regard, while in the signalized intersections, the intelligent vehicle could communicate with the intelligent traffic assets such as traffic lights to optimize its journey. There are many scientific attempts which have been analyzing different aspects of those new technologies, see, for example, Refs. [64–66]. In today's modern cities, ITSs are welcomed and used in transportation, traffic, and urban management. ITS has also been recognized as the main requirement for moving toward resilient and smart cities.

The quality of ITS services could significantly be improved by AI algorithms. For example, sensors and cameras attached to roads and streets can collect a lot of information from traffic details. These data are sent to the center and then, using AI and bulk data analysis programs, the traffic pattern is analyzed and examined. By processing these data, valuable information such as traffic forecasts, accidents, shortest routes, and roadblocks can be provided. In this regard, using AI in the ITS and traffic management can reduce traffic volume, improve road safety, and reduce greenhouse gases. Over time, various criteria and subcriteria have been considered for each human activity for its optimal performance. These criteria for locating in GIS software have been applied in different ways in the past years. AI is an advanced tool that has recently been widely used in conjunction with GIS and ITS. The entry of AI into complex problem-solving has made it possible to combine GIS spatial capabilities and advanced AI features to enable people to achieve the best option to achieve their goals. This prevents a lot of vehicle traffic from finding the desired destination, reducing overall emissions.

The use of intelligent geo-sensing in urban mobility purposes could solve problems such as failure to calculate fares accurately based on distance traveled, time-consuming processes to monitor the performance of drivers and vehicles using manpower, influential factors in improving the satisfaction of the citizens, etc. Also, transportation management uses positioning systems to comprehensively control everything from electronic fare payment to location, taxi positioning, travel routing, and even the application of traffic restrictions, speed control, and fuel consumption management. The automatic vehicle location (AVL) is an ITS used to obtain instantaneous information about the location of the vehicle in the management of the general freight and passenger transport fleet. AVL is a fleet management system based on tracking using GPS and GPRS/SMS technologies. In this system, each vehicle is equipped with a tracker.

In the fleet management system, which is done with a GPS tracking device, the control of the moving subject is done in a control center. The needs of organizations and companies to manage location and track motion, track visitors on a map, and view the route accurately on urban, regional, national, and even global maps in detail force them to use these systems. With this advanced system, control of the vehicle is possible, and it is possible to apply correct management methods, careful planning, and security measures. Also, the use of this type of system is expected today in advanced societies and provides outstanding solutions for managers and officials of organizations and governments to reduce fuel consumption and greenhouse gases.

7. Conclusion

The transportation system is an integral part of the economy. In today's world, communities' benefit from the effects of transportation, and as cities grow and develop, the need for public services and facilities has increased. Since economic development depends on the evolution of transportation systems, transportation systems have developed in tangent with economic growth. This has created problems such as an increase in the number of vehicles in metropolitan areas and negative consequences for human health and the environment; therefore, public concern surrounding these impacts has risen accordingly. The advancement in computing and positioning technologies could open a new window for moving toward less carbon, low emission, and sustainable transport. Geo-sensing technologies could provide a reliable and continuous position of all mobile and stationary objects of our planet; new advancements in processing technologies could develop efficient management systems such as ITS which could manage the flow of thousands of moving vehicles; and the intelligent algorithms such as AI could boost the quality of decisions made. Eco-driving, eco-routing, IoT, anonymous vehicles, signalized intersections, and many other technologies and services are all born from the combination of geo-sensing and processing technologies.

References

[1] M.J. Bruton, Introduction to Transportation Planning, Routledge, 2021.

[2] H. Suzuki, R. Cervero, K. Iuchi, Transforming Cities with Transit: Transit and Land-Use Integration for Sustainable Urban Development, World Bank Publications, 2013.

[3] S. Guan, Q. Zhang, Estimating the contribution of intelligent transportation to urban economic growth, in: 2020 13th International Symposium on Computational Intelligence and Design (ISCID), IEEE, 2020.

[4] R. Mammadov, The importance of transportation in tourism sector, in: 7th Silk Road International Conference "Challenges and Opportunities of Sustainable Economic Development in Eurasian Countries", 2012.

[5] S. Sarjana, N. Khayati, L. Warini, P. Wiyati, The importance of transportation management in optimizing supply chain management at industrial estate, J. Transp. Multimoda 18 (1) (2020).

[6] K. Eyuboglu, U. Uzar, The impact of tourism on CO_2 emission in Turkey, Curr. Issues Tourism 23 (13) (2020) 1631–1645.

[7] M. Mohsin, Q. Abbas, J. Zhang, M. Ikram, N. Iqbal, Integrated effect of energy consumption, economic development, and population growth on CO_2 based environmental degradation: a case of transport sector, Environ. Sci. Pollut. Control Ser. 26 (32) (2019) 32824–32835.

[8] S. Wang, C. Li, H. Zhou, Impact of China's economic growth and energy consumption structure on atmospheric pollutants: based on a panel threshold model, J. Clean. Prod. 236 (2019) 117694.

[9] O. Ghaffarpasand, M.R. Talaie, H. Ahmadikia, A.T. Khozani, M.D. Shalamzari, A high-resolution spatial and temporal on-road vehicle emission inventory in an Iranian metropolitan area, Isfahan, based on detailed hourly traffic data, Atmos. Pollut. Res. 11 (9) (2020) 1598–1609.

[10] G. Duranton, Growing through cities in developing countries, World Bank Res. Obs. 30 (1) (2015) 39–73.

[11] J.B.H. Yap, I.N. Chow, K. Shavarebi, Criticality of construction industry problems in developing countries: analyzing Malaysian projects, J. Manag. Eng. 35 (5) (2019) 04019020.
[12] O. Ghaffarpasand, D.C.S. Beddows, K. Ropkins, F.D. Pope, Real-world assessment of vehicle air pollutant emissions subset by vehicle type, fuel and EURO class: new findings from the recent UK EDAR field campaigns, and implications for emissions restricted zones, Sci. Total Environ. 734 (2020) 139416.
[13] J. Guerrero-Ibáñez, S. Zeadally, J. Contreras-Castillo, Sensor technologies for intelligent transportation systems, Sensors 18 (4) (2018) 1212.
[14] A. Sładkowski, W. Pamuła, Intelligent Transportation Systems-Problems and Perspectives, Springer, 2016.
[15] F. Zhu, Z. Li, S. Chen, G. Xiong, Parallel transportation management and control system and its applications in building smart cities, IEEE Trans. Intell. Transport. Syst. 17 (6) (2016) 1576–1585.
[16] Y. Huang, E.C.Y. Ng, J.L. Zhou, N.C. Surawski, E.F.C. Chan, G. Hong, Eco-driving technology for sustainable road transport: a review, Renew. Sustain. Energy Rev. 93 (2018) 596–609.
[17] J. Hong, Z. Chu, Q. Wang, Transport infrastructure and regional economic growth: evidence from China, Transportation 38 (5) (2011) 737–752.
[18] B. Herrendorf, J. Schmitz, A. James, A. Teixeira, The role of transportation in US economic development: 1840–1860, Int. Econ. Rev. 53 (3) (2012) 693–716.
[19] F. Nistor, C.C. Popa, The Role of Transport in Economic Development, 2014.
[20] G. Huang, J. Zhang, J. Yu, X. Shi, Impact of transportation infrastructure on industrial pollution in Chinese cities: a spatial econometric analysis, Energy Econ. 92 (2020) 104973.
[21] T. Münzel, M. Sørensen, A. Daiber, Transportation noise pollution and cardiovascular disease, Nat. Rev. Cardiol. (2021) 1–18.
[22] A. Wang, C. Stogios, Y. Gai, J. Vaughan, G. Ozonder, S. Lee, I.D. Posen, E.J. Miller, M. Hatzopoulou, Automated, electric, or both? Investigating the effects of transportation and technology scenarios on metropolitan greenhouse gas emissions, Sustain. Cities Soc. 40 (2018) 524–533.
[23] WHO, Air Pollution, 2019. https://www.who.int/data/gho/data/themes/theme-details/GHO/air-pollution.
[24] D. Manzoor, R. Kohanhoosh Nejad, A comparative review of global transportation energy outlook, Iran. J. Energy 20 (2) (2017) 51–65.
[25] W. Tjan, R.R. Tan, D.C. Foo, A graphical representation of carbon footprint reduction for chemical processes, J. Clean. Prod. 18 (9) (2010) 848–856.
[26] T. Bektaş, G. Laporte, The pollution-routing problem, Transp. Res. Part B Methodol. 45 (8) (2011) 1232–1250.
[27] T. Stephenson, J.E. Valle, X. Riera-Palou, Modeling the relative GHG emissions of conventional and shale gas production, Environ. Sci. Technol. 45 (24) (2011) 10757–10764.
[28] T. Wiedmann, J. Minx, A definition of 'carbon footprint, Ecol. Econ. Res. Trends 1 (2008) 1–11.
[29] I. Williams, S. Kemp, J. Coello, D.A. Turner, L.A. Wright, A beginner's guide to carbon footprinting, Carbon Manag. 3 (1) (2012) 55–67.
[30] O. Edenhofer, Climate Change 2014: Mitigation of Climate Change, Cambridge University Press, 2015.

[31] R.K. Pachauri, M.R. Allen, V.R. Barros, J. Broome, W. Cramer, R. Christ, J.A. Church, L. Clarke, Q. Dahe, P. Dasgupta, Climate change 2014: synthesis report, in: Contribution of Working Groups I, II and III to the Fifth Assessment Report of the Intergovernmental Panel on Climate Change, IPCC, 2014.

[32] F. Tubiello, M. Salvatore, R. Cóndor Golec, A. Ferrara, S. Rossi, R. Biancalani, S. Federici, H. Jacobs, A. Flammini, Agriculture, Forestry and Other Land Use Emissions by Sources and Removals by Sinks, 2014. Rome, Italy.

[33] EPA, Sources of Greenhouse Gas Emissions, 2019.

[34] P. Chastenay, M. Riopel, Development and validation of the moon phases concept inventory for middle school, Phys. Rev. Phys. Educ. Res. 16 (2) (2020) 020107.

[35] K. Subramaniam, S. Padalkar, Visualisation and reasoning in explaining the phases of the moon, Int. J. Sci. Educ. 31 (3) (2009) 395–417.

[36] I. Testa, S. Galano, S. Leccia, E. Puddu, Development and validation of a learning progression for change of seasons, solar and lunar eclipses, and moon phases, Phys. Rev. ST Phys. Educ. Res. 11 (2) (2015) 020102.

[37] N. Dai, Compass. Thirty Great Inventions of China, Springer, 2020, pp. 663–683.

[38] J.J. Gudmens, Staff Ride Handbook for the Attack on Pearl Harbor, 7 December 1941: A Study of Defending America, DIANE Publishing, 2005.

[39] A.K. Brown, Receiver autonomous integrity monitoring using a 24-satellite GPS constellation, in: Institute of Navigation, Technical Meeting, 1987.

[40] J. Spilker, B.W. Parkinson, Overview of GPS operation and design, Theory Appl. 1 (1996) 29–55.

[41] G. Xu, Y. Xu, GPS, Springer, 2007.

[42] Ali, M. E. Geographic Information System (GIS): Definition, Development, Applications & Components.

[43] B.F. Khashoggi, A. Murad, Issues of healthcare planning and GIS: a review, ISPRS Int. J. Geo-Inf. 9 (6) (2020) 352.

[44] S. Teixeira, Qualitative geographic information systems (GIS): an untapped research approach for social work, Qual. Soc. Work 17 (1) (2018) 9–23.

[45] V. Čabanová, M. Miterpáková, M. Druga, Z. Hurníková, D. Valentová, GIS-based environmental analysis of fox and canine lungworm distribution: an epidemiological study of Angiostrongylus vasorum and Crenosoma vulpis in red foxes from Slovakia, Parasitol. Res. 117 (2) (2018) 521–530.

[46] Z. Liang, D. Qiao, T. Sung, Research on 3D virtual simulation of geology based on GIS, Arab. J. Geosci. 14 (5) (2021) 1–8.

[47] N. Neofitou, K. Papadimitriou, C. Domenikiotis, L. Tziantziou, P. Panagiotaki, GIS in environmental monitoring and assessment of fish farming impacts on nutrients of Pagasitikos Gulf, Eastern Mediterranean, Aquaculture 501 (2019) 62–75.

[48] V. Raimondi, C. Falco, D. Curzi, A. Olper, Trade effects of geographical indication policy: the EU case, J. Agric. Econ. 71 (2) (2020) 330–356.

[49] R. Sharma, S.S. Kamble, A. Gunasekaran, Big GIS analytics framework for agriculture supply chains: a literature review identifying the current trends and future perspectives, Comput. Electron. Agric. 155 (2018) 103–120.

[50] G. Yang, C. Zhou, W. Wang, S. Ma, H. Liu, Y. Liu, Z. Zhao, Recycling sustainability of waste paper industry in Beijing City: an analysis based on value chain and GIS model, Waste Manag. 106 (2020) 62–70.

[51] F. Zhang, N. Cao, Application and research progress of geographic information system (GIS) in agriculture, in: 2019 8th International Conference on Agro-Geoinformatics (Agro-Geoinformatics), IEEE, 2019.
[52] J.D. Lippincott, "Jack Dangermond." View, vol. 19, 2019, pp. 46–48.
[53] F. Haque, M.A. Abas, Review of driving behavior towards fuel consumption and road safety, Jurnal Mekanikal 41 (1) (2018).
[54] M. Afify, A.S. Abuabed, N. Alsbou, Smart engine speed control system with ECU system interface, in: 2018 IEEE International Instrumentation and Measurement Technology Conference (I2MTC), IEEE, 2018.
[55] P. Kochar, M. Supriya, Vehicle speed control using Zigbee and GPS, in: International Conference on Smart Trends for Information Technology and Computer Communications, Springer, 2016.
[56] S. Hatagale, R. Manatkar, Fleet maintenance using IOT technology, Psychol. Educ. J. 57 (9) (2020) 6251–6253.
[57] The Ultimate Car Maintenance Checklist, 2021 from, https://www.bridgestonetire.com/tread-and-trend/drivers-ed/ultimate-car-maintenance-checklist#.
[58] A. Soltani-Sobh, K. Heaslip, R. Bosworth, R. Barnes, Effect of improving vehicle fuel efficiency on fuel tax revenue and greenhouse gas emissions, Transp. Res. Rec. 2502 (1) (2015) 71–79.
[59] P. Barla, M. Gilbert-Gonthier, M.A. Lopez Castro, L. Miranda-Moreno, Eco-driving training and fuel consumption: impact, heterogeneity and sustainability, Energy Econ. 62 (2017) 187–194.
[60] N. Chakraborty, A. Mondal, S. Mondal, Intelligent charge scheduling and eco-routing mechanism for electric vehicles: a multi-objective heuristic approach, Sustain. Cities Soc. 69 (2021) 102820.
[61] L. Greer, J.L. Fraser, D. Hicks, M. Mercer, K. Thompson, Intelligent Transportation Systems Benefits, Costs, and Lessons Learned: 2018 Update Report, Dept. of Transportation. ITS Joint Program Office, United States, 2018.
[62] A.T. Proper, R.P. Maccubbin, L.C. Goodwin, Intelligent Transportation Systems Benefits: 2001 Update, United States, Federal Highway Administration, 2001.
[63] J.S. Sussman, Perspectives on Intelligent Transportation Systems (ITS), Springer Science & Business Media, 2008.
[64] A.D. Boursianis, M.S. Papadopoulou, P. Diamantoulakis, A. Liopa-Tsakalidi, P. Barouchas, G. Salahas, G. Karagiannidis, S. Wan, S.K. Goudos, Internet of Things (IoT) and agricultural unmanned aerial vehicles (UAVs) in smart farming: a comprehensive review, IoT (2020) 100187.
[65] N. Hussein, R. Hassan, M.T. Fahey, Effect of pavement condition and geometrics at signalised intersections on casualty crashes, J. Saf. Res. 76 (2021) 276–288.
[66] A.N. Stephens, V. Beanland, N. Candappa, E. Mitsopoulos-Rubens, B. Corben, M.G. Lenné, A driving simulator evaluation of potential speed reductions using two innovative designs for signalised urban intersections, Accid. Anal. Prev. 98 (2017) 25–36.

Chapter 4

Language of response surface methodology as an experimental strategy for electrochemical wastewater treatment process optimization

A. Yagmur Goren[1], Yaşar K. Recepoğlu[2], Alireza Khataee[3,4]
[1]*Department of Environmental Engineering, Izmir Institute of Technology, Urla, Izmir, Turkey;*
[2]*Department of Chemical Engineering, Izmir Institute of Technology, Urla, Izmir, Turkey;*
[3]*Department of Environmental Engineering, Gebze Technical University, Gebze, Turkey;*
[4]*Research Laboratory of Advanced Water and Wastewater Treatment Processes, Department of Applied Chemistry, Faculty of Chemistry, University of Tabriz, Tabriz, Iran*

1. Introduction

The increasing water demand with the population growth causes migration to drought-prone regions, industrial development, and climate change, leading to changing weather patterns in populated areas, along with increased pressures on water resources. Hence, several regions of the world face water scarcity, suggesting that water resource requirement by sectors cannot be met entirely due to the effect of excessive water use on water quality or supply [1,2]. In particular, water resource consumption has increased swiftly with the accelerated technological growth and consequently causes the formation of a considerable amount of wastewater effluents, including various pollutants, resulting in severe environmental disorders and pollution problems [3]. For instance, the textile production sector consumes enormous amount of water in its overall process and produces vast amounts of dye wastewater [4]. The food industry sector generates considerable amounts of organic pollutants containing wastewater effluents such as olive oil mills [5,6]. Therefore, water recycling, reclamation, and reuse are strictly needed to address wastewater management by solving water resource concerns and producing novel sources

Artificial Intelligence and Data Science in Environmental Sensing
https://doi.org/10.1016/B978-0-323-90508-4.00009-5

of high-quality water sources [7]. The current research status and development of affordable treatment methods are rising daily for various industrial wastewaters. Different physical and chemical treatment technologies have been reviewed for the treatment of wastewater, such as adsorption [8–10], biological degradation [11], precipitation with chemical agents [12], coagulation-flocculation [13], ion exchange [14], membrane processes [15–18], etc. All these treatment methods have different performance characteristics and different direct impacts on the environment. Accordingly, electrochemical processes have attracted significant attention for the treatment of industrial wastewater, since they can degrade a wide variety of contaminants [19,20]. They provide numerous advantages compared with other technologies, such as ambient pressure and temperature, along with robust performance and capability to adjust to variations in the flow rate and inlet composition. They do produce relatively less waste and generally require no auxiliary chemicals. Due to other technologies and the adaptation to other applications, the small footprint of electrochemical processes or their modular design makes them attractive methods. Nevertheless, toxic by-products in the treated water and relatively high-cost electrodes have decelerated the electrochemical treatment [21]. Instead, response surface methodology (RSM) is preferred to develop mathematical equations to overcome these drawbacks in the pollutant elimination, quantitative evaluation, and optimization of electrochemical processes. This is because this approach enables reductions in both the cost and the consumption of precious resources, such as materials and energy, by giving the identical amount of data in fewer experiments, assessing interactions between the development of empirical models and factors [22–24].

In this chapter, a summary and a discussion are reviewed elaborately for the RSM approach utilized in the EWT processes to overcome process challenges by optimizing parameters. This review clarifies that the RSM approach is used in EWT processes reported in the literature and summarizes the optimum operational variables determined by the RSM approach.

2. Strategy of response surface methodology

RSM is a statistical approach to design experimental runs, evaluate the self and interactive effects of independent operating variables, and optimize the process with few experiments [25–27]. The main concept of the method is the development of mathematical models fitted to the experimental data from the designed set of experiments and confirmation of the model obtained by the statistical analysis [28]. The optimum operational conditions for a system are first determined by screening independent factors and selecting desired responses, followed by choosing the experimental design strategy and running the experiments to obtain the results. After that, the model is confirmed by utilizing response graphs and analysis of variance, and optimal conditions are determined in the end [29]. Among all the consecutive steps, assessing independent factors and selecting dependent responses are the most vital RSM action. It may lead to unpredictable outcomes if the factors are not specified correctly. Besides, the selection of a suitable experimental design strategy is of importance to estimate and evaluate surface responses by applying one of the design strategies proposed in the literature, such as central composite design (CCD), Box–Behnken design (BBD), Doehlert design (DD), and full factorial design (FFD) [30–36].

BBD consists of three interlocked 2^2 factorial designs and a center point so that it is a three-level factorial design reducing the number of required experiments. BBD does not include points where all factors are simultaneously at their lowest or highest levels, which is advantageous for avoiding experiments performed under extreme conditions [37,38]. BBD has been widely applied in the experimental design for agrochemicals, bioprocessing, food engineering, pharmaceuticals, and other industries to extract biological active compounds, such as phenolic compounds, polysaccharides, and proteins, from various sources for human use [39–41]. CCD is a widely studied design model to create a second-order response surface model in environmental processes [42–45]. Although CCD needs three types of independent operating variable levels (axial, center, and cube levels) from the factorial design, it presents equal information as the three-level FFD with substantially few experimental tests. Furthermore, CCD produces a beneficial estimation of linear and quadratic interaction impacts of factors affecting the selected process [46,47]. A factorial design is another statistical technique useful in planning experiments to control few factors whose effects are investigated at two or more levels. A general FFD is applicable if multiple levels of several factors are dealt with. The significant advantage of a FFD in the modeling of complex systems is the capability of defining one factor at a time over other traditional techniques for modeling a multivariable system [48,49]. On the other hand, the DD has high flexibility compared to other designs because it can be utilized to produce response surfaces with the right prediction of the mathematical model variables with three operating variables at various levels. Meanwhile, all experimental domains are explored with a minimum number of experiments, following a sequential approach, first studying only two factors, three factors, and so on [50]. After deciding on an appropriate experimental design strategy, the obtained mathematical model is fitted to experimental data. First, mathematical models are coded, such as -1, 0, and $+1$. Then, the coded data are fitted to the model using a regression in which it widely uses the coefficient of determination (R^2) as the exclusive validation value for any model. However, sometimes a high R^2 does not necessarily mean that the model is well fitted. As the number of operating factors increases, the model may produce a deceptively high R^2 value due to incorporating the random noise in the data, which lessens the ability to make estimations. Hence, adjusted R^2 and predicted R^2 are reported, and the R^2 and residual plots to overcome such problems [51]. Other confirming tools to illustrate the responses using an appropriate software are contour or three-dimensional surface plots that provide a piece of knowledge about the individual effects of operating variables and their relations by having a rough prediction of the optimal design response. Furthermore, the analysis of variance methods identifies statistically significant parameters that are essential to determine the most critical factor in multiparameter models by providing information about the fit accuracy. In this method, the F-value is considered as the quotient of the mean square and residual mean square while the model is being examined. The *P*-value can also be calculated based on the marginal significance level within a statistical hypothesis test, representing the occurrence probability of a given event. Even though *P*-values $<.05$ may be considered for the threshold of significance, the factor *P*-value is lower than .05, which can be considered a significant factor for higher precision of modeling [52,53].

Finding out optimal conditions is the final step in RSM where operating variables in any process are adjusted to determine the appropriate levels that profit the most promising results. The optimum conditions can either be acquired using a single target to efficiency maximization as the primary target by keeping the specified values for factors or making the optimization process be a multiobjective optimization targeted for higher efficiency and lower operating cost [54,55].

3. Practical application of RSM in electrochemical processes for wastewater treatment

3.1 Electrocoagulation

In recent years, electrocoagulation (EC) as good water and wastewater treatment technology has attained significant attention [56,57]. The EC process is a widely used electrochemical treatment process that removes organic pollutants, heavy metals, dyes, suspended solids, anions, oils, and other pollutants from water and wastewater using electricity. A simple EC system involves an anode electrode (i.e., iron, aluminum, and iron-aluminum hybrid) and a cathode electrode (i.e., aluminum, iron, stainless steel, and titanium) in solution; electrolytic oxidation of anode electrode and reduction reactions at the cathode electrode take place when the current or voltage is supplied to the electrode materials. Through the dissolution of the anode electrode, the metal ions (Al^{3+}, Fe^{2+}, and Fe^{3+}) react with OH^- ions produced by the electrolysis of water on the cathode electrode to produce the corresponding monomeric form and polymeric metallic (oxyhydr)oxide complex species ($Me_n(OH)_m^{3n-m}$, transformed finally into solid $Me(OH)_3$), which enhances the production of the flocks by the destabilization of pollutants [58,59].

The EC is an applicable and environmentally friendly wastewater treatment technology compared with other conventional technologies due to its high removal performance, low sludge formation, comfortable and compact operation, no need for additional chemicals, and moderately low operating costs. Effectiveness of the EC process is significantly related to operating parameters, such as electrode material, electrode distance, initial pH, aeration rate, supplied current, pollutant concentrations, and operating time. However, the proper selection of these operating conditions is one of the critical challenges in applying the EC process for wastewater treatment. Commonly, the experimental runs are performed by analyzing one factor, while other parameters are sustained continuously. This procedure has several disadvantages, such as the need for a significant number of experimental runs resulting in high operating costs and waste of time and being insufficient for the study of the combined effect of operating variables. On the other hand, the statistical analysis of data with experimental design methodologies is a promising procedure to collect a significant amount of information from extracted data with a small number of experimental runs. Furthermore, the assessment of statistical significance and accuracy for the effects of the studied operating parameters and the determination of combined effects of parameters can be evaluated using mathematical models through an experimental design. Therefore, the RSM has attracted considerable interest in optimizing, modeling, and experimental design of treatment processes. According to the literature, RSM can effectively optimize and design the EC technology for wastewater treatment [60].

The literature on wastewater treatment using the EC process with RSM for optimization is summarized in Table 4.1. Several water types were studied to optimize the wastewater removal performance of the EC process with RSM. These include licorice processing, textile, pharmaceutical, metal plating, landfill leachate, cold meat industry, mill industry, metalworking industry, sugar industry, hospital, power plant, instant coffee production, cheese whey, swine slaughterhouse, tannery, bilge, urban, gray, biodiesel, palm oil mill effluent, and baker's yeast wastewater. The most used RSM model and statistical software are CCD and Design-Expert, respectively. Furthermore, the EC process is essential because the operating time, current, pH, mixing rate, electrolyte concentration, and electrode distance affect the system removal performance and operational cost.

Hendaoui et al. [60] applied RSM for optimizing the removal of indigo dyeing wastewater by continuous EC process. In the CCD model, the researchers identified pH, voltage, and air flow rate as three practical operational factors to optimize the color, COD, and conductivity removal efficiency as three responses. They specified two desired targets in the optimization process. The CCD was conducted without specifying any operating parameter value for 95% color and COD removal efficiency and 35% conductivity removal efficiency at the minimum total cost. Using this target, they achieved color removal efficiency of 90.5%, COD removal efficiency of 76.3, the conductivity removal efficiency of 30.1%, and total cost of $0.532/m^3$, with a solution having the pH value of 7.2, the flow rate of 1.1 L/min, and voltage of 65.79 V.

TABLE 4.1 Summary of optimization of electrocoagulation technology for wastewater treatment with response surface methodology (RSM).

Water type	RSM					Optimum conditions			References
	Design model	Independent variables	Dependent responses	Statistical software	Run number	Target of model	Independent parameters	Dependent parameters	
Licorice processing wastewater	CCD	T_{EC}, min j, A/m^2 $C_{,NaCl}$, mg/L M_i, rpm	$\%Re_{,color}$ $\%Re_{,COD}$ $\%Re_{,turbidity}$ $\%Re_{,alkalinity}$ ENC, kWh/m^3	Design-Expert	30	Maximum Re for COD, color, and turbidity and minimum ENC	$T_{EC} = 81.98$ min $J = 351.02\ A/m^2$ $C_{,NaCl} = 300$ mg/L $M_i = 45$ rpm	$\%Re_{,color} = 90.9$ $\%Re_{,COD} = 91.6\%$ $Re_{,turbidity} = 82.9\%$ $Re_{,alkalinity} = 74.4$ $ENC = 2\ kWh/m^3$	[61]
Textile wastewater	CCD	T_{EC}, min j, A/m^2 $C_{,Na2SO4}$, mg/L pH	$\%Re_{,color}$	Design-Expert	30	Maximum Re for color	$T_{EC} = 50$ min $J = 10\ A/m^2$ $C_{,Na2SO4} = 78.8$ mg/L pH = 5.7	$\%Re_{,color} = 97.68$	[62]
Pharmaceutical wastewater	CCD	T_{EC}, min j, mA/cm^2 $C_{,MNZ}$, mg/L ED, cm pH	$\%Re_{,MNZ}$	NA	50	$Re_{,MNZ}$ greater than 98%	$T_{EC} = 14.6$ min $J = 6.0\ mA/cm^2$ $C_{,MNZ} = 21.6$ mg/L ED = 3 cm pH = 8.2	$\%Re_{,MNZ} = 100$	[63]
Metal plating wastewater	CCD	T_{EC}, min j, mA/cm^2 pH	$\%Re_{,color}$ $\%Re_{,COD}$ $\%Re_{,chromium}$ $\%Re_{,nickel}$ $\%Re_{,zinc}$	Statgraphics Centurion	15	NA	$T_{EC} = 30$ min $J = 30\ mA/cm^2$ pH = 5	$\%Re_{,color} = 99.9$ $\%Re_{,COD} = 76.2\%$ $Re_{,chromium} = 98.9$ $\%Re_{,nickel} = 96.3$ $\%Re_{,zinc} = 99.8$	[64]

Landfill leachate wastewater	CCD	j, A/dm^2 ED, cm pH	%$Re_{,color}$ %$Re_{,TOC}$ ENC	Design-Expert	20	Maximum Re for color and TOC and minimum power consum-ption	$J = 5.25\ A/dm^2$ $ED = 1$ cm $pH = 7.83$	%$Re_{,color} = 74.57$ %$Re_{,TOC} = 51.74$ $ENC = 14.8\ kWh/m^3$	[65]
Cold meat industry wastewater	BBD	T_{EC}, min j, mA/cm^2 pH	%$Re_{,COD}$	Design-Expert	17	Maximum Re for COD	$T_{EC} = 35.12$ min $J = 5.4\ mA/cm^2$ $pH = 8.2$	%$Re_{,COD} = 88.91$	[66]
Sugar industry wastewater	–		%$Re_{,color}$ %$Re_{,COD}$	Design-Expert	23	Maximum Re for color and COD	$T_{EC} = 60$ min $i = 1$ A $ED = 1.5$ mm $pH = 7$	%$Re_{,color} = 88.6$ %$Re_{,COD} = 69.2$	[72]
Instant coffee production wastewater	CCD	T_{EC}, min j, A/m^2 pH	%$Re_{,COD}$	Minitab	17	100% Re for COD	$T_{EC} = 10$ min $J = 108.3\ A/m^2$ $pH = 7$	%$Re_{,COD} = 93.3$	[73]
Hospital wastewater	CCD	j, mA/cm^2 pH	%$Re_{,TOC}$	NA	13	Maximum Re for TOC	$j = 4.46\ mA/cm^2$ $pH = 5.45$	%$Re_{,TOC} = 99.9$	[74]
Power plant wastewater	CCD	T_{EC}, min i, A ED, cm pH	%$Re_{,nickel}$ %$Re_{,iron}$	Minitab	31	Maximum Re for nickel and iron	$T_{EC} = 18$ min $i = 1.5$ A $ED = 1$ cm $pH = 8.1$	%$Re_{,nickel} = 99.9$ %$Re_{,iron} = 99.8$	[75]
Swine slaughterhouse wastewater	CCD	T_{EC}, min j, A/m^2 pH	%$Re_{,COD}$	Minitab	20	Maximum Re for COD	$T_{EC} = 9.7$ min $j = 126.2\ A/m^2$ $pH = 8.36$	%$Re_{,COD} = 99.9$	[76]

Continued

TABLE 4.1 Summary of optimization of electrocoagulation technology for wastewater treatment with response surface methodology (RSM).—cont'd

Water type	RSM					Optimum conditions			References
	Design model	Independent variables	Dependent responses	Statistical software	Run number	Target of model	Independent parameters	Dependent parameters	
Hospital wastewater	BBD	T_{EC}, min V, volt $C_{,electrolyte}$, M	$\%Re_{,COD}$	Design-Expert	17	Maximum Re for COD	$T_{EC} = 34.26$ min $V = 12$ V $C_{,electrolyte} = 0.38$ M	$\%Re_{,COD} = 65.0$	[77]
Cheese whey wastewater	BBD	T_{EC}, min j, mA/cm^2 pH	$\%Re_{,COD}$	Minitab	15	Maximum Re for COD	$T_{EC} = 20$ min $j = 60\ mA/cm^2$ $pH = 4.54$	$\%Re_{,COD} = 86.4$	[78]
Bilge wastewater	CCD	T_{EC}, min j, mA/cm^2 T, °C	$\%Re_{,COD}$ $\%Re_{,oil\text{-}grease}$	NA	20	Maximum Re for both COD and oil and grease	$T_{EC} = 13$ min $j = 9.87\ mA/cm^2$ $T = 29\ °C$	$\%Re_{,COD} = 90.3$ $\%Re_{,oil\text{-}grease} = 81.7$	[79]
Tannery wastewater	CCD	T_{EC}, min j, mA/cm^2 pH	$\%Re_{,COD}$	Statgraphics Centurion	20	Maximum Re for COD	$T_{EC} = 32.9$ min $j = 110\ mA/cm^2$ $pH = 3$	$\%Re_{,COD} = 86.51$	[80]
Urban wastewater	BBD	T_{EC}, min j, mA/cm^2 pH	$\%Re_{,COD}$ $\%Re_{,turbidity}$ $\%Re_{,phosphate}$	Design-Expert	17	Maximum Re for COD, turbidity, and phosphate	$T_{EC} = 17.9$ min $j = 14.9\ mA/cm^2$ $pH = 7.07$	$\%Re_{,COD} = 69.1$ $\%Re_{,turbidity} = 100$ $\%Re_{,phosphate} = 100$	[81]

Landfill leachate wastewater	CCD	$C_{,NaCl}$, g/L ED, cm pH	$\%Re_{,color}$ $\%Re_{,COD}$	Design-Expert	20	Maximum Re for both color and COD	$C_{,NaCl} = 2$ g/L ED = 1.16 cm pH = 7.73	$\%Re_{,color} = 90.2$ $\%Re_{,COD} = 46.1$	[82]
Metalworking industry wastewater	CCD	T_{EC}, min j, mA/cm^2 pH	$\%Re_{,COD}$ $\%Re_{,turbidity}$	Statgraphics Centurion	15	Maximum Re for COD and turbidity	$T_{EC} = 21.78$ min $j = 55$ mA/cm^2 pH = 6	$\%Re_{,COD} = 76.72$ % $Re_{,turbidity} = 99.97$	[83]
Gray wastewater	BBD	T_{EC}, min j, mA/cm^2 ED, cm pH	$\%Re_{,COD}$ $\%Re_{,TS}$ $\%Re_{,FC}$	Design-Expert	29	Maximum Re for COD, TS, and FC	$T_{EC} = 20$ min $j = 20$ mA/cm^2 ED = 5 cm pH = 7	$\%Re_{,COD} = 95.47$ $\%Re_{,TS} = 99.87$ $\%Re_{,FC} = 97.15$	[84]
Metal cutting wastewaters	CCD	T_{EC}, min j, A/m^2 pH	$\%Re_{,COD}$ $\%Re_{,turbidity}$ $\%Re_{,TOC}$	Design-Expert	17	Maximum Re for COD, turbidity, and TOC	$T_{EC} = 20.6$ min $j = 66.39$ A/m^2 pH = 7.03	$\%Re_{,COD} = 88.43$ % $Re_{,turbidity} = 66.39$ $\%Re_{,TOC} = 79.1$	[85]
Palm oil mill effluent wastewater	BBD	T_{EC}, hour V, volt $C_{,NaCl}$, g/L	$\%Re_{,COD}$ $\%Re_{,turbidity}$ $\%Re_{,Fe}$ $\%Re_{,Mg}$ $\%Re_{,Ca}$	Minitab	NA	Maximum Re for COD, turbidity, Fe, mg, and Ca	$T_{EC} = 6$ h V = 4 V $C_{,NaCl} = 0$ g/L	$\%Re_{,COD} = 42.94$ % $Re_{,turbidity} = 83.16$ $\%Re_{,Fe} = 23.62$ $\%Re_{,Mg} = 27.56$ $\%Re_{,Ca} = 47.83$	[86]

Continued

TABLE 4.1 Summary of optimization of electrocoagulation technology for wastewater treatment with response surface methodology (RSM).—cont'd

	RSM					Optimum conditions			
Water type	Design model	Independent variables	Dependent responses	Statistical software	Run number	Target of model	Independent parameters	Dependent parameters	References
Baker's yeast wastewater	CCD	T_{EC}, min j, A/m^2 pH	$\%Re_{,COD}$ $\%Re_{,color}$ $\%Re_{,TOC}$ ENC ELC OC W_{sludge}	Design-Expert	20	Maximum Re for COD, color, and TOC, and minimum OC and W_{sludge} (ELC and ENC in range)	$T_{EC} = 22$ min $j = 50$ A/m^2 $pH = 4$	$\%Re_{,COD} = 48$ $\%Re_{,color} = 88$ $\%Re_{,TOC} = 49$ $ENC = 1.08$ kWh/m^3 $ELC = 0.208$ $kgAl/m^3$ $OC = 0.4$ €/m^3 $W_{sludge} = 1.3$ kg/m^3	[87]
Galvanic by-product wastewater	CCD	T_{EC}, min j, A/m^2 M_i, rpm pH	$\%Re_{,COD}$ $\%Re_{,turbidity}$ $\%Re_{,color}$ $\%Re_{,TSS}$ $\%Re_{,chromium}$ $\%Re_{,nickel}$ $\%Re_{,zinc}$ $\%Re_{,copper}$	Statistica	28	Maximum Re for all responses	$T_{EC} = 35$ min $j = 97.7$ A/m^2 $M_i = 150$ rpm $pH = 6.5$	$\%Re_{,COD} = 90$ $\%Re_{,turbidity} = 100$ $\%Re_{,color} = 100$ $\%Re_{,TSS} = 90$ $\%Re_{,chromium} = 100$ $\%Re_{,nickel} = 100$ $\%Re_{,zinc} = 99$ $\%Re_{,copper} = 99$	[88]
Biodiesel wastewater	BBD	T_{EC}, min V, volt pH	$\%Re_{,COD}$ $\%Re_{,oil-grease}$ $\%Re_{,SS}$	Minitab	15	Maximum Re for all responses	$T_{EC} = 23.5$ min $V = 18.2$ V $pH = 6.06$	$\%Re_{,COD} = 55.43$ $\%Re_{,oil-grease} = 98.4$ $\%Re_{,SS} = 96.6$	[89]

Mill industry wastewater	CCD	T_{EC}, min j, mA/cm^2 pH	%Re$_{,COD}$	Statgraphics Centurion	15	Maximum Re for COD	$T_{EC} = 27.55$ min $j = 96$ mA/cm^2 pH = 9	%Re$_{,COD} = 34.70$	[90]
Mill industry wastewater	CCD	T_{EC}, min j, mA/cm^2 pH	%Re$_{,phenol}$	Statgraphics Centurion	15	Maximum Re for phenol	$T_{EC} = 30$ min $j = 91.15$ mA/cm^2 pH = 8.57	%Re$_{,phenol} = 92.32$	
Mill industry wastewater	CCD	T_{EC}, min j, mA/cm^2 pH	%Re$_{,calcium}$	Statgraphics Centurion	15	Maximum Re for calcium	$T_{EC} = 30$ min $j = 95.27$ mA/cm^2 pH = 9	%Re$_{,calcium} = 33.36$	
CNC (metalworking fluid) industry wastewater	CCD	T_{EC}, min j, mA/cm^2 pH	%Re$_{,COD}$	Statgraphics Centurion	15	Maximum Re for COD	$T_{EC} = 30$ min $j = 48$ mA/cm^2 pH = 4	%Re$_{,COD} = 98.69$	[91]
CNC (metalworking fluid) industry wastewater	CCD	T_{EC}, min j, mA/cm^2 pH	%Re$_{,copper}$	Statgraphics Centurion	15	Maximum Re for copper	$T_{EC} = 30$ min $j = 40$ mA/cm^2 pH = 8	%Re$_{,copper} = 100$	
CNC (metalworking fluid) industry wastewater	CCD	T_{EC}, min j, mA/cm^2 pH	%Re$_{,nickel}$	Statgraphics Centurion	15	Maximum Re for nickel	$T_{EC} = 30$ min $j = 38$ mA/cm^2 pH = 10	%Re$_{,nickel} = 94.56$	

BBD, Box–Behnken design; $C_{,NaCl}$, NaCl concentration; *CCD*, central composite design; *ED*, electrode distance; *ELC*, electrode consumption; *ENC*, energy consumption; *FC*, fecal coliform; *i*, applied current; *j*, current density; *Mi*, mixing intensity; *MNZ*, metronidazole; *NA*, not available; *OC*, operating cost; *Re*, removal efficiency; *SS*, suspended solid; *T*, temperature; T_{EC}, electrolysis time; *TOC*, total organic carbon; *TS*, total solids; *TSS*, total suspended solids; *V*, voltage; W_{sludge}, amount of sludge.

In the second target, they neglected the conductivity and conducted CCD to achieve 95% color and COD removal efficiency. For this target, they observed the following independent and dependent parameters: solution pH = 7.6, flow rate = 0.99 L/min, and voltage = 100.8 V, which resulted in the target COD and color removal efficiency of 95%. In a research, the treatment of molasses wastewater was investigated using the EC process and optimization of the process through RSM. The BBD method considered reaction time, current density, and wastewater dilution rate as the independent operating parameters for maximizing the COD, ammonium, and nitrate removal efficiencies and minimizing the operating cost. The maximum COD, ammonium, and nitrate removal efficiencies were found to be 49.7%, 51.4%, and 51.7%, respectively, and the minimum operating cost was 0.982 €/m^3 at an operating time of 3.5 h, current density of 32.6 mA/cm^2, and water dilution of 45% [67].

In another study, Refs. [68,69] reported the optimization of chicken industry wastewater removal and the production of H_2 gas using the EC process with aluminum electrodes. They utilized the Design-Expert software to use the BBD with three independent operating parameters, viz. operating time, electrode surface area, and current density, to optimize the COD removal efficiency and H_2 gas production as a dependent parameter. At an operating time of 30 min, a current density of 15 A/m^2, and an electrode surface area of 5 m^2, the maxima of COD removal efficiency and gas production were found to be 99% and 0.8 mL/L, respectively.

The removal of anaerobically treated wastewater by a continuous EC reactor (Fig. 4.1A) was optimized using RSM by Ref. [70]. They used the BBD model to optimize three factors with three initial COD concentrations (200, 275, and 350 mg/L), current densities (2, 5, and 8 mA/cm^2), and operating times (5, 10, and 15 min), as well as their responses as effluent COD and phosphate concentration and turbidity (Fig. 4.1B–D). As a target of the model, they proposed that the current density and operating time were the minimum values. These parameters control the removal efficiency and total cost of the process. Furthermore, the initial COD concentration was specified in the range of 200–350 mg/L for a response effluent COD concentration of 100 mg/L and effluent turbidity and phosphate values in the design matrix range. Based on this target, the optima of effluent COD, phosphate, and turbidity values were found to be 87 mg/L, 0.59 mg/L, and 12.6 NTU at an initial COD concentration of 274 mg/L, current density of 2 mA/cm^2, and operating time of 5 min, respectively.

The optimization of dye bath wastewater treatment by the EC process with aluminum electrodes using RSM-CCD was studied by Ref. [71]. The independent operating variables were solution pH (3.3–6.7), current density (29.7–105.3 A/m^2), and operating time (4.8–55.2 min), while color removal, COD removal, and operating cost were the dependent parameters. The individual and combined effects of independent operating parameters on the responses are presented. The optimum independent operating parameters were

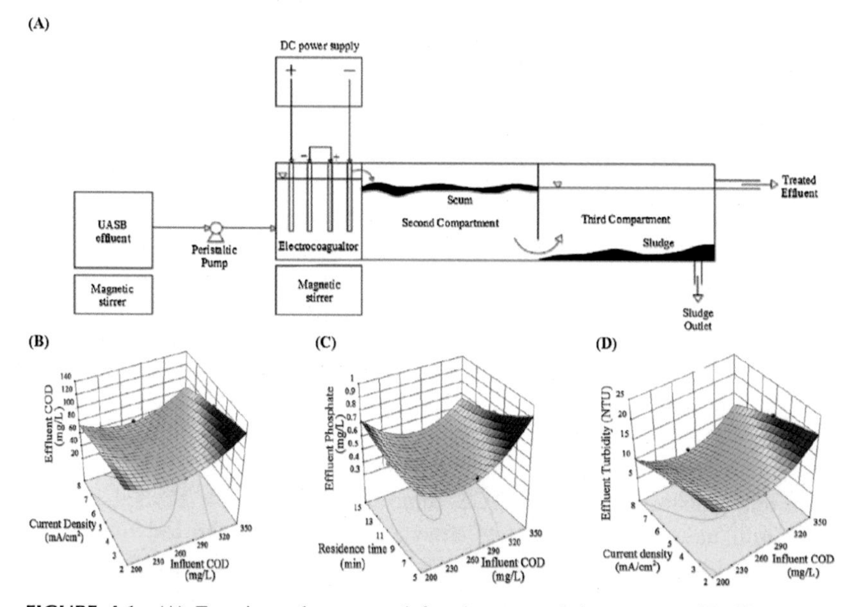

FIGURE 4.1 (A) Experimental setup used for electrocoagulation process, (B) 3D response surface plot for effluent COD concentration at operating time of 5 min, (C) 3D plot for effluent phosphate concentration at current density of 2 mA/cm^2, and (D) 3D plot for effluent turbidity operating time of 5 min. *Reproduced with modification from A.R. Makwana, M.M. Ahammed, Continuous electrocoagulation process for the post-treatment of anaerobically treated municipal wastewater, Process Saf. Environ. Protect. 102 (2016), 724–733 with permission from Elsevier (License Number: 5017680271193).*

reported to be an initial pH of 5.01, a current density of 64 A/m^2, and an operating time of 28.5 min for maximum color removal efficiency of 85.8%, COD removal percentage of 76.9%, and operating cost of 1.84 €/m^3. The most important remark of this study indicated that the majority of operating cost (by 70%) in the process considerably resulted from high acid consumption to adjust pH due to the high alkalinity of dye bath wastewater, which is usually neglected in many studies.

3.2 Electro-Fenton

Electro-Fenton (EF) technology is another emerging electrochemical process for the treatment of wastewater and water. The EF process can be described as the continuous electro-formation of H_2O_2 at a cathode surface with air or O_2 supply at the presence of iron catalyst to the treated solution for the generation of hydroxyl radicals at the bulk through Fenton's reaction [92]. The most critical affecting parameters in the EF process are ferrous/ferric concentration, hydrogen peroxide (H_2O_2) concentration, pollutant concentration, initial pH, electrolyte type, operation mode, current density, temperature, operating time, and electrode material [93]. On the other hand, slow H_2O_2 production, the need for pH adjustment, and high sludge formation are significant drawbacks of EF processes. Therefore, several researches have been conducted on optimizing operating parameters using RSM to overcome these disadvantages and enhance the EF process performance.

The studies on optimization of the EF process for wastewater treatment are summarized in Table 4.2. Various wastewaters from agro-food, pharmaceutical, petrochemical, and textile industries were treated using the EF process. Optimization of the process was performed using both BBD and CCD models to obtain maximum removal efficiency and minimum operating cost. Moreover, the operating time, current density, pH, and H_2O_2/Fe^{2+} molar ratio were widely studied as independent operating parameters, while removal of COD, total suspended solid (TSS), and color were selected as dependent parameters.

Zhu et al. [26] used RSM with the BBD model to optimize the EF process performance for the treatment of cooking wastewater. Optimum parameters were found as 1.8 h electrolysis time, 0.6 mM of Fe^{2+}, pH 4, and 3.7 mA/cm^2 of current density with 55.8% TOC removal. Refs. [83,90] studied EO for paper mill industry wastewater treatment and evaluated the individual and combined effects of the process variables according to the RSM approach. The performance of the process was optimized as a 30 min run, a pH of 2, a current density of 96 mA/cm^2, and H_2O_2/COD molar ratio 2.0 for 74.31% of COD removal. The operating cost was calculated to be 7.02 €/m^3 depending on the electrode and energy consumption at optimum conditions.

Akkaya et al. [99] employed an RSM approach using the BBD to develop a mathematical model and optimize EF process parameters for the treatment of dairy wastewater. Optimum conditions were determined via ANOVA statistics as H_2O_2/COD ratio of 2, a current density of 32 mA/cm^2, a pH of 2.4, and a reaction time of 45 min. The removal efficiencies were over 88% for orthophosphate, suspended solid, and color, where the *P*-value was more significant than 1.7×10^{-3} with a 95% confidence level for all responses for EF processes.

Kaur et al. [100] investigated the degradation mechanisms of textile industry wastewater contaminants by the EF process and system performance optimization using RSM. The CCD model was conducted for the design of experiments. Effects of operating time, current, and ferrous concentrations were investigated on COD and color removal efficiencies and energy consumption. The CCD model showed that COD and color removal efficiencies increased with increasing the supplied current and operating time, while the energy consumption increased with increasing the two variables. An operating time of 175 min, a current of 2.72 A, and a ferrous concentration of 1 mM led to the maximum color and COD removal efficiencies of 100% with minimum specific energy consumption of 0.025 kWh/kg of COD.

Davarnejad and Nasiri [101] conducted RSM with the BBD model to investigate the treatment of slaughterhouse wastewater by the EF process. The independent operating variables for the study were operating time

TABLE 4.2 Summary of optimization of electro-Fenton process for wastewater treatment with response surface methodology (RSM).

Water type	RSM					Optimum conditions			References
	Design model	Independent variables	Dependent responses	Statistical software	Run number	Target of model	Independent parameters	Dependent parameters	
Cheese whey wastewater	CCD	$C_{,FeSO47H2O}$, mol $C_{,H2O2/FesSO47H2O7H2O}$, mol/mol i, A	$\%Re_{,COD}$ $\%Re_{,current}$ ENC, kWh/kgCOD	Minitab	20	Maximum Re for COD, and current and minimum ENC	$MR_{FeSO47H2O} = 0.625$ $C_{,H2O2/FesSO47H2O7H2O} = 14.48$ mol/mol $i = 1.22$ a	$\%Re_{,COD} = 92.73$ $\%Re_{,current} = 51.37$ ENC = 0.13 kWh/kgCOD	[94]
Carbonated soft drink industry wastewater	BBD	T_{EC}, min j, mA/cm^2 pH $MR_{H2O2/Fe2+}$, mol $VR_{H2O2/SDW}$, mL/L	$\%Re_{,COD}$	Design-Expert	46	Maximum Re for COD	$T_{EC} = 41.55$ min $j = 46.12$ mA/cm^2 pH = 4.14 $MR_{H2O2/Fe2+} = 0.98$ mol $VR_{H2O2/SDW} = 2.74$ mL/L	$\%Re_{,COD} = 73.1$	[95]
Food industry wastewater	CCD	T_{EC}, min j, mA/cm^2 pH $R_{H2O2/COD}$	$\%Re_{,COD}$	Statgraphics Centurion	30	Maximum Re for COD	$T_{EC} = 27.11$ min $j = 86$ mA/cm^2 pH = 2.38 $R_{H2O2/COD} = 2$	$\%Re_{,COD} = 59.1$	[96]
			$\%Re_{,TSS}$			Maximum Re for TSS	$T_{EC} = 19.26$ min $j = 84.36$ mA/cm^2 pH = 2 $R_{H2O2/COD} = 0.5$	$\%Re_{,TSS} = 99.96$	

Continued

TABLE 4.2 Summary of optimization of electro-Fenton process for wastewater treatment with response surface methodology (RSM).—cont'd

	RSM					Optimum conditions			
Water type	Design model	Independent variables	Dependent responses	Statistical software	Run number	Target of model	Independent parameters	Dependent parameters	References
Alcoholic wastewater	NA	T_{EC}, min j, mA/cm^2 pH $MR_{H2O2/Fe2+}$ VF_{H2O2}, mL/L	$\%Re_{,COD}$	Design-Expert	46	Maximum Re for COD	$T_{EC} = 69.29$ min $j = 51.54$ mA/cm^2 pH = 2.84 $MR_{H2O2/Fe2+} = 3.7$ $VF_{H2O2} = 1.62$ mL/L	$\%Re_{,COD} = 67.65$	[97]
Palm oil mill wastewater	BBD	T_{EC}, min V, volt $C_{,NaCl}$, M VF_{H2O2}, mL	$\%Re_{,COD}$	NA	29	Maximum Re for COD	$T_{EC} = 35.92$ min V = 15.78 V $C_{,NaCl} = 0.06$ M $VF_{H2O2} = 14.79$ mL	$\%Re_{,COD} = 99.56$	[98]
Dairy wastewater	BBD	T_{EC}, min j, mA/cm^2 pH $MR_{H2O2/Fe2+}$ $VR_{H2O2/DW}$, mL/L	$\%Re_{,COD}$	Design-Expert	46	Maximum Re for COD and color	$T_{EC} = 90$ min $j = 56$ mA/cm^2 pH = 7.52 $MR_{H2O2/Fe2+} = 3.97$ $VR_{H2O2/DW} = 0.898$ mL/L	$\%Re_{,COD} = 93.93$	[104]
			$\%Re_{,color}$				$T_{EC} = 86$ min $j = 55.1$ mA/cm^2 pH = 7.48 $MR_{H2O2/Fe2+} = 3.99$ $VR_{H2O2/DW} = 0.907$ mL/L	$\%Re_{,color} = 97.32$	

Landfill leachate wastewater	CCD	T_{EC}, min j, mA/cm^2 pH $MR_{H2O2/Fe2+}$	$\%Re_{,COD}$ $\%Re_{,color}$	Design-Expert	30	Maximum Re for COD and color	$T_{EC} = 43$ min $j = 49$ mA/cm^2 pH = 3 $MR_{H2O2/Fe2+} = 1$	$\%Re_{,COD} = 94.16$ $\%Re_{,color} = 97.07$	[105]
Coking wastewater	CCD	j, mA/cm^2 pH $C_{,Fe2+}$,mg/L	$\%Re_{,TOC}$	Minitab	20	Maximum Re for TOC	$j = 30.9$ mA/cm^2 pH = 4.05 $C_{,Fe2+} = 0.35$ mg/L	$\%Re_{,TOC} = 73.8$	[106]
Biotreated coking wastewater	BBD	V, volt ED, cm pH	$\%Re_{,COD}$	Design-Expert	17	Maximum Re for COD	V = 10 V ED = 1 cm pH = 3.42	$\%Re_{,COD} = 106.9$	[107]
Poultry wastewater	BBD	T_{EC}, min j, mA/cm^2 pH $VR_{H2O2/PW}$, mL/L	$\%Re_{,COD}$ $\%Re_{,turbidity}$ ENC, kWh/L	Design-Expert	29	Maximum Re for COD and turbidity and minimum energy consumption	$T_{EC} = 30.52$ min $j = 10.21$ mA/cm^2 pH = 3.25 $VR_{H2O2/PW} = 20.32$ mL/L	$\%Re_{,COD} = 97.11$ $\%Re_{,turbidity} = 93.23$ ENC = 0.0091 kWh/L	[68]
Medical laboratory wastewater	CCD	T_{EC}, min i, A pH $R_{H2O2/COD}$	$\%Re_{,COD}$	Statgraphics Centurion	30	Maximum Re for COD	$T_{EC} = 33.4$ min i = 3 A pH = 3.4 $R_{H2O2/COD} = 1.1$	$\%Re_{,COD} = 55.1$	[108]
			$\%Re_{,BOD5}$			Maximum Re for BOD_5	$T_{EC} = 39.1$ min i = 3.7 A pH = 3.3 $R_{H2O2/COD} = 1.1$	$\%Re_{,BOD5} = 42.5$	
			$\%Re_{,toxicity}$			Maximum Re for toxicity	$T_{EC} = 33.6$ min i = 3.9 A pH = 2.6 $R_{H2O2/COD} = 1.06$	$\%Re_{,toxicity} = 99.7$	

Continued

TABLE 4.2 Summary of optimization of electro-Fenton process for wastewater treatment with response surface methodology (RSM).—cont'd

	RSM					Optimum conditions			
Water type	Design model	Independent variables	Dependent responses	Statistical software	Run number	Target of model	Independent parameters	Dependent parameters	References
Composting leachate wastewater	NA	T_{EC}, min i, A pH C_{H2O2}, M	$\%Re_{,COD}$	Design-Expert	NA	Maximum Re for COD	$T_{EC} = 100$ min $i = 3$ A $pH = 3$ $C_{H2O2} = 0.25$ M	$\%Re_{,COD} = 63.4$	[109]
Soybean oil plant wastewater	BBD	T_{EC}, min j, mA/cm^2 pH $MR_{H2O2/Fe2+}$ $VR_{H2O2/WW}$, mL/L	$\%Re_{,Oil}$	Design-Expert	33	Maximum Re for oil	$T_{EC} = 86.69$ min $j = 60.38$ mA/cm^2 $pH = 3.3$ $MR_{H2O2/Fe2+} = 3.69$ $VR_{H2O2/WW} = 2.09$ mL/L	$\%Re_{,Oil} = 93.46$	[110]
Pharmaceutical wastewater	BBD	T_{EC}, min j, mA/cm^2 $MR_{H2O2/Fe2+}$ $VR_{H2O2/WW}$, mL/L	$\%Re_{,COD}$	Design-Expert	46	Maximum Re for COD	$T_{EC} = 58.49$ min $j = 42.9$ mA/cm^2 $pH = 2.96$ $MR_{H2O2/Fe2+} = 3.78$ $VR_{H2O2/WW} = 1.37$ mL/L	$\%Re_{,COD} = 93$	[111]
Petrochemical wastewater	CCD	T_{EC}, min j, mA/cm^2 $MR_{H2O2/Fe2+}$ $VR_{H2O2/WW}$, mL/L	$\%Re_{,COD}$ $\%Re_{,color}$	Design-Expert	47	Maximum Re for COD and color	$T_{EC} = 78.97$ min $j = 68.65$ mA/cm^2 $pH = 3.06$ $MR_{H2O2/Fe2+} = 4.99$ $VR_{H2O2/WW} = 2.14$ mL/L	$\%Re_{,COD} = 53.94$ $\%Re_{,color} = 67.35$	[112]

Coke plant wastewater	BBD	T_{EC}, h j, mA/cm^2 pH	%Re,$_{COD}$	NA	17	Maximum Re for COD	$T_{EC} = 2$ h $j = 3.2$ mA/cm^2 pH = 2.6	%Re,$_{COD}$ = 70	[113]
Gray wastewater	BBD	T_{EC}, min j, mA/cm^2 $MR_{H2O2/Fe2+}$ pH	%Re,$_{COD}$ %Re,$_{TSS}$	Design-Expert	29	Maximum Re for COD and TSS	$T_{EC} = 13.59$ min $j = 9.76$ mA/cm^2 $MR_{H2O2/Fe2+} = 0.71$ pH = 3.94	%Re,$_{COD}$ = 89.92 %Re,$_{TSS}$ = 84.33	[114]
Textile wastewater	BBD	T_{EC}, min i, A C_{FeSO4}, mM	%Re,$_{COD}$ %Re,$_{color}$ ENC,Wh	Design-Expert	17	Maximum Re for COD and color and minimum energy consumption	$T_{EC} = 90$ min $i = 0.32$ A $C_{FeSO4} = 0.53$ mM	%Re,$_{COD}$ = 100 %Re,$_{color}$ = 90.3 ENC = 1.27 Wh	[115]
Oilfield produced wastewater	CCD	T_{EC}, min C_{Fe2+},mM i, mA	%Re,$_{COD}$ ENC,kWh/ kg COD	Minitab	20	Maximum Re for COD and minimum energy consumption	$T_{EC} = 81$ min $C_{Fe2+} = 0.306$ mM $i = 156.6$ mA	%Re,$_{COD}$ = 73.33 ENC = 0.901 kWh/ kg COD	[116]
Olefin plant spent caustic wastewater	BBD	T_{EC}, min j, mA/cm^2 $MR_{H2O2/Fe2+}$ $VR_{H2O2/WW}$, mL/ L pH	%Re,$_{COD}$	Design-Expert	46	Maximum Re for COD	$T_{EC} = 70.67$ min $j = 58.85$ mA/cm^2 pH = 2.96 $MR_{H2O2/Fe2+} = 3.74$ $VR_{H2O2/WW} = 1.59$ mL/L pH	%Re,$_{COD}$ = 81.2	[117]

BBD, Box–Behnken design; *CCD*, central composite design; *DW*, dairy wastewater; *ED*, electrode distance; *ENC*, energy consumption; *i*, applied current; *j*, current density; *MR*, molar ratio; *NA*, not available; *PW*, poultry wastewater; *R*, ratio; *Re*, removal efficiency; *SDW*, soft drink wastewater; *SS*, suspended solids; T_{EC}, electrolysis time; *TSS*, total suspended solids; *V*, voltage; *VF*, volume fraction; *VR*, volume ratio; *WW*, wastewater.

(10−60 min), initial pH (2−8), the molar ratio of H_2O_2/Fe^{2+} (0.5−5), current density (20−80 mA/cm^2), and the ratio of H_2O_2/wastewater (0.3−2.14 mL/L), while color and COD removals were the dependent parameters. The optimization of color and COD removal efficiencies was separately specified as a target of the model. The model determined an optimum COD removal efficiency of 94.4% at an operating time of 55.6 min, an initial pH of 4.4, an H_2O_2/Fe^{2+} molar ratio of 3.7, a current density of 74.1 mA/cm^2, and an H_2O_2/wastewater ratio of 1.63 mL/L, while maximum color removal efficiency of 89.2% was achieved at optimized operating parameters (operating time of 49.2 min, initial pH of 3.4, H_2O_2/Fe^{2+} molar ratio of 3.6, a current density of 67.9 mA/cm^2, and H_2O_2/wastewater ratio of 1.44 mL/L).

Refs. [102,103] tested the photo-EF technology for the efficient treatment of textile wastewater (Fig. 4.2A). At optimum operational conditions obtained using RSM, namely $j = 40$ mA/cm^2, pH = 4, $\sigma = 5768$ μS/cm, and $Fe^{2+} = 0.3$ mM, the solar-driven process achieved complete color removal, TOC mineralization of 70%, and COD reduction of 83%, after an electrolysis time of 15 min (Fig. 4.2B−D). The process yielded a biocompatible (BOD5/COD>0.4) and highly oxidized (AOS = 2.24) effluent. The Pareto analysis was also used to determine factors having the most significant cumulative effect on response variables (Fig. 4.2E). The operating costs were estimated as 1.56 USD/m^3. The outcomes of this study demonstrated that photo-EF process could be a promising treatment method for industrial wastewater effluent to achieve allowable discharge criteria recommended by environmental authorities.

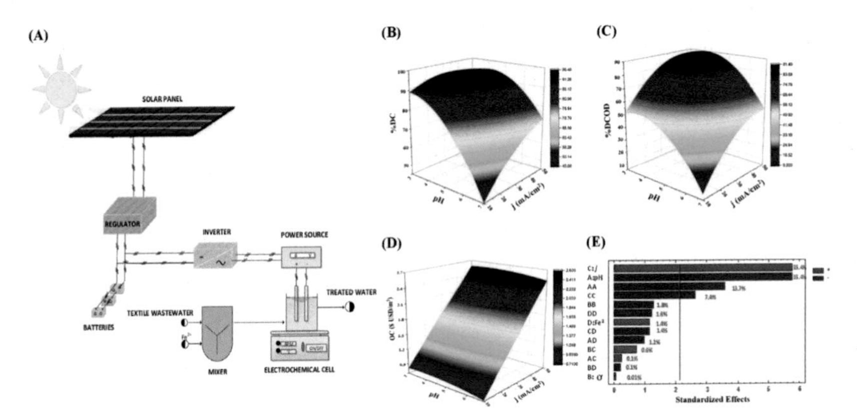

FIGURE 4.2 (A) Diagram of the photo-electro-Fenton process, (B) 3D response plots for the combined effect of applied current and pH on %DC, (C) 3D response plots for %DCOD, (D) 3D response plots for OC, (E) Pareto diagram for %DCOD. *Reproduced with modification from E. GilPavas, I. Dobrosz-Gomez, M.Á. Gómez-García, Optimization of solar-driven photo-electro-Fenton process for the treatment of textile industrial wastewater, J. Water Process Eng. 24 (2018) 49−55 and E. GilPavas, I. Dobrosz-Gómez, M.Á. Gómez-García, Optimization of sequential chemical coagulation-electro-oxidation process for the treatment of an industrial textile wastewater, J. Water Process Eng. 22 (2018), 73−79 with permission from Elsevier (License Number: 5017680840926).*

3.3 Electro-oxidation

Electro-oxidation (EO) is a process that decreases the concentration of contaminants in sludge and wastewater by the mineralization of organic pollutants. In the EO process, a DC is supplied in an EO system, combustion or electrochemical alteration is improved, and free radicals are produced that degrade organic contaminants [24,118]. The primary mechanism of EO is reported as one of the easiest methods for mineralizing pollutants since stable anode materials, such as boron-doped diamond electrodes as the most active anode for the oxidation of different pollutants, are employed in this process [119].

Piya-Areetham et al. [120] investigated COD and color removal from distillery wastewater by EO processes. The kinetics of COD decline was the pseudo−first-order reaction with a fast rate constant of 6.78 min^{-1} at a pH of 1 with the highest oxidation efficiency. The energy consumption was found in the range of 24.08−28.07 kWh/m^3 or 2.82−4.83 kWh/kgCOD wastewater based on the concentration of additives that promoted the reduction of color and COD (92.24% and 89.62%, respectively). Refs. [105,121] applied the CCD model to optimize the EO process for landfill leachate wastewater. The maximum COD and color removal efficiencies were 48.7% and 58.2% at an operating time of 240 min, initial electrolyte concentration of 3000 mg/L, and current density 80 mA/cm^2 without optimization. On the other hand, the COD and color removal efficiencies increased from 48.7% to 49.3% and 59.2%, respectively, by optimizing the operating parameters with the CCD. Titanium coated with iridium dioxide (IrO_2) and ruthenium dioxide (RuO_2) containing EO of mature landfill leachate pretreated by a sequencing batch reactor of BBD and RSM were used by Ref. [122]. Of 50 detected pollutants, 20 organics were entirely removed after the EO. The ANOVA showed that neither the quadratic effect nor the interaction effect was considerable on ammonia nitrogen removal, while both effects were remarkable on COD removal. Kaur et al. [34] performed the EO containing RuO_2-coated Ti electrode (Ti/RuO_2) to treat actual textile wastewater. At the optimum condition using the BBD and data analysis, the actual values of COD and color removal and energy consumption were found to be 80.0%, 97.25%, and 0.679 Wh, respectively. COD removal was found to be faster than that of color, revealing rate constant and R^2 values to be 0.025 min^{-1} and 0.015 min^{-1} and 0.988 and 0.986, respectively. It was also observed that most of the organics were oxidized entirely and eliminated during the process.

Kaur et al. [123] used RSM to study textile wastewater treatment by the EO process by a CCD model with operating time, pH, applied current, and retention time as independent operating variables and degradation efficiency as a dependent variable. Furthermore, the combined effects of independent variables on the degradation efficiency were presented by 3D response plots. An optimized degradation efficiency of 86% was obtained at an initial pH of 5.4, operating time of 130 min, a retention time of 143 min, and an applied current

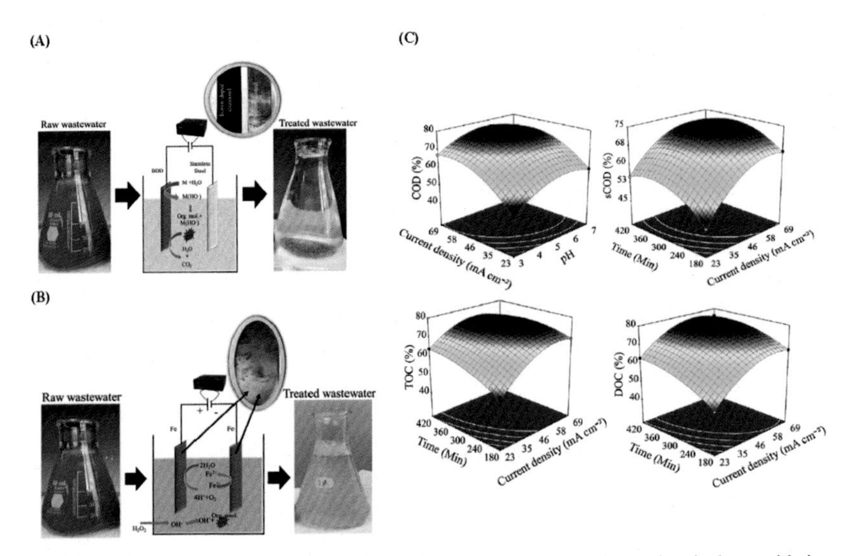

FIGURE 4.3 (A) Schematic diagram of the electro-oxidation, (B) electrochemical peroxidation process, and (C) 3D response plots for combined effects of current density, operating time, and initial pH on COD, sCOD, TOC, and DOC removal efficiency. *Reproduced with modification from S. Sharma, H. Simsek, Sugar beet industry process wastewater treatment using electrochemical methods and optimization of parameters using response surface methodology, Chemosphere, 238 (2020) 124669 with permission from Elsevier (License Number: 5017681326185).*

of 1.4 A. The authors reported that the EO process showed significant performance for textile wastewater treatment with the optimization of parameters using the CCD model. Recently, sugar beet wastewater with a high concentration of COD has been treated via the EO (Fig. 4.3A) and electrochemical peroxidation (ECP) (Fig. 4.3B) for organic removal in a laboratory-scale study. The treatment performances of both processes were optimized using the BBD and compared depending on energy consumption and removal efficiency during the operation. While EO could remove 75% of organics at optimum conditions of operation time (393 min), current density (48.5 mA/cm^2) (Fig. 4.3C) and pH (5.3), total and soluble COD and dissolved and total organic carbon were removed by 65%, 64%, 63%, and 66%, respectively, by the ECP at optimum conditions of operation time (361 min), current density (48 mA/cm^2) and H_2O_2 dosage (21 mL/L) [124]. A focused literature summary on the EO for the treatment of real wastewaters, including RSM studies, is listed in Table 4.3. There are fewer studies implementing the EO method for wastewater treatment than other electrochemical processes. Optimization of the EO process performance for textile, landfill leachate, slaughterhouse, gold leaching, and pharmaceutical wastewater treatment applications was analyzed using RSM with the BBD model. Independent variables were commonly specified as operating time and current density to maximize the removal of COD and minimize energy consumption.

TABLE 4.3 Summary of optimization of electro-oxidation process for wastewater treatment with response surface methodology (RSM).

	RSM					Optimum conditions			
Water type	Design model	Independent variables	Dependent responses	Statistical software	Run number	Target of model	Independent parameters	Dependent parameters	References
Pharmaceutical wastewater	BBD	T_{EC}, min j, mA/cm^2 pH	$\%Re_{,COD}$	Design-Expert	15	Maximum Re for COD minimum applied current and operating time	$T_{EC} = 86.89$ min $j = 76.1$ mA/cm^2 $pH = 6.56$	$\%Re_{,COD} = 30.9$	[125]
Landfill leachate wastewater	BBD	T_{EC}, min j, mA/cm^2 $C_{,Na2SO4}$, mol/L	$\%Re_{,COD}$ $\%Re_{,NH3-N}$	Design-Expert	15	Maximum Re for COD and NH_3-N	$T_{EC} = 37.1$ min $j = 242.84$ mA/cm^2 $C_{,Na2SO4} = 0.07$ mol/L	$\%Re_{,COD} = 54.99$ $\%Re_{,NH3-N} = 71.07$	[126]
Textile dye wastewater	BBD	T_{EC}, h j, A/dm^2 V_R, mL	$\%Re_{,COD}$	Minitab	15	Maximum Re for COD	$T_{EC} = 6$ h $j = 4.0$ A/dm^2 $V_R = 300$ mL	$\%Re_{,COD} = 97.17$	[127]
Gold leaching wastewater	BBD	j, mA/cm^2 $C_{,CN}$, mg/L M_i, rpm	$\%Re_{,COD}$ X_{CN} ENC, kWh/kg	Statgraphics Centurion	15	Maximum Re for COD, and maximum X_{CN}, and minimum ENC	$j = 100$ mA/cm^2 $C_{,CN} = 1000$ mg/L $M_i = 750$ rpm	$\%Re_{,COD} = 52.63$ $X_{CN} = 0.65$ $ENC = 38.2$ kWh/kg	[128]

Continued

TABLE 4.3 Summary of optimization of electro-oxidation process for wastewater treatment with response surface methodology (RSM).—cont'd

	RSM					Optimum conditions			
Water type	Design model	Independent variables	Dependent responses	Statistical software	Run number	Target of model	Independent parameters	Dependent parameters	References
Slaughterhouse wastewater	NA	T_{EC}, min j, mA/cm^2 C_{COD}, mg/L	$\%Re_{,COD}$ $\%Re_{,BOD}$ $\%Re_{,color}$	Design-Expert	32	Maximum Re for COD, BOD, and color	$T_{EC} = 55$ min $j = 30$ mA/cm^2 $C_{COD} = 220$ mg/L	$\%Re_{,COD} = 82.2$ $\%Re_{,BOD} = 85.5$ $\%Re_{,color} = 96.4$	[129]
Landfill leachate wastewater	NA	T_{EC}, h j, mA/cm^2 C_{COD}, mg/L	$\%Re_{,COD}$ $\%Re_{,BOD}$ $\%Re_{,color}$ $R_{BOD:COD}$ pH	Design-Expert	32	Maximum Re for COD, BOD, and color	$T_{EC} = 4$ h $j = 79.9$ mA/cm^2 $C_{COD} = 1414$ mg/L	$\%Re_{,COD} = 68.5$ $\%Re_{,BOD} = 73.9$ $\%Re_{,color} = 79.9$ $R_{BOD:COD} = 0.036$ pH = 9.24	[130]

BBD, Box–Behnken design; *CN*, cyanide; C_{Na2SO4}, Na_2SO_4 concentration; j, current density; *Mi*, mixing intensity; *NA*, not available; *R*, ratio; T_{EC}, electrolysis time; V_R, volume of reactor; X_{CN}, cyanide conversion.

3.4 Hybrid processes

The existing methods alone may not be sufficient for producing effluents of reusable quality and safe in terms of performance and cost. Two or more technologies can be integrated to obtain better process intensification. Daghrir et al. [131] studied the performance of a hybrid process combining EC and EO to treat domestic wastewater loaded with organic matter. Al (or Fe) as bipolar and sacrificial and graphite as monopolar electrodes including process were investigated in terms of simultaneously producing a coagulant and an oxidant agent. The FFD and CCD methodologies were successively used to define the optimal operating conditions for the COD removal. The treatment using Al electrode and a current intensity of 0.7 A during 39 min was determined to be the optimal conditions in terms of cost/effectiveness, which resulted in 78% of COD removal for a total cost of 0.78 US $/m^3, revealing that EC and EO could form the basis of a process capable of removing COD from domestic, municipal, and industrial effluents. Borba et al. [132] conducted a study to determine optimal experimental conditions for the photo-peroxi-EC (PPEC) hybrid process to treat tannery industrial wastewater using RSM. The optimization analysis presented the optimum conditions of the PPEC process with a pH of 4, a current density of 34.2 mA/cm^2, 6 g of H_2O_2/L, and an electrolysis time of 120 min, suggesting satisfactory results for the PPEC reaction in terms of both efficient chromium and COD removal and low operating cost. The chemical coagulation (CC)-EO hybrid process was reported as another integrated method for treating wastewater produced from the textile industry. First, the best dosage of the coagulant and the pH of the process were determined as 600 mg/L of $Al_2(SO_4)_3$ and 9.3, respectively, using the jar test for the CC, yielding 93% of turbidity, 53% of COD, and 24% of TOC removal. Then, CC effluent was treated by the EO having optimum operational conditions of conductivity (4.7 mS/cm), current density (15 mA/cm^2), and pH (5.6), determined by the BBD and RSM. At these conditions, the sequential CC-EO process produced 100% of color, 93.5% of COD, and 75% of TOC removal efficiencies after 45 min of reaction time with an estimated operating cost of 6.91 USD/m^3 [102,103]. Sharma and Simsek [133] conducted a study on combined EC and EO and the ECP processes for the treatment of canola oil refinery (COR) wastewaters. For the EC and EO process, the optimized values were a current density of 13.27 mA/cm^2 and an electrolysis time of 403 min for soluble COD (sCOD) removal efficiency of 98.6%, with DOC removal efficiency of 95.3% at 10.83 mA/cm^2 of current density and 420 min electrolysis time. This study revealed that the application of combined EC and EO process had a significant potential in treating canola oil refinery wastewater, handling the issues related to biological processes. The combined EC and EO process has recently been compared with the ECP to treat sunflower oil refinery wastewater. Optimized conditions of the EC process using BBD were determined with pH (6), applied current density (9.16 mA/cm^2), and operation time (19.06 min), achieving >75% of organics degradation. The treated effluent from the EC was then sent to the EO process and organics were removed (90%–95%) from the sunflower oil refinery wastewater under optimized pH (5.8), current density (8.58 mA/cm^2), and operation time (243 min). It was observed that the combination of EC and EO methods was successful and more efficient than the ECP process in removing higher organic pollutants with less energy consumption [134]. Ibarra-Taquez et al. [135] optimized the EC-EO hybrid process with the BBD model for coffee effluent wastewater treatment. The effects of independent operating variables such as initial pH (2.5–8) and the current densities of EC and EO (50–150 and 200–500 A/m^2, respectively) were investigated on the COD removal efficiency as a response. An optimum COD removal efficiency of 74.1% was achieved at an initial pH of 7.98 and EC and EO current densities of 149.23 and 500 A/m^2, respectively. Similarly, a continuous multistage EC-EO hybrid process was used for textile wastewater treatment [136]. The RSM with BBD model was conducted to determine the optimum levels of operating variables with minimizing operational cost for maximum COD removal efficiency. A maximized COD removal efficiency of 71.44% and minimized operating cost of 1.47 USD/m^3 at a pH of 4, a current density of 4.1 mA/cm^2, and a conductivity of 3.7 mS/cm were predicted by the BBD model.

4. Merits and demerits of RSM

RSM estimates a response based on a combination of factor levels and determining the optimum operating conditions to maximize a system's performance. One of the primary advantages of RSM is that a large amount of information is obtained from a limited number of experiments. Building models and graphical illustrations can study the main effects of variables and their interaction on the response. It determines the factor levels providing the optimum response and optimum conditions resulting from multiple responses. The duration of the project can be estimated owing to well-planned experiments. Besides these advantages, it also has some limitations. One of the most important limitations is to fit data only to first- or second-order polynomials that restrict the behavior of responses so that it cannot explain all systems containing the curvature. Another shortcoming is that as the number of independent variables increases, the number of experiments also increases, which exhibits low prediction capability outside the experimental domain; therefore, the factors of interest should vary continuously over the experimental domain. Although the model can tell what happens at different conditions, it cannot explain the process mechanism [22]. For future studies, the number of applied scenarios can be considered to better understand the logic of optimization in several scopes. Considering realistic process conditions, both independent and dependent parameters can be varied to enhance the optimization. Besides the removal efficiency, other important parameters, such as operating cost, energy, and electrode consumption, can be highly encouraged based on the process of interest. Instead of using single RSM model, multimodels can also be used and compared to determine meaningful outcomes.

5. Conclusions

In recent years, electrochemical treatment methods have gained increasing interest due to their superior technical merits for the effective removal of a broad range of pollutants in wastewaters rarely eliminated by conventional methods. This chapter has reviewed and verified the application, usability, and limitations of RSM modeling to analyze and optimize different electrochemical processes used in various industrial wastewater treatments. In most of the related reviewed articles, diagnostic and statistical analyses showed that RSM could be a reliable tool to optimize treatment process parameters by significantly improving the removal efficiency and operational cost reduction. However, there was inadequate information on the RSM implementation in some for optimization on the preliminary work regarding the range of process variables.

The optimization of electrochemical treatment technologies using RSM for wastewater treatment has been widely implemented in EC, EO, EF, and their combinations. The EC process is the most reported technique due to its relatively low cost, low sludge, and almost no toxic by-product formation, no need for pH adjustment, ease of operation and maintenance, and effectiveness. RSM with the CCD model was generally applied to optimize independent operating variables such as initial pH, current density, and electrolysis time, which resulted in maximum removal efficiency and minimum operating cost of the EC process. The BBD model was also preferred in some EC studies. Moreover, the EO process removal performance was commonly optimized using RSM with the BBD using the applied current and electrolysis time, while both BBD and CCD models were well fitted for the EF process optimization considering H_2O_2 concentrations, in addition to other parameters. To conclude, RSM can be implemented in pilot and/or industrial-scale wastewater treatment processes to determine its applicability in real life. It is unfortunately limited by lab-scale studies, which resulted in a reduction in the operating and fixed costs and an improvement in the process performance as the promising outcomes.

References

[1] T.W. Seow, C.K. Lim, M.H.M. Nor, M.F.M. Mubarak, C.Y. Lam, A. Yahya, Z. Ibrahim, Review on wastewater treatment technologies, Int. J. Appl. Environ. Sci. 11 (1) (2016) 111–126.

[2] J. Liu, H. Yang, S.N. Gosling, M. Kummu, M. Flörke, S. Pfister, N. Hanasaki, Y. Wada, X. Zhang, C. Zheng, Water scarcity assessments in the past, present, and future, Earths Future 5 (6) (2017) 545–559.

[3] H. Tahir, M. Sultan, N. Akhtar, U. Hameed, T. Abid, Application of natural and modified sugar cane bagasse for the removal of dye from aqueous solution, J. Saudi Chem. Soc. 20 (2016) S115–S121.

[4] H. Yin, P. Qiu, Y. Qian, Z. Kong, X. Zheng, Z. Tang, H. Guo, Textile wastewater treatment for water reuse: a case study, Processes 7 (1) (2019) 34.

[5] M. Ahmadi, F. Vahabzadeh, B. Bonakdarpour, E. Mofarrah, M. Mehranian, Application of the central composite design and response surface methodology to the advanced treatment of olive oil processing wastewater using Fenton's peroxidation, J. Hazard Mater. 123 (1−3) (2005) 187−195.

[6] D. Cecconet, D. Molognoni, A. Callegari, A.G. Capodaglio, Agro-food industry wastewater treatment with microbial fuel cells: energetic recovery issues, Int. J. Hydrog. Energy 43 (1) (2018) 500−511.

[7] G.W. Miller, Integrated concepts in water reuse: managing global water needs, Desalination 187 (1−3) (2006) 65−75.

[8] M. Dakiky, M. Khamis, A. Manassra, M. Mer'Eb, Selective adsorption of chromium (VI) in industrial wastewater using low-cost abundantly available adsorbents, Adv. Environ. Res. 6 (4) (2002) 533−540.

[9] Y.K. Recepoğlu, N. Kabay, İ. Yılmaz-İpek, M. Arda, M. Yüksel, K. Yoshizuka, S. Nishihama, Deboronation of geothermal water using N-methyl-D-glucamine based chelating resins and a novel fiber adsorbent: batch and column studies, J. Chem. Technol. Biotechnol. 92 (7) (2017) 1540−1547.

[10] H. Karimi-Maleh, M. Shafieizadeh, M.A. Taher, F. Opoku, E.M. Kiarii, P.P. Govender, S. Ranjbari, M. Rezapour, Y. Orooji, The role of magnetite/graphene oxide nano-composite as a high-efficiency adsorbent for removal of phenazopyridine residues from water samples, an experimental/theoretical investigation, J. Mol. Liq. 298 (2020) 112040.

[11] S. Sivaprakasam, S. Mahadevan, S. Sekar, S. Rajakumar, Biological treatment of tannery wastewater by using salt-tolerant bacterial strains, Microb. Cell Factories 7 (1) (2008) 1−7.

[12] T. Zhang, L. Ding, H. Ren, X. Xiong, Ammonium nitrogen removal from coking wastewater by chemical precipitation recycle technology, Water Res. 43 (20) (2009) 5209−5215.

[13] S. Verma, B. Prasad, I.M. Mishra, Pretreatment of petrochemical wastewater by coagulation and flocculation and the sludge characteristics, J. Hazard Mater. 178 (1−3) (2010) 1055−1064.

[14] P. Peralta-Zamora, E. Esposito, R. Pelegrini, R. Groto, J. Reyes, N. Durán, Effluent treatment of pulp and paper, and textile industries using immobilised horseradish peroxidase, Environ. Technol. 19 (1) (1998) 55−63.

[15] E. Hassanzadeh, M. Farhadian, A. Razmjou, N. Askari, An efficient wastewater treatment approach for a real woolen textile industry using a chemical assisted NF membrane process, Environ. Nanotechnol. Monit. Manag. 8 (2017) 92−96.

[16] F. Ebrahimi, Y. Orooji, A. Razmjou, Applying membrane distillation for the recovery of nitrate from saline water using pvdf membranes modified as superhydrophobic membranes, Polymers 12 (12) (2020) 2774.

[17] V. Vatanpour, S. Khorshidi, Surface modification of polyvinylidene fluoride membranes with ZIF-8 nanoparticles layer using interfacial method for BSA separation and dye removal, Mater. Chem. Phys. 241 (2020) 122400.

[18] A. Karimi, A. Khataee, A. Ghadimi, V. Vatanpour, Ball-milled Cu_2S nanoparticles as an efficient additive for modification of the PVDF ultrafiltration membranes: application to separation of protein and dyes, J. Environ. Chem. Eng. 9 (2) (2021) 105115.

[19] C. Magro, E.P. Mateus, J.M. Paz-Garcia, A.B. Ribeiro, Emerging organic contaminants in wastewater: understanding electrochemical reactors for triclosan and its by-products degradation, Chemosphere 247 (2020) 125758.

[20] J. Treviño-Reséndez, A. Medel, Y. Meas, Electrochemical technologies for treating petroleum industry wastewater, Curr. Opin. Electrochem. 27 (2021) 100690.

[21] J. Radjenovic, D.L. Sedlak, Challenges and opportunities for electrochemical processes as next-generation technologies for the treatment of contaminated water, Environ. Sci. Technol. 49 (19) (2015) 11292–11302.

[22] A.T. Nair, A.R. Makwana, M.M. Ahammed, The use of response surface methodology for modelling and analysis of water and wastewater treatment processes: a review, Water Sci. Technol. 69 (3) (2014) 464–478.

[23] R. Keyikoglu, O. Karatas, H. Rezania, M. Kobya, V. Vatanpour, A. Khataee, A review on treatment of membrane concentrates generated from landfill leachate treatment processes, Separ. Purif. Technol. 259 (2020) 118182.

[24] M. Moradi, Y. Vasseghian, A. Khataee, M. Kobya, H. Arabzade, E.-N. Dragoi, Service life and stability of electrodes applied in electrochemical advanced oxidation processes: a comprehensive review, J. Ind. Eng. Chem. 87 (2020) 18–39.

[25] A.R. Khataee, M.B. Kasiri, L. Alidokht, Application of response surface methodology in the optimization of photocatalytic removal of environmental pollutants using nanocatalysts, Environ. Technol. 32 (15) (2011) 1669–1684.

[26] X. Zhu, J. Tian, R. Liu, L. Chen, Optimization of fenton and electro-fenton oxidation of biologically treated coking wastewater using response surface methodology, Separ. Purif. Technol. 81 (3) (2011) 444–450.

[27] M. Safarpour, J. Barzin, A. Khataee, Z. Kordkatooli, Two-stage phase separation of cellulose acetate membranes modified with plasma-treated natural zeolite: response surface modeling, Polym. Adv. Technol. 30 (4) (2019) 889–901.

[28] S. Karimifard, M.R.A. Moghaddam, Application of response surface methodology in physicochemical removal of dyes from wastewater: a critical review, Sci. Total Environ. 640 (2018) 772–797.

[29] L. Zhang, Y. Zeng, Z. Cheng, Removal of heavy metal ions using chitosan and modified chitosan: a review, J. Mol. Liq. 214 (2016) 175–191.

[30] A.R. Khataee, M. Fathinia, S. Aber, M. Zarei, Optimization of photocatalytic treatment of dye solution on supported TiO_2 nanoparticles by central composite design: intermediates identification, J. Hazard Mater. 181 (1–3) (2010a) 886–897.

[31] A.R. Khataee, G. Dehghan, E. Ebadi, M. Pourhassan, Central composite design optimization of biological dye removal in the presence of macroalgae Chara sp, Clean 38 (8) (2010b) 750–757.

[32] K. Cruz-González, O. Torres-Lopez, A.M. García-León, E. Brillas, A. Hernández-Ramírez, J.M. Peralta-Hernández, Optimization of electro-Fenton/BDD process for decolorization of a model azo dye wastewater by means of response surface methodology, Desalination 286 (2012) 63–68.

[33] S. Ellouze, S. Kessemtini, D. Clematis, G. Cerisola, M. Panizza, S.C. Elaoud, Application of Doehlert design to the electro-Fenton treatment of Bismarck Brown Y, J. Electroanal. Chem. 799 (2017) 34–39.

[34] P. Kaur, J.P. Kushwaha, V.K. Sangal, Evaluation and disposability study of actual textile wastewater treatment by electro-oxidation method using Ti/RuO_2 anode, Process Saf. Environ. Protect. 111 (2017) 13–22.

[35] E. Sharifpour, H. Haddadi, M. Ghaedi, Optimization of simultaneous ultrasound assisted toxic dyes adsorption conditions from single and multi-components using central composite design: application of derivative spectrophotometry and evaluation of the kinetics and isotherms, Ultrason. Sonochem. 36 (2017) 236–245.

[36] L. Alidokht, S. Oustan, A. Khataee, CrVI reductive transformation process by humic acid extracted from bog peat: effect of variables and multi-response modeling, Chemosphere 263 (2021) 128221.

[37] S.A. Maruyama, S.V. Palombini, T. Claus, F. Carbonera, P.F. Montanher, N. E. de Souza, J.V. Visentainer, S. Gomes, M. Matsushita, Application of box-behnken design to the study of fatty acids and antioxidant activity from enriched white bread, J. Braz. Chem. Soc. 24 (9) (2013) 1520–1529.

[38] A. Ahmad, M.U. Rehman, A.F. Wali, H.A. El-Serehy, F.A. Al-Misned, S.N. Maodaa, H.M. Aljawdah, T.M. Mir, P. Ahmad, Box–Behnken response surface design of polysaccharide extraction from rhododendron arboreum and the evaluation of its antioxidant potential, Molecules 25 (17) (2020) 3835.

[39] R. Ragonese, M. Macka, J. Hughes, P. Petocz, The use of the Box–Behnken experimental design in the optimisation and robustness testing of a capillary electrophoresis method for the analysis of ethambutol hydrochloride in a pharmaceutical formulation, J. Pharmaceut. Biomed. Anal. 27 (6) (2002) 995–1007.

[40] S.A. Pasma, R. Daik, M.Y. Maskat, O. Hassan, Application of Box-Behnken design in optimization of glucose production from oil palm empty fruit bunch cellulose, Int. J. Polym. Sci. 2013 (2013), 104502.

[41] S. Jibril, N. Basar, H.M. Sirat, R.A. Wahab, N.A. Mahat, L. Nahar, S.D. Sarker, Application of Box–Behnken design for ultrasound-assisted extraction and recycling preparative HPLC for isolation of anthraquinones from *Cassia singueana*, Phytochem. Anal. 30 (1) (2019) 101–109.

[42] A.R. Khataee, A. Karimi, R.D.C. Soltani, M. Safarpour, Y. Hanifehpour, S.W. Joo, Europium-doped ZnO as a visible light responsive nanocatalyst: sonochemical synthesis, characterization and response surface modeling of photocatalytic process, Appl. Catal. Gen. 488 (2014) 160–170.

[43] A. Khataee, S. Bozorg, B. Vahid, Response surface optimization of heterogeneous Fenton-like degradation of sulfasalazine using Fe-impregnated clinoptilolite nanorods prepared by Ar-plasma, Res. Chem. Intermed. 43 (7) (2017) 3989–4005.

[44] M. Isam, L. Baloo, S.R.M. Kutty, S. Yavari, Optimisation and modelling of pb (ii) and cu (ii) biosorption onto red algae (*gracilaria changii*) by using response surface methodology, Water 11 (11) (2019) 2325.

[45] A.J. Sisi, A. Khataee, M. Fathinia, B. Vahid, Y. Orooji, Comparative study of sonocatalytic process using MOF-5 and peroxydisulfate by central composite design and artificial neural network, J. Mol. Liq. 316 (2020) 113801.

[46] G.E.P. Box, K.B. Wilson, On the experimental attainment of optimum conditions, J. Roy. Stat. Soc. B 13 (1) (1951) 1–38.

[47] A.N.Z. Alshehria, K.M. Ghanem, S.M. Al-Garni, Application of a five level central composite design to optimize operating conditions for electricity generation in a microbial fuel cell, J. Taibah Univ. Sci. 10 (6) (2016) 797–804.

[48] M. Khaoula, B. Wided, H. Chiraz, H. Béchir, Boron removal by electrocoagulation using full factorial design, J. Water Resour. Protect. 5 (2013) 867–875.

[49] S.A. Kordkandi, M. Forouzesh, Application of full factorial design for methylene blue dye removal using heat-activated persulfate oxidation, J. Taiwan Inst. Chem. Eng. 45 (5) (2014) 2597–2604.

[50] B. Zhou, Y. Li, J. Gillespie, G.-Q. He, R. Horsley, P. Schwarz, Doehlert matrix design for optimization of the determination of bound deoxynivalenol in barley grain with trifluoroacetic acid (TFA), J. Agric. Food Chem. 55 (25) (2007) 10141–10149.

[51] A. Witek-Krowiak, K. Chojnacka, D. Podstawczyk, A. Dawiec, K. Pokomeda, Application of response surface methodology and artificial neural network methods in modelling and optimization of biosorption process, Bioresour. Technol. 160 (2014) 150–160.

[52] R.A. Fisher, Statistical methods for research workers, in: Breakthroughs in Statistics, Springer, 1992, pp. 66–70.

[53] A.Y. Goren, M. Kobya, Arsenic removal from groundwater using an aerated electrocoagulation reactor with 3D Al electrodes in the presence of anions, Chemosphere 263 (2021) 128253.

[54] E. Sık, M. Kobya, E. Demirbas, M.S. Oncel, A.Y. Goren, Removal of As(V) from groundwater by a new electrocoagulation reactor using Fe ball anodes: optimization of operating parameters, Desalin. Water Treat. 56 (5) (2015) 1177–1190.

[55] E. Demirbas, M. Kobya, M.S. Oncel, E. Sık, A.Y. Goren, Arsenite removal from groundwater in a batch electrocoagulation process: optimization through response surface methodology, Separ. Sci. Technol. 54 (5) (2019) 775–785.

[56] J.N. Hakizimana, B. Gourich, M. Chafi, Y. Stiriba, C. Vial, P. Drogui, J. Naja, Electrocoagulation process in water treatment: a review of electrocoagulation modeling approaches, Desalination 404 (2017) 1–21.

[57] D.T. Moussa, M.H. El-Naas, M. Nasser, M.J. Al-Marri, A comprehensive review of electrocoagulation for water treatment: potentials and challenges, J. Environ. Manag. 186 (2017) 24–41.

[58] A.Y. Goren, M. Kobya, M.S. Oncel, Arsenite removal from groundwater by aerated electrocoagulation reactor with Al ball electrodes: human health risk assessment, Chemosphere 251 (2020) 126363.

[59] M. Kobya, P.I. Omwene, Z. Ukundimana, Treatment and operating cost analysis of metalworking wastewaters by a continuous electrocoagulation reactor, J. Environ. Chem. Eng. 8 (2) (2020) 103526.

[60] K. Hendaoui, F. Ayari, I.B. Rayana, R.B. Amar, F. Darragi, M. Trabelsi-Ayadi, Real indigo dyeing effluent decontamination using continuous electrocoagulation cell: study and optimization using response surface methodology, Process Saf. Environ. Protect. 116 (2018) 578–589.

[61] S. Abbasi, M. Mirghorayshi, S. Zinadini, A.A. Zinatizadeh, A novel single continuous electrocoagulation process for treatment of licorice processing wastewater: optimization of operating factors using RSM, Process Saf. Environ. Protect. 134 (2020) 323–332.

[62] M.R. Samarghandi, A. Dargahi, A. Shabanloo, H.Z. Nasab, Y. Vaziri, A. Ansari, Electrochemical degradation of methylene blue dye using a graphite doped PbO_2 anode: optimization of operational parameters, degradation pathway and improving the biodegradability of textile wastewater, Arab. J. Chem. 13 (8) (2020) 6847–6864.

[63] S. Ahmadzadeh, M. Dolatabadi, Electrochemical treatment of pharmaceutical wastewater through electrosynthesis of iron hydroxides for practical removal of metronidazole, Chemosphere 212 (2018) 533–539.

[64] M.K. Oden, H. Sari-Erkan, Treatment of metal plating wastewater using iron electrode by electrocoagulation process: optimization and process performance, Process Saf. Environ. Protect. 119 (2018) 207–217.

[65] P. Asaithambi, D. Beyene, A.R.A. Aziz, E. Alemayehu, Removal of pollutants with determination of power consumption from landfill leachate wastewater using an electrocoagulation process: optimization using response surface methodology (RSM), Appl. Water Sci. 8 (2) (2018) 1–12.

[66] J. Morales-Rivera, B. Sulbarán-Rangel, K.J. Gurubel-Tun, J. del Real-Olvera, V. Zúñiga-Grajeda, Modeling and optimization of COD removal from cold meat industry wastewater by electrocoagulation using computational techniques, Processes 8 (9) (2020) 1139.

[67] C. Tsioptsias, D. Petridis, N. Athanasakis, I. Lemonidis, A. Deligiannis, P. Samaras, Post-treatment of molasses wastewater by electrocoagulation and process optimization through response surface analysis, J. Environ. Manag. 164 (2015) 104–113.

[68] K. Thirugnanasambandham, S. Kandasamy, V. Sivakumar, R. Mohanavelu, Modeling of by-product recovery and performance evaluation of Electro-Fenton treatment technique to treat poultry wastewater, J. Taiwan Inst. Chem. Eng. 46 (2015) 89–97.

[69] K. Thirugnanasambandham, V. Sivakumar, J.P. Maran, Optimization of process parameters in electrocoagulation treating chicken industry wastewater to recover hydrogen gas with pollutant reduction, Renew. Energy 80 (2015) 101–108.

[70] A.R. Makwana, M.M. Ahammed, Continuous electrocoagulation process for the post-treatment of anaerobically treated municipal wastewater, Process Saf. Environ. Protect. 102 (2016) 724–733.

[71] A. Aygun, B. Nas, M.F. Sevimli, Treatment of reactive dyebath wastewater by electrocoagulation process: optimization and cost-estimation, Kor. J. Chem. Eng. 36 (9) (2019) 1441–1449.

[72] O. Sahu, B. Mazumdar, P.K. Chaudhari, Electrochemical treatment of sugar industry wastewater: process optimization by response surface methodology, Int. J. Environ. Sci. Technol. 16 (3) (2019) 1527–1540.

[73] H.M. Bui, Optimization of electrocoagulation of instant coffee production wastewater using the response surface methodology, Pol. J. Chem. Technol. 19 (2) (2017) 67–71.

[74] S. Veli, A. Arslan, D. Bingöl, Application of response surface methodology to electrocoagulation treatment of hospital wastewater, Clean 44 (11) (2016) 1516–1522.

[75] Z. Beiramzadeh, M. Baqersad, M. Aghababaei, Application of the response surface methodology (RSM) in heavy metal removal from real power plant wastewater using electrocoagulation, Eur. J. Environ. Civil Eng. 23 (2019) 1–19.

[76] H.M. Bui, Applying response surface methodology to optimize the treatment of swine slaughterhouse wastewater by electrocoagulation, Pol. J. Environ. Stud. 27 (5) (2018).

[77] D. Ekawati, R. Nadira, Application of response surface methodology (RSM) for wastewater of hospital by using electrocoagulation, IOP Conf. Ser. Mater. Sci. Eng. 345 (1) (2018) 12011.

[78] U.T. Un, A. Kandemir, N. Erginel, S.E. Ocal, Continuous electrocoagulation of cheese whey wastewater: an application of response surface methodology, J. Environ. Manag. 146 (2014) 245–250.

[79] K. Ulucan, H.A. Kabuk, F. Ilhan, U. Kurt, Electrocoagulation process application in bilge water treatment using response surface methodology, Int. J. Electrochem. Sci 9 (5) (2014) 2316.

[80] G. Varank, H. Erkan, S. Yazycy, A. Demır, G. Engin, Electrocoagulation of tannery wastewater using monopolar electrodes: process optimization by response surface methodology, Int. J. Environ. Res. 8 (1) (2014) 165–180.

[81] A.R. Makwana, M.M. Ahammed, Electrocoagulation process for the post-treatment of anaerobically treated urban wastewater, Separ. Sci. Technol. 52 (8) (2017) 1412–1422.

[82] N. Huda, A.A.A. Raman, M.M. Bello, S. Ramesh, Electrocoagulation treatment of raw landfill leachate using iron-based electrodes: effects of process parameters and optimization, J. Environ. Manag. 204 (2017) 75–81.

[83] S.Y. Guvenc, Y. Okut, M. Ozak, B. Haktanir, M.S. Bilgili, Process optimization via response surface methodology in the treatment of metal working industry wastewater with electrocoagulation, Water Sci. Technol. 75 (4) (2017) 833–846.

[84] T. Karichappan, S. Venkatachalam, P.M. Jeganathan, Optimization of electrocoagulation process to treat grey wastewater in batch mode using response surface methodology, J. Environ. Health Sci. Eng. 12 (1) (2014) 1–8.

[85] M. Kobya, E. Demirbas, M. Bayramoglu, M.T. Sensoy, Optimization of electrocoagulation process for the treatment of metal cutting wastewaters with response surface methodology, Water Air Soil Pollut. 215 (1) (2011) 399–410.

[86] A. Nasution, B.L. Ng, E. Ali, Z. Yaakob, S.K. Kamarudin, Electrocoagulation of palm oil mill effluent for treatment and hydrogen production using response surface methodology, Pol. J. Environ. Stud. 23 (5) (2014) 1669–1677.

[87] E. Gengec, M. Kobya, E. Demirbas, A. Akyol, K. Oktor, Optimization of baker's yeast wastewater using response surface methodology by electrocoagulation, Desalination 286 (2012) 200–209.

[88] F.R. Espinoza-Quiñones, A.N. Módenes, P.S. Theodoro, S.M. Palácio, D.E.G. Trigueros, C.E. Borba, M.M. Abugderah, A.D. Kroumov, Optimization of the iron electro-coagulation process of Cr, Ni, Cu, and Zn galvanization by-products by using response surface methodology, Separ. Sci. Technol. 47 (5) (2012) 688–699.

[89] O. Chavalparit, M. Ongwandee, Optimizing electrocoagulation process for the treatment of biodiesel wastewater using response surface methodology, J. Environ. Sci. 21 (11) (2009) 1491–1496.

[90] S.Y. Guvenc, H.S. Erkan, G. Varank, M.S. Bilgili, G.O. Engin, Optimization of paper mill industry wastewater treatment by electrocoagulation and electro-Fenton processes using response surface methodology, Water Sci. Technol. 76 (8) (2017) 2015–2031.

[91] M.K. Oden, Treatment of CNC industry wastewater by electrocoagulation technology: an application through response surface methodology, Int. J. Environ. Anal. Chem. 100 (1) (2020) 1–19.

[92] E. Brillas, I. Sirés, M.A. Oturan, Electro-Fenton process and related electrochemical technologies based on Fenton's reaction chemistry, Chem. Rev. 109 (12) (2009) 6570–6631.

[93] H. He, Z. Zhou, Electro-Fenton process for water and wastewater treatment, Crit. Rev. Environ. Sci. Technol. 47 (21) (2017) 2100–2131.

[94] Camcıoğlu, B. Özyurt, S. Sengül, H. Hapoğlu, Evaluation of electro-fenton method on cheese whey treatment: optimization through response surface methodology, Desalin. Water Treat. 172 (2019) 270–280. October 2018.

[95] R. Davarnejad, J. Azizi, A. Joodaki, S. Mansoori, Optimization of electro-Fenton oxidation of carbonated soft drink industry wastewater using response surface methodology, Maced. J. Chem. Chem. Eng. 39 (2) (2020) 129–137.

[96] G. Varank, S. Yazici Guvenc, A. Demir, A comparative study of electrocoagulation and electro-Fenton for food industry wastewater treatment: multiple response optimization and cost analysis, Separ. Sci. Technol. 53 (17) (2018) 2727–2740.

[97] R. Davarnejad, J. Azizi, Alcoholic wastewater treatment using electro-Fenton technique modified by Fe_2O_3 nanoparticles, J. Environ. Chem. Eng. 4 (2) (2016) 2342–2349.

[98] A. Chairunnisak, B. Arifin, H. Sofyan, M.R. Lubis, Comparative study on the removal of COD from POME by electrocoagulation and electro-Fenton methods: process optimization, IOP Conf. Ser. Mater. Sci. Eng. 334 (1) (2018) 12026.

[99] G.K. Akkaya, H.S. Erkan, E. Sekman, S. Top, H. Karaman, M.S. Bilgili, G.O. Engin, Modeling and optimizing Fenton and electro-Fenton processes for dairy wastewater treatment using response surface methodology, Int. J. Environ. Sci. Technol. 16 (5) (2019) 2343–2358.

[100] P. Kaur, J.P. Kushwaha, V.K. Sangal, Transformation products and degradation pathway of textile industry wastewater pollutants in Electro-Fenton process, Chemosphere 207 (2018) 690–698.

[101] R. Davarnejad, S. Nasiri, Slaughterhouse wastewater treatment using an advanced oxidation process: optimization study, Environ. Pollut. 223 (2017) 1–10.

[102] E. GilPavas, I. Dobrosz-Gomez, M.Á. Gómez-García, Optimization of solar-driven photo-electro-Fenton process for the treatment of textile industrial wastewater, J. Water Process Eng. 24 (2018) 49–55.

[103] E. GilPavas, I. Dobrosz-Gómez, M.Á. Gómez-García, Optimization of sequential chemical coagulation-electro-oxidation process for the treatment of an industrial textile wastewater, J. Water Process Eng. 22 (2018) 73–79.

[104] R. Davarnejad, M. Nikseresht, Dairy wastewater treatment using an electrochemical method: experimental and statistical study, J. Electroanal. Chem. 775 (2016) 364–373.

[105] S. Mohajeri, H.A. Aziz, M.H. Isa, M.A. Zahed, M.N. Adlan, Statistical optimization of process parameters for landfill leachate treatment using electro-Fenton technique, J. Hazard Mater. 176 (1–3) (2010) 749–758.

[106] B. Zhang, J. Sun, Q. Wang, N. Fan, J. Ni, W. Li, Y. Gao, Y.-Y. Li, C. Xu, Electro-Fenton oxidation of coking wastewater: optimization using the combination of central composite design and convex optimization method, Environ. Technol. 38 (19) (2017) 2456–2464.

[107] Y. Wang, X. Zhou, N. Jiang, G. Meng, J. Bai, Y. Lv, Treatment of biotreated coking wastewater by a heterogeneous electro-Fenton process using a novel Fe/activated carbon/Ni composite cathode, Int. J. Electrochem. Sci 15 (2020) 4567–4585.

[108] I. Basturk, G. Varank, S. Murat Hocaoglu, S. Yazici Guvenc, Medical Laboratory Wastewater Treatment by Electro-fenton Process: Modeling and Optimization Using Central Composite Design, Water Environment Research, 2020, pp. 1–16.

[109] G. Khajouei, S. Mortazavian, A. Saber, N.Z. Meymian, H. Hasheminejad, Treatment of composting leachate using electro-Fenton process with scrap iron plates as electrodes, Int. J. Environ. Sci. Technol. 16 (8) (2019) 4133–4142.

[110] R. Davarnejad, M. Sabzehei, F. Parvizi, S. Heidari, A. Rashidi, Study on soybean oil plant wastewater treatment using the electro-fenton technique, Chem. Eng. Technol. 42 (12) (2019) 2717–2725.

[111] R. Behfar, R. Davarnejad, Pharmaceutical wastewater treatment using UV-enhanced electro-Fenton process: comparative study, Water Environ. Res. 91 (11) (2019) 1526–1536.

[112] R. Davarnejad, M. Mohammadi, A.F. Ismail, Petrochemical wastewater treatment by electro-Fenton process using aluminum and iron electrodes: statistical comparison, J. Water Process Eng. 3 (2014) 18–25.

[113] X. Zhou, Z. Hou, L. Lv, J. Song, Z. Yin, Electro-Fenton with peroxi-coagulation as a feasible pre-treatment for high-strength refractory coke plant wastewater: parameters optimization, removal behavior and kinetics analysis, Chemosphere 238 (2020) 124649.

[114] K. Thirugnanasambandham, V. Sivakumar, Optimization of treatment of grey wastewater using Electro-Fenton technique—Modeling and validation, Process Saf. Environ. Protect. 95 (2015) 60–68.

[115] P. Kaur, V.K. Sangal, J.P. Kushwaha, Parametric study of electro-Fenton treatment for real textile wastewater, disposal study and its cost analysis, Int. J. Environ. Sci. Technol. 16 (2) (2019) 801–810.

[116] R.Q. Al-Khafaji, A.H.A.K. Mohammed, Optimization of continuous electro-fenton and photo electro-fenton processes to treat iraqi oilfield produced water using surface response methodology, IOP Conf. Ser. Mater. Sci. Eng. 518 (6) (2019) 62007.

[117] R. Davarnejad, M. Bakhshandeh, Olefin plant spent caustic wastewater treatment using electro-Fenton technique, Egypt. J. Pet. 27 (4) (2018) 573–581.

[118] A. Turan, R. Keyikoglu, M. Kobya, A. Khataee, Degradation of thiocyanate by electrochemical oxidation process in coke oven wastewater: role of operative parameters and mechanistic study, Chemosphere 255 (2020) 127014.

[119] J.A. Barrios, E. Becerril, C. De León, C. Barrera-Díaz, B. Jiménez, Electrooxidation treatment for removal of emerging pollutants in wastewater sludge, Fuel 149 (2015) 26–33.

[120] P. Piya-Areetham, K. Shenchunthichai, M. Hunsom, Application of electrooxidation process for treating concentrated wastewater from distillery industry with a voluminous electrode, Water Res. 40 (15) (2006) 2857–2864.

[121] S. Mohajeri, H.A. Aziz, M.H. Isa, M.A. Zahed, M.J.K. Bashir, M.N. Adlan, Application of the central composite design for condition optimization for semi-aerobic landfill leachate treatment using electrochemical oxidation, Water Sci. Technol. 61 (5) (2010) 1257–1266.

[122] H. Zhang, X. Ran, X. Wu, D. Zhang, Evaluation of electro-oxidation of biologically treated landfill leachate using response surface methodology, J. Hazard Mater. 188 (1–3) (2011) 261–268.

[123] P. Kaur, M.A. Imteaz, M. Sillanpää, V.K. Sangal, J.P. Kushwaha, Parametric optimization and MCR-ALS kinetic modeling of electro oxidation process for the treatment of textile wastewater, Chemometr. Intell. Lab. Syst. 203 (2020) 104027.

[124] S. Sharma, H. Simsek, Sugar beet industry process wastewater treatment using electrochemical methods and optimization of parameters using response surface methodology, Chemosphere 238 (2020) 124669.

[125] A.M. Deshpande, Ramakant, S. Satyanarayan, Treatment of pharmaceutical wastewater by electrochemical method: optimization of operating parameters by response surface methodology, J. Hazard. Toxic Radioact. Waste 16 (4) (2012) 316–326.

[126] J.E. Silveira, J.A. Zazo, G. Pliego, E.D. Bidóia, P.B. Moraes, Electrochemical oxidation of landfill leachate in a flow reactor: optimization using response surface methodology, Environ. Sci. Pollut. Control Ser. 22 (8) (2015) 5831–5841.

[127] J. Sendhil, P.K.A. Muniswaran, C.A. Basha, Real textile dye wastewater treatment by electrochemical oxidation: application of response surface methodology (Rsm), Int. J. ChemTech Res. 7 (6) (2015) 2681–2690.

[128] I. Dobrosz-Gómez, M.Á.G. García, G.H. Gaviria, E. GilPavas, Mineralization of cyanide originating from gold leaching effluent using electro-oxidation: multi-objective optimization and kinetic study, J. Appl. Electrochem. 50 (2) (2020) 217–230.

[129] Z.B. Awang, M.J.K. Bashir, S.R.M. Kutty, M.H. Isa, Post-treatment of slaughterhouse wastewater using electrochemical oxidation, Res. J. Chem. Environ. 15 (2) (2011) 229–237.

[130] M.J.K. Bashir, M.H. Isa, S.R.M. Kutty, Z.B. Awang, H.A. Aziz, S. Mohajeri, I.H. Farooqi, Landfill leachate treatment by electrochemical oxidation, Waste Manag. 29 (9) (2009) 2534–2541.

[131] R. Daghrir, P. Drogui, F. Zaviska, Effectiveness of a hybrid process combining electro-coagulation and electro-oxidation for the treatment of domestic wastewaters using response surface methodology, J. Environ. Sci. Health A 48 (3) (2013) 308–318.

[132] F.H. Borba, D. Seibert, L. Pellenz, F.R. Espinoza-Quiñones, C.E. Borba, A.N. Módenes, R. Bergamasco, Desirability function applied to the optimization of the Photoperoxi-Electrocoagulation process conditions in the treatment of tannery industrial wastewater, J. Water Process Eng. 23 (2018) 207–216.

[133] S. Sharma, H. Simsek, Treatment of canola-oil refinery effluent using electrochemical methods: a comparison between combined electrocoagulation+ electrooxidation and electrochemical peroxidation methods, Chemosphere 221 (2019) 630–639.

[134] S. Sharma, A. Aygun, H. Simsek, Electrochemical treatment of sunflower oil refinery wastewater and optimization of the parameters using response surface methodology, Chemosphere 249 (2020) 126511.

[135] H.N. Ibarra-Taquez, E. GilPavas, E.R. Blatchley III, M.-Á. Gómez-García, I. Dobrosz-Gómez, Integrated electrocoagulation-electrooxidation process for the treatment of soluble coffee effluent: optimization of COD degradation and operation time analysis, J. Environ. Manag. 200 (2017) 530–538.

[136] E. GilPavas, P. Arbeláez-Castaño, J. Medina, D.A. Acosta, Combined electrocoagulation and electro-oxidation of industrial textile wastewater treatment in a continuous multi-stage reactor, Water Sci. Technol. 76 (9) (2017) 2515–2525.

Chapter 5

Artificial intelligence and sustainability: solutions to social and environmental challenges

Firouzeh Taghikhah[1,2], Eila Erfani[2], Ivan Bakhshayeshi[3], Sara Tayari[2], Alexandros Karatopouzis[2], Bavly Hanna[2]
[1]*College of Asia and the Pacific, Australian National University, Canberra, ACT, Australia;* [2]*Faculty of Engineering and Information Technology, University of Technology Sydney, Sydney, NSW, Australia;* [3]*Faculty of Science and Engineering, Macquarie University, Sydeny, NSW, Australia*

1. Introduction

People worldwide are experiencing significant environmental, health, and economic problems caused by anthropogenically induced changes in the biophysical environment, loss of biodiversity, depletion of natural resources, and climate change. The planet is warming, glaciers are melting, cloud forests are drying, plastic pollution is choking the oceans, biodiversity is rapidly declining, and extreme weather conditions are becoming more frequent. Recent catastrophic bushfires in Australia, which devastated more than 5 million hectares of land and killed an estimated 500 million animals, are a good example of Earth's ever-more frequent environmental crises [1]. There is an urgent need to create solutions to these daunting socioenvironmental challenges.

How to achieve environmental sustainability, defined as meeting the resource and services needs of current and future generations without compromising the health of global ecosystems, is a highly complex question that requires an equally sophisticated response. Artificial intelligence (AI) technologies offer considerable promise as the means to generate the necessary solutions. AI, in which machines can "learn from experience, adjust to new inputs, and perform human-like tasks," can reveal insights and patterns in unstructured data (data generated from videos, images, social media content) and combine computational resources to solve complex problems. AI enables

Artificial Intelligence and Data Science in Environmental Sensing
https://doi.org/10.1016/B978-0-323-90508-4.00006-X

large-scale pattern recognition informed by vast amounts of objective data, free of bias of human emotions. It has the potential to help decision-makers devise science-based solutions and policies for environmental sustainability [2–5]. AI can play a vital role in the development of operative environmental governance, the management of natural resources, and inform innovative and science-based solutions, policies, and principles for the use of natural resources. In resolving the climate crisis, both adaptation and mitigation are essential. Mitigation refers to preventive actions designed to avoid contributing to environmental degradation, while adaptation means becoming more responsive and resilient to the unavoidable negative impacts. AI is playing a critical role in forming and driving climate change adaptation and mitigation policies [6]. AI-based adaptation tools assist with designing effective behavioral change interventions and improving disaster management and the resilience of socioecological systems [7]. AI-based mitigation tools have the potential to find optimal solutions for reducing the environmental impacts of industries, cities, transport, energy, and agriculture.

AI capabilities can be categorized into data analysis (machine learning [ML] and computer vision), human cognition and emotions (natural language processing [NLP], affective computing), and thinking and decision-making (decision support and self-learning) (Russell and Norvig, 2009). In some cases, ensemble models—combinations of multiple AI methods, for example, ML and NLP—are used to deal with highly complex problems. Such AI-based decision-making models and systems are already being applied in many fields. For example, in the health domain, deep learning (part of a broader family of ML methods based on artificial neural networks [ANNs] with representation learning) technology has been combined with electrocardiographic to detect patients with reduced left ventricular ejection fraction [8], enabling early and more successful treatment of this serious condition. In agriculture, ML has optimized irrigation planning and water management [9] and enabled accurate estimations of land-related service provision and land use change [10]. It is also used in developing tools for automatic detection of plant diseases [11]. AI has advanced traditional spatial modeling by collecting environmental data at very high temporal and spatial resolutions, helping with the automation of precision agricultural operations, afforestation, and managing bushfires and deforestation [12]. In the energy conservation and renewable energy domain, decision support systems that leverage AI provide more accurate probabilistic estimates of energy production and distribution, energy planning and design, and operation and maintenance [13]. NLP has been used to model social media data to predict public opinion for biodiversity conservation [14]. Promoting adaptive, resilient living requires a more inclusive concept of governance and collaboration between governmental and nongovernmental actors, commercial and not-for-profit actors at the individual, local, and national levels. Consequently, policymakers often deploy AI-powered tools to analyze how environmental policy alternatives will change behaviors [15].

This chapter highlights the intersection of AI and sustainability. In the next two sections, we focus on how AI can assist in addressing social and environmental problems by presenting two case studies. The first case involves the application of unsupervised learning algorithms, a subset of ML, to data about recycling behavior, for the purpose of informing new policies or improving the effectiveness of existing programs. The next section provides an overview of how AI can be used to determine how to reduce bushfire risk and improve the management of renewable energy. Finally, we discuss the practical implications of deploying AI-based technology and its potential positive and negative impacts.

2. AI and social change: the case of food and garden waste management

Food waste is a major environmental issue in highly developed countries. According to the Australian Bureau of Statistics, in 2018–19, the industrial sector generated an estimated 2.2 million tonnes of organic waste, while households produced 6.4 million tonnes.[1] In 2017, organic waste accounted for more than 5% of Australia's greenhouse gas emissions. The Australian government spends about $20 billion annually on food waste management. In 2019, the Department of Agriculture, Water, and the Environment in New South Wales (NSW) reported that approximately 40% of food waste went into household garbage bins [16].

Local councils play an indispensable role in combating food waste. In March 2021, several NSW local governments introduced a food organics and garden organics (FOGO) collection service to tackle this fast-growing issue. The FOGO program involves collection of specific waste streams and composting them to produce a nutrient-rich product that can be sold or used on public land to improve soil quality and water retention. The program reduces the amount of agricultural and organic waste that ends up in landfill, thereby reducing emissions of methane and other greenhouse gases and improving urban sustainability. Moreover, it decreases total waste collection costs because industrial composting is less expensive than landfilling.

The FOGO program's effectiveness depends heavily on the concept of "high participation; low pollution." It requires clear and organized coordination, as well as close surveillance, to raise community awareness, interest, and dedication in order to reduce the pollution rates. The purpose of this case study was to find evidence-based strategies to improve the effectiveness of the FOGO program. The main research question was: what are the determinants of citizen participation in waste management and adoption of FOGO bins?

1. https://www.abs.gov.au/statistics/environment/environmental-management/waste-account-australia-experimental-estimates/latest-release.

2.1 AI-powered analysis of FOGO survey data

Many Australian councils provide FOGO services to residents, and some collect high-quality data from them about those services. In 2020, in the City of Wollongong, NSW, 1000 residents responded to a survey about their household's waste management. The data included information about neighborhood influence, environmental beliefs, awareness of the FOGO program, attitude toward using FOGO bins, organic waste collection habits, social media influence, and demographics such as age and level of education. These data were provided to the authors of this chapter to discover the behavioral patterns of residents.

In analyzing the survey data, AI algorithms were used to automatically detect FOGO bin users' and nonusers' behavioral patterns. Following Taghikhah et al. [17], after preprocessing the data, the DBSCAN algorithm (an unsupervised ML method) was applied. Density-based clustering algorithms such as DBSCAN automatically detect the number of clusters in the data and are suitable for cases involving clusters that are not compact and/or distinct. In contrast to ad hoc methods that divide records based on one attribute, this method includes all attributes to identify the cohort outliers. Partitioning methods (e.g., K-means) and hierarchical clustering find spherical-shaped clusters or convex clusters, while DBSCAN identifies arbitrary-shaped clusters under fewer restrictions.

The DBSCAN algorithm detected heterogeneous clusters of residents. Around 21% of households were labeled as noise, as expected, while the remainder were categorized into two clusters: (1) those households who do not use the FOGO bin or use it rarely (38%) and (2) those households using the FOGO bin frequently (41% of total). A principal component reduction approach was used to visualize the identified clusters (Fig. 5.1). The variables used to segregate clusters are presented in Fig. 5.2 to facilitate interpretation.

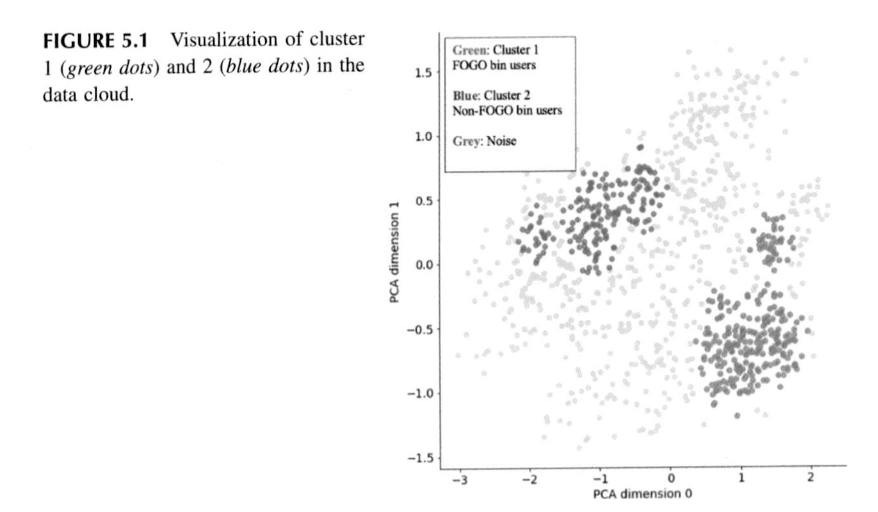

FIGURE 5.1 Visualization of cluster 1 (*green dots*) and 2 (*blue dots*) in the data cloud.

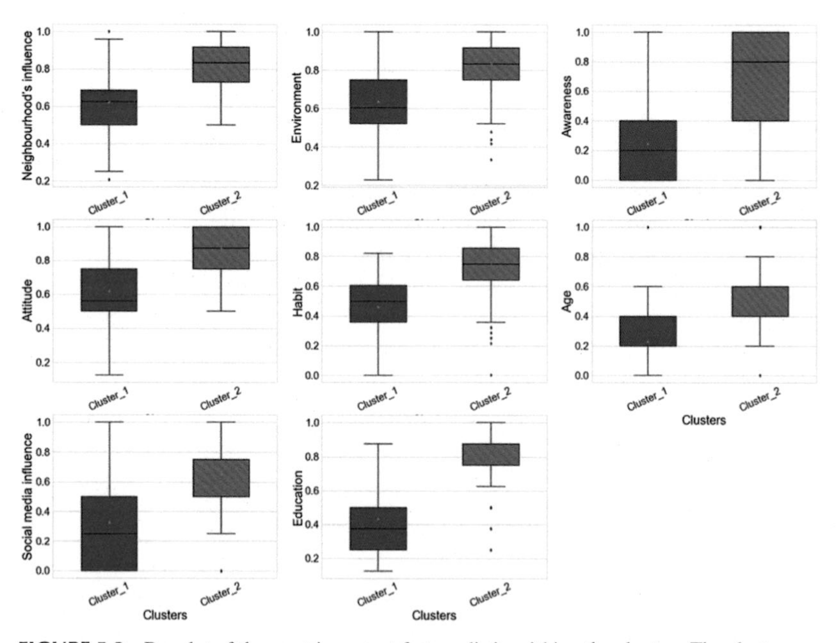

FIGURE 5.2 Boxplot of the most important factors distinguishing the clusters. The clusters are clearly different with respect to most variables, but some overlaps exist.

The residents in cluster 1 had lower awareness about the FOGO program, were not social media users, and cared less about environmental issues. The results also show that the younger people separated their waste less often than older people and were less likely to report willingness to take part in the FOGO program. Awareness of the project and its benefits was the critical factor driving the behavior of cluster 2 households. Social media, including online platforms (Twitter, Facebook), as well as traditional approaches such as posters and brochures, significantly affected the behavior of this group. Additionally, people in this cluster were highly concerned about environmental issues (more than 70%), and their participation rate in implementing the FOGO program is the highest. Neighborhood effects and social learning drove the behavior of cluster 2 residents, indicating the presence of norms, peer pressure, and networks. More highly educated people were more motivated to use FOGO bins.

Analysis of the characteristics of both clusters resulted in the conclusion that awareness is the most important factor driving residents' decisions about using FOGO bins. Therefore, raising awareness is crucial to improve the motivation and participation of residents in the FOGO program and presumably other environmental activities. Social media influences, environmental beliefs, neighborhood influences, education, habits, attitudes, and age were the other factors that affected residents' behavior.

These findings are in line with those of Hasan [18] and Islam et al. [19], who reported that raising awareness is the best approach to reducing household waste. Similarly, Debrah et al. [20] reported the literature about attitudes, awareness, and environmental knowledge related to solid waste management collected between 2010 and 2019. They concluded that weak ecological knowledge in less developed countries leads to environmental and waste management problems. Studies in Australia [21], Thailand [22], South Africa [23], and the United States [24] suggest that sustainability awareness is the main driving factor of people's recycling behavior. Barbosa and Veloso [25] showed that environmentally unfriendly attitude and lack of awareness about sustainability have roots in cultural beliefs. The United Nations Environment Program also emphasized that raising public knowledge about sustainability and reducing environmental emissions are key factors in successfully managing the climate change crisis [26].

2.2 Using AI insights to improve waste management

It can be concluded from the results presented above that improving the performance of Wollongong's FOGO program requires consistent teamwork and engagement with the community. Driving widespread adoption of sustainable practices needs a long-term education strategy. Local governments should design strategies based on the sociobehavioural characteristics of their residents to increase their motivation for participation. For example, councils' activities can be adjusted according to the most important factors (raising awareness and social media influences). Currently, Randwick council mainly motivates residents by organizing community engagement events and providing information; but without properly assessing the heterogeneity of the communities, it cannot correctly determine the appropriate communication channels necessary for raising knowledge across their entirety. For instance, Australians spend an average of 5.5 h per day using digital devices for internet access, emailing, and using social media[2]; therefore, awareness programs should include online information campaigns and digital nudging methods as well as traditional "physical" events. Based on the survey analysis, Wollongong council could focus on educating younger residents especially about the importance of recycling practices by spreading FOGO-related information through social media accounts via influencers.

This case study shows how AI can be employed to shape evidence-based policies designed to deliver higher-quality services and improve environmental efficiency and performance. Governments that embed AI-enabled processes in policymaking empower their citizens to have closer and better interactions and integration with all agencies affecting their lives. AI should

2. https://www.reviews.org/au/mobile/aussie-screentime-in-a-lifetime/.

not replace policymakers, but can help them take more rigorous and adaptive approaches to decision-making [26a]. Future research should investigate the role of AI in facilitating the participation of citizens and collective intelligence for designing and implementing policies.

3. AI and ecosystem services: insights into bushfire management and renewable energy production

3.1 AI role in predicting bushfire occurrence and spread

Wildfires are an essential element of many ecosystems, playing critical roles in ecosystem dynamics and the survival of species that have developed in response to fire [27]. Bushfires are a natural occurrence in Australia, a frequent feature of the landscape, and well-integrated into Indigenous Australians' rich mythology. However, they also pose considerable threats to human welfare and production systems and are major contributors to the carbon emissions that drive climate change [28]. In recent years, the risk of bushfires in Australia and elsewhere has risen as a result of expanding the urbanization and land use changes, in tandem with climate change. Research from the Australian Bureau of Meteorology and the Bushfire Cooperative Research Center found that increases in temperature, wind speed, and dryness increase bushfire hazards [29].

Property loss and deaths due to bushfires have increased due to urban expansion into formerly forested areas. In Australia, nearly 265,000 ha were burned between 1990 and 1999, an average of 26,500 ha per year, increasing to 60,000 ha per year between 2000 and 2013 [30]. Australia's bushfires smoke-related health cost is estimated around $1.95 billion [31], and since 1850 the bushfires have claimed the lives of more than 800 people [32]. The Black Saturday fires in the state of Victoria burned 245,000 ha in 2009, and approximately 2000 homes were damaged or destroyed [33]. During the summer of 2019–20, Australia had numerous heatwaves and major bushfires that generated dense smoke [34], resulting in 417 fatalities [35]. Learning to coexist with bushfires while safeguarding human lives and property, as well as the ecosystems that sustain us, is an increasingly critical issue.

Uncontrolled fires alter landscape ecology, causing irreversible harm to natural resources [28]. Increasing loss of natural environments, life, and property due to wildfires worldwide has driven a growing understanding of the importance of fuel reduction burning to reduce the risk and severity of wildfires [36]. However, this technique has downsides, such as reduced air quality and the potential of fire escape that endangers humans, wildlife, and infrastructure [37]. Another way to reduce fire risk is mechanical fuel load reduction (thinning of vegetation), but this method creates higher forest floor fuel loads than controlled burning, and heavy vehicles compact the soil and damage flora and animal habitat [38].

Climate change is making bushfires more challenging and unpredictable, and the traditional approach to bushfire control may no longer be adequate to deal with the rising complexity of these disasters [39]. The need to develop and implement an efficient wildfire management system capable of dealing with new and unforeseen hazards has prompted academics and practitioners to turn to AI. AI has been utilized in bushfire management since the 1990s, initially via applications such as neural networks and expert systems [40]. Considerable advances in blaze monitoring and observation have been accomplished, owing largely to the increased availability and capabilities of remote sensing technology. Satellites such as NASA TERRA, AQUA, and GOES include onboard fire detection sensors, and others, such as the LANDSAT series, monitor vegetation distribution and change [40]. Furthermore, AI-based advances in weather prediction and climate models provide finer geographical resolution and longer-range forecasts, improving the ability to anticipate and plan for intense fire weather [41]. A data-centric approach to wildfire modeling has become standard, and in recent years ML has been used widely in bushfire management [40]. AI can be used to assess and map bushfire vulnerability using predictive modeling approaches and may also provide a toolset for the creation of an effective fuel treatment system. For example, Bates et al. [42] conducted an ML and statistical classification study of dry and wet thunderstorm days, based on daily data from ground-based sensors and gauges for lightning flash count and precipitation, as well as a complete dataset of atmospheric variables over 2004–13 in Australia.

Fire management agencies use fire hazard indices to analyze fire weather conditions and give public warnings. The McArthur Fire Forest Hazard Index and the Grassland Fire Danger Index are the most frequently used fire danger indicators in Australia [43]. These indices are computed at weather stations using weather variable readings and fuel load information [44]. When estimating the danger of catastrophic fire weather occurrences in a large nation like Australia, it is also necessary to compute the geographical distribution of these indicators. Zhang et al. [45] created a fire occurrence probability map for south-eastern Australia, using AI hotspots identified using a moderate resolution imaging spectroradiometer. The results showed that wildfires are more likely to occur in hilly areas, forests, savannas, and places with dense vegetation and less likely to occur in grasslands and shrublands. Wildfires are also more likely to occur near human infrastructure. The identification of environmental and socioeconomic variables that strongly predict wildfire occurrence, as well as the geographical patterns of wildfires, can help fire managers adopt suitable management activities. Fig. 5.3 shows the fire occurrence probability map produced by applying the model to raster layers.

3.2 Artificial intelligence for energy conservation and renewable energy

Improving the reliability of energy systems cost-effectively is a topic of growing importance. Though the use of renewable energy resources has been

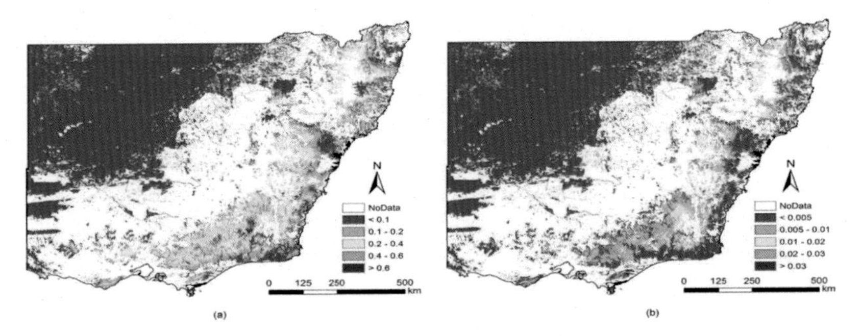

FIGURE 5.3 Maps depicting (A) the projected chance of wildfire occurrence in South-Eastern Australia and (B) the standard deviation of the probability maps generated by 10-fold method. *Retrieved from Zhang, Y., Lim, S., Sharples, J.J., 2016. Modelling spatial patterns of wildfire occurrence in South-Eastern Australia. Geomat. Nat. Hazards Risk 7 (6), 1800–1815.*

increasing worldwide, optimizing power generation, planning energy storage, and forecasting their demand have become more challenging due to increasing variability in the characteristics of production and distribution systems. The integration of fast-acting supply and demand management is critical to provide stability and efficiency in the operation of the energy value chain. Since electrical grid operation entered the digital era, massive amount of data about the energy production process and consumption patterns are generated. If these big data are processed correctly, they can inform managers' decisions rapidly and efficiently. Bringing AI and big data together enables businesses to develop intelligent tools to support real-time decisions, especially related to energy production, distribution and conservation, and system operation and maintenance.

Wind, solar, hydropower, and biomass are major sources of renewable energy generation. Wind energy has some advantages over other sources of renewable energy. It has cheaper installation costs, and the generated energy is more economically efficient [46]. ML algorithms, such as adaptive neuro-fuzzy systems, neural networks, and pattern recognition, are extensively used in developing predictive models of wind power generation at different time scales. This is a complicated task due to the stochastic and intermittent nature of wind and the power generated by turbines [47]. There are four categories for wind power prediction in terms of timing: very short-term prediction, short-term prediction, medium-term prediction, and long-term prediction. It is important to use the proper technique for each time scale and consider the limitations of the wind turbines for optimization, control, and modeling. Accurate prediction of energy generation can assist power system management, turbine control load tracking and trading energy, and monitoring energy systems (Colak et al., 2012).

On the energy demand side, AI can be used to identify conservative consumers, learn their attributes and preferences, and develop strategies to motivate behavior change for greater energy efficiency. Multiple researchers have reviewed the AI techniques used to categorize customers and forecast their consumption. In particular, Antonopoulos et al. [48] presented a comprehensive review of the AI applications in energy demand side response. For example, ANNs are commonly used for forecasting energy prices and energy lead in the short term. Unsupervised ML is mostly used for predicting consumption patterns and clustering consumers so that targeted reward and penalty schemes can be designed and implemented [49].

The use of batteries in electric vehicles and for energy storage in grids is increasing rapidly worldwide. Recent developments in ML methods have meant that ML strategies have become dominant in the battery management domain [50]. These methods are mainly used in macroscopic performance, including battery state estimation [51,52], lifetime and storage mode prediction [53,54], fault detection [55], balancing cells [56], and efficient thermal and charging management [57,58].

Environmental concerns about the carbon emissions associated with the extraction of lithium, the key material in modern lithium-ion batteries, are well recognized. The lithium-ion battery production process is complicated and energy-consuming. The materials and components, as well as the energy consumption, carbon intensity, and cost, affect the quality and specifications of the final product. Schnell and Reinhart [59] used AI methods to analyze failure modes and parametric effects to improve the quality and control over the battery production chain. Adopting decision tree techniques, Turetskyy et al. [60] presented a predictive model to estimate the maximum battery capacity. They demonstrated that data mining techniques not only improve the product quality but also help to reduce energy demand, environmental impacts, and production costs. Cunha et al. [61] used supervised learning to classify electrode descriptors and find interactions between electrode manufacturing features with high accuracy. Their method reduced the time and cost of lithium-ion batteries manufacturing process. Schnell et al. [62] relied on data mining to identify and monitor lithium cell production processes in an attempt to reduce the scrap rate and cost while maximizing quality. Wessel et al. [63] developed a data mining approach which can deal with noise effect and provide accurate and robust estimations of battery power capacity.

Despite the recent utilization of AI in battery production studies, there is room for improvement. For example, AI can accurately predict the capacity of a battery before the formation process, which is usually expensive. Data mining methods and image recognition techniques can also identify defects in the battery supply chain. Such analysis enables engineers to save costs and optimize battery production to enhance final product performance at lower environmental cost. While different types of AI techniques in renewable energy management have been applied, more experimental studies need to be conducted to detect and validate the optimal solutions in real-life trials. Additional research, along with industrial projects and large-scale experimentation, is necessary to drive the emergence of more accurate models and AI solutions and allows AI/ML techniques to become mainstream in the energy sector.

4. Challenges of using AI to achieve sustainability

Although AI is a promising and powerful technology that can be harnessed in efforts to achieve socioenvironmental sustainability, the increased use of AI technologies in organizational process and humane practices faces some challenges. In leveraging AI technologies, ethical issues such as responsibility, accountability, and liability, as well as data accuracy, privacy, and accessibility, must be considered.

The accuracy, authenticity, and correctness of predictions made using AI-based models and techniques, working from patterns available in historical datasets, have been questioned in some contexts [64]. In a highly dynamic climatic system, predictions can be inaccurate because historical datasets lack data related to human-induced climate effects, which are evolving and have issues of variance–bias trade-offs. Deep learning and neural network techniques that simulate the human decision-making process are limited to exploring relationships between variables. AI techniques combine datasets collected from multiple sources and analyze vast amounts of data. This raises potential privacy problems and can increase the chances of cybersecurity threats and risks to data confidentiality. Applying AI technologies in fields such as healthcare requires considerable financial investment, presenting an initial hurdle to adoption. In addition, AI techniques can create unintended adverse consequences, such as the use of massive datasets and big data centers producing a large carbon footprint. Another challenge of predictive model development is that there should be a standard database to test the model properly. This database should cover all the different data types and datasets, input parameters, error metrics, and standard time scales.

5. Implications and conclusion

This chapter highlights AI potentials to produce new insights and large efficiency gains in relation to environmental sustainability and building climate change resilience and mitigation capabilities. AI can be a powerful enabler of the global effort to minimize the negative impacts of human production and consumption on societies, governance systems, and the environment. The first case study exemplifies how AI is opening new avenues in designing interventions. It shows how AI can contribute to the development of policies and mechanisms that can discourage unsustainable production and encourage people to make greener choices so that citizens become agents of change. Governments' incentivization of communities, industries, and individuals to adopt environmentally friendly practices can be accelerated by AI.

The second case study demonstrated how AI's applications in wildfire control have resulted in numerous innovative techniques for mitigating the danger of catastrophic fires. The development and deployment of AI-powered technologies can give insight into complex aspects of fire behavior. The increasing threat of bushfires and other natural hazards emphasizes the necessity to handle this complexity with innovative approaches involving the next generation of AI technologies.

There are several potential future directions for the use of AI in environmental and social management. One such innovative application is the An Eye on Recovery project, a collaboration between WWF-Australia and Conservation International that is using sensor cameras to monitor the recuperation of animals affected by wildfires.[3] In this project, 600 cameras are being installed to monitor wildlife in areas devastated by the Black Summer bushfires. The findings will reveal how and when many animal species repopulate regions damaged by fire. Another innovative AI-powered technology is a citizen science app for individuals who live in fire-prone areas. Once a fire has started, the AI model can forecast the direction, extent, severity, and boundaries of the fire rapidly and accurately, allowing for targeted and strategic interventions, and helping communities to respond effectively.

There is a need to develop a comprehensive set of design principles for developing AI applications for sustainability that take into account the concerns, interests, and values of users. In addition, appropriate frameworks and regulations are needed to enable regulators to govern effectively with the assistance of AI. Similarly, the legal and ethical aspects of the collection, analysis, and transfer of vast amounts of structured and unstructured data used in AI need to be considered.

Because AI-powered systems are based on probability, the confidence level of AI models' predictions varies, and this can affect users' trust in the results. Timely and appropriate explanation of AI systems' capabilities and limits will help users to understand how AI can assist them to accomplish their goals. Future studies should focus on developing ways to enhance AI literacy, optimize understanding, and foster trust. Interdependencies among various factors such as psychological, sociological, and organizational and economic factors that contribute to successful implementations of AI should be investigated.

In this chapter, we described AI-related research and practical solutions to environmental problems, using two case studies. We hope it will motivate researchers and practitioners to use AI to make further contributions to greater global environmental sustainability. We argue that AI can support the development of culturally appropriate organizational processes and individual practices to reduce the natural resource, energy, and greenhouse gas intensity

3. https://www.wwf.org.au/what-we-do/species/eye-on-recovery#gs.6w9vr9.

of human activities. AI researchers and practitioners should advance the theory and practice for leveraging AI for sustainability. They must employ AI to capture the complexity of the real world and investigate the interdependencies among various psychological, sociological, organizational, and economic factors that contribute to successful implementations of AI. System dynamic method and design thinking approaches should be considered for stakeholder-centric environmental governance; and the economic feasibility and value of AI for sustainability should be investigated.

References

[1] D. Celermajer, R. Lyster, G.M. Wardle, R. Walmsley, E. Couzens, The Australian bushfire disaster: how to avoid repeating this catastrophe for biodiversity, Wiley Interdiscip. Rev. Clim. Change 12 (3) (2021) e704.

[2] M. Asadnia, L.H. Chua, X. Qin, A. Talei, Improved particle swarm optimization–based artificial neural network for rainfall-runoff modeling, J. Hydrol. Eng. 19 (7) (2014) 1320–1329.

[3] M. Asadnia, A.M. Khorasani, M.E. Warkiani, An accurate PSO-GA based neural network to model growth of carbon nanotubes, J. Nanomater. 2017 (2017).

[4] M. Asadnia, M.S. Yazdi, A.M. Khorasani, An improved particle swarm optimization based on neural network for surface roughness optimization in face milling of 6061-T6 aluminum, Mech. Eng. 5 (19) (2010) 3191–3201.

[5] M. Farahnakian, M.R. Razfar, M. Moghri, M. Asadnia, The selection of milling parameters by the PSO-based neural network modeling method, Int. J. Adv. Manuf. Syst. 57 (1) (2011) 49–60.

[6] D. Rolnick, et al., Tackling Climate Change with Machine Learning, 2019 arXiv preprint arXiv:1906.05433.

[7] F. Taghikhah, A. Voinov, N. Shukla, T. Filatova, M. Anufriev, Integrated modeling of extended agro-food supply chains: a systems approach, Eur. J. Oper. Res. 288 (3) (2021) 852–868.

[8] M. Anderson, S.L. Anderson, How should AI be developed, validated, and implemented in patient care? AMA J. Ethics 21 (2) (2019) 125–130.

[9] T. Hill, How Artificial Intelligence is Reshaping the Water Sector, in: Water Fin. & Mgmt, March 5, 2018 http://waterfm.com/artificial…,2019.

[10] W. Chen, L. Zhao, Q. Kang, F. Di, Systematizing heterogeneous expert knowledge, scenarios and goals via a goal-reasoning artificial intelligence agent for democratic urban land use planning, Cities 101 (2020) 102703.

[11] S. Brewster, Wilting Shrubs? Diagnose Plant Diseases with an App, MIT Technology Review, 2016.

[12] T. Liu, et al., Unmanned aerial vehicle and artificial intelligence revolutionizing efficient and precision sustainable forest management, J. Clean. Prod. (2021) 127546.

[13] A. Youssef, M. El-Telbany, A. Zekry, The role of artificial intelligence in photo-voltaic systems design and control: a review, Renew. Sustain. Energy Rev. 78 (2017) 72–79.

[14] T. Toivonen, et al., Social media data for conservation science: a methodological overview, Biol. Conserv. 233 (2019) 298–315.

[15] F. Taghikhah, A. Voinov, N. Shukla, Extending the supply chain to address sustainability, J. Clean. Prod. 229 (2019) 652–666.

[16] D. Donnelly, E. Wu, Food Waste Australian Household Attitudes and Behaviours: National Benchmarking Study, 2019. Available: https://fightfoodwastecrc.com.au/wp-content/uploads/2019/11/Summary-Report_final.pdf.

[17] F. Taghikhah, A. Voinov, N. Shukla, T. Filatova, Shifts in consumer behavior towards organic products: theory-driven data analytics, J. Retailing Consum. Serv. 61 (2021) 102516.

[18] S. Hasan, Public awareness is key to successful waste management, J. Environ. Sci. Heal. A 39 (2) (2004) 483–492.

[19] M.T. Islam, P. Dias, N. Huda, Young consumers'e-waste awareness, consumption, disposal, and recycling behavior: a case study of university students in Sydney, Australia, J. Clean. Prod. 282 (2021) 124490.

[20] J.K. Debrah, D.G. Vidal, M.A.P. Dinis, Raising awareness on solid waste management through formal education for sustainability: a developing countries evidence review, Recycling 6 (1) (2021) 6.

[21] T. Issa, T. Issa, V. Chang, Sustainability and green IT education: practice for incorporating in the Australian higher education curriculum, Int. J. Sustain. Educ. 9 (2) (2014) 19–30.

[22] N. Phoorisart, Sustainability awareness in Thailand, in: Sustainability, Green IT and Education Strategies in the Twenty-First Century, Springer, 2017, pp. 103–147.

[23] W.F. Strydom, Applying the theory of planned behavior to recycling behavior in South Africa, Recycling 3 (3) (2018) 43.

[24] M.J. Clarke, J.A. Maantay, Optimizing recycling in all of New York City's neighborhoods: using GIS to develop the REAP index for improved recycling education, awareness, and participation, Resour. Conserv. Recycl. 46 (2) (2006) 128–148.

[25] L. Barbosa, L. Veloso, Consumption, domestic life and sustainability in Brazil, J. Clean. Prod. 63 (2014) 166–172.

[26] P. Sayeg, H. Lubis, United Nations Environment Program, 2014.

[26a] M. Anjum, V. Alexey, T. Firouzeh, F.P. Salvatore, Discussoo: towards an intelligent tool for multi-scale participatory modeling, Environ. Model. Softw. 140 (2021) 105044.

[27] J.G. Pausas, J.E. Keeley, A burning story: the role of fire in the history of life, Bioscience 59 (7) (2009) 593–601.

[28] D. Vakalis, H. Sarimveis, C. Kiranoudis, A. Alexandridis, G. Bafas, A GIS based operational system for wildland fire crisis management I. Mathematical modelling and simulation, Appl. Math. Model. 28 (4) (2004) 389–410.

[29] C. Lucas, K. Hennessy, G. Mills, J. Bathols, Bushfire Weather in Southeast Australia: Recent Trends and Projected Climate Change Impacts, 2007.

[30] D.o. t. Environment, C.o.A. Energy, Australian Greenhouse Emissions Information System, AGEIS), 2014.

[31] F.H. Johnston, et al., Unprecedented health costs of smoke-related PM 2.5 from the 2019–20 Australian megafires, Nat. Sustain. 4 (1) (2021) 42–47.

[32] L. Hughes, J. Fenwick, The Burning Issue: Climate Change and the Australian Bushfire Threat, 2015.

[33] B. Teague, S. Pascoe, R. McLeod, The 2009 Victorian Bushfires Royal Commission Final Report: Summary, 2010.

[34] J.J. Sharples, S.C. Lewis, S.E. Perkins-Kirkpatrick, Modulating influence of drought on the synergy between heatwaves and dead fine fuel moisture content of bushfire fuels in the Southeast Australian region, Weather Clim. Extremes 31 (2021) 100300.

[35] N. Borchers Arriagada, A.J. Palmer, D.M. Bowman, G.G. Morgan, B.B. Jalaludin, F.H. Johnston, Unprecedented smoke-related health burden associated with the 2019–20 bushfires in eastern Australia, Med. J. Aust. 213 (6) (2020) 282–283.

[36] J. Russell-Smith, R. Thornton, Perspectives on prescribed burning, Front. Ecol. Environ. 11 (s1) (2013).

[37] F. Ximenes, et al., Mechanical fuel load reduction in Australia: a potential tool for bushfire mitigation, Aust. For. 80 (2) (2017) 88–98.

[38] L. Volkova, C.J. Weston, Effect of thinning and burning fuel reduction treatments on forest carbon and bushfire fuel hazard in Eucalyptus sieberi forests of South-Eastern Australia, Sci. Total Environ. 694 (2019) 133708.

[39] M.K. Linnenluecke, A. Griffiths, The 2009 Victorian bushfires: a multilevel perspective on organizational risk and resilience, Organ. Environ. 26 (4) (2013) 386–411.

[40] P. Jain, S.C. Coogan, S.G. Subramanian, M. Crowley, S. Taylor, M.D. Flannigan, A review of machine learning applications in wildfire science and management, Environ. Rev. 28 (4) (2020) 478–505.

[41] P. Bauer, A. Thorpe, G. Brunet, The quiet revolution of numerical weather prediction, Nature 525 (7567) (2015) 47–55.

[42] B.C. Bates, A.J. Dowdy, R.E. Chandler, Classification of Australian thunderstorms using multivariate analyses of large-scale atmospheric variables, J. Appl. Meteorol. Climatol. 56 (7) (2017) 1921–1937.

[43] A. Hadisuwito, F. Hassan, A comparative study of drought factors in the Mcarthur Forest Fire Danger Index in Indonesian Forest, Ecol. Environ. Conserv. 27 (2021) 202–206.

[44] L. Sanabria, X. Qin, J. Li, R. Cechet, C. Lucas, Spatial interpolation of McArthur's forest fire danger index across Australia: observational study, Environ. Model. Software 50 (2013) 37–50.

[45] Y. Zhang, S. Lim, J.J. Sharples, Modelling spatial patterns of wildfire occurrence in South-Eastern Australia, Geomatics Nat. Hazards Risk 7 (6) (2016) 1800–1815.

[46] S.K. Jha, J. Bilalovic, A. Jha, N. Patel, H. Zhang, Renewable energy: present research and future scope of Artificial Intelligence, Renew. Sustain. Energy Rev. 77 (2017) 297–317.

[47] I. Colak, G. Fulli, S. Bayhan, S. Chondrogiannis, S. Demirbas, Critical aspects of wind energy systems in smart grid applications, Renew. Sustain. Energy Rev. 52 (2015) 155–171.

[48] I. Antonopoulos, et al., Artificial intelligence and machine learning approaches to energy demand-side response: a systematic review, Renew. Sustain. Energy Rev. 130 (2020) 109899.

[49] R. Raza, N.U. Hassan, C. Yuen, Determination of consumer behavior based energy wastage using IoT and machine learning, Energy Build. 220 (2020) 110060.

[50] Z. Zhou, et al., An efficient screening method for retired lithium-ion batteries based on support vector machine, J. Clean. Prod. 267 (2020) 121882.

[51] Z. Wei, J. Zhao, R. Xiong, G. Dong, J. Pou, K.J. Tseng, Online estimation of power capacity with noise effect attenuation for lithium-ion battery, IEEE Trans. Ind. Electron. 66 (7) (2018) 5724–5735.

[52] H. Tian, P. Qin, K. Li, Z. Zhao, A review of the state of health for lithium-ion batteries: research status and suggestions, J. Clean. Prod. 261 (2020) 120813.

[53] K. Liu, X. Hu, Z. Wei, Y. Li, Y. Jiang, Modified Gaussian process regression models for cyclic capacity prediction of lithium-ion batteries, IEEE Trans. Transp. Electr. 5 (4) (2019) 1225–1236.

[54] X. Tang, K. Liu, X. Wang, B. Liu, F. Gao, W.D. Widanage, Real-time aging trajectory prediction using a base model-oriented gradient-correction particle filter for Lithium-ion batteries, J. Power Sources 440 (2019) 227118.

[55] R. Yang, R. Xiong, H. He, Z. Chen, A fractional-order model-based battery external short circuit fault diagnosis approach for all-climate electric vehicles application, J. Clean. Prod. 187 (2018) 950–959.

[56] L.H. Saw, Y. Ye, A.A. Tay, Integration issues of lithium-ion battery into electric vehicles battery pack, J. Clean. Prod. 113 (2016) 1032–1045.

[57] L.K. Maia, L. Drünert, F. La Mantia, E. Zondervan, Expanding the lifetime of Li-ion batteries through optimization of charging profiles, J. Clean. Prod. 225 (2019) 928–938.

[58] R. Xiong, H. Chen, C. Wang, F. Sun, Towards a smarter hybrid energy storage system based on battery and ultracapacitor-A critical review on topology and energy management, J. Clean. Prod. 202 (2018) 1228–1240.

[59] J. Schnell, G. Reinhart, Quality management for battery production: a quality gate concept, Procedia CIRP 57 (2016) 568–573.

[60] A. Turetskyy, S. Thiede, M. Thomitzek, N. von Drachenfels, T. Pape, C. Herrmann, Toward data-driven applications in lithium-ion battery cell manufacturing, Energy Technol. 8 (2) (2020) 1900136.

[61] R.P. Cunha, T. Lombardo, E.N. Primo, A.A. Franco, Artificial intelligence investigation of NMC cathode manufacturing parameters interdependencies, Batter. Supercaps 3 (1) (2020) 60–67.

[62] J. Schnell, et al., Data mining in lithium-ion battery cell production, J. Power Sources 413 (2019) 360–366.

[63] J. Wessel, A. Turetskyy, O. Wojahn, C. Herrmann, S. Thiede, Tracking and tracing for data mining application in the lithium-ion battery production, Procedia CIRP 93 (2020) 162–167.

[64] R. Nishant, M. Kennedy, J. Corbett, Artificial intelligence for sustainability: challenges, opportunities, and a research agenda, Int. J. Inf. Manag. 53 (2020) 102104.

Chapter 6

Application of multi-criteria decision-making tools for a site analysis of offshore wind turbines

Mohammad Yazdi[1,3], Arman Nedjati[2], Esmaeil Zarei[1], Rouzbeh Abbassi[3]

[1]*Centre for Risk, Integrity, and Safety Engineering (C-RISE), Faculty of Engineering and Applied Science, Memorial University of Newfoundland, St. John's, NL, Canada;* [2]*Industrial Engineering Department, Quchan University of Technology, Quchan, Iran;* [3]*School of Engineering, Faculty of Science and Engineering, Macquarie University, Sydney, NSW, Australia*

1. Decision-making in renewable energy investments

Decision-making analysis was initially used to investigate oil and gas exploration difficulties in the 1960s [1]. It experienced an increasing demand in the public domain and then applied in energy planning and policy analysis to explore the energy—economy relationships in the 1970s. Single criterion approaches were frequently utilized to estimate the most efficient solutions [2]. Therefore, several influencing factors draw attention to using multicriteria methods to reach a set of best decisions. As a result, the importance and suitability of these techniques in dealing with strategic and policy decisions, especially under uncertainty and multiple conflicting criteria, are substantially acknowledged. Considering renewable energy investments, numerous diverse factors and stakeholders' preferences should prevent infant mortality in such new markets. In other words, various groups (i.e., local communities, authorities, potential investors, environments, government, academic sector, and customers) bring different interests and concerns requiring a multi-criteria decision-making approach to reach a mutual comprehension and concessions [3].

Renewable energy systems have substantially stimulated funds worldwide in recent decades, adhering to global climate change concerns. This energy can be generated through natural resources such as wind, rain, biomass, geothermal heat, tides, and solar. However, sustainable development, meeting

Artificial Intelligence and Data Science in Environmental Sensing
https://doi.org/10.1016/B978-0-323-90508-4.00008-3

the current demands without compromising the capability of future generations to meet their own needs [4], of renewable energy systems face technical, environmental, social, economic, and academic barriers and many factors in these obstacles are inextricably intertwined.

Strantzali E and Aravossis K [4] reviewed decision-making approaches in renewable energy systems from 1983 to 2014. They concluded that multi-criteria decision-making (MCDM) techniques [5–11] have been mainly applied in six areas, including regional planning to meet the energy demand (21%), power generation technologies and projects evaluation (20%), energy strategy and management (19%), national planning (17%), environmental impact assessment (11%), and planning of electrical network (4%). MCDM techniques such as analytic hierarchy process (AHP), analytic network process (ANP), the preference ranking organization method for enrichment evaluation (PROMETHEE), multiattribute utility theory (MAUT), the elimination and choice translating reality (ELECTRE), the technique for order preference by similarity to ideal solutions (TOPSIS), fuzzy, and their hybrid forms are frequently applied in this domain during the mentioned timeframe. They categorized influencing criteria used to assess energy planning projects into four main groups (dominant subfactors): (1) technical criteria (e.g., efficiency [31%], reliability [20%], resource availability [18%], nominal power capacity [17%], and maturity [16%] and safety [10%]), (2) economic criteria (e.g., investment cost [52%], operation and maintenance cost [34%], energy cost [23%], payback period [16%]), (3) environmental criteria (CO_2 emissions [52%], land use [33%], impact on ecosystems [31%], NO_x emissions [22%], and SO_2 emission [17%]), and (4) social criteria (e.g., job creation [46%], social acceptability [28%], social benefits [15%], visual impact [14%], local development [13%], impact on health [10%]) [4].

Considering the advantages and drawbacks of each MCDM method [12–16], the selection of methods often is based on the preference of analysts and decision-makers, context, and sensitivity of decisions. However, it should be considered that the validity, reliability, and suitability of the applied techniques need to be explored in each project. Hobbs and Horn [17] indicated that a method could significantly affect the final decisions. They argued that a method change often results in more difference than a change of the decision-maker applying the method. It is believed that ideally, more than one technique should be used in a decision process. It seems that fuzzy theory and fuzzy-based extensions of conventional methods are becoming more popular to deal with uncertainty due to data scarcity and lack or insufficient knowledge regarding the influencing criteria [18–22]. Take another example of a new hybrid model, Zarei et al. [23] integrated logic-based safety assessment technique (i.e., bow-tie technique) with improved D numbers theory and best-worst method to deal with epistemic uncertainty in the operational safety management of the green hydrogen infrastructure [23].

2. Decision-making tools on the development and design of offshore wind power farms

Wind power farms (offshore/onshore) play an incredible renewable energy source and have an important role in eco-friendly and continuously optimizing the energy domain and dealing with climate change by reducing greenhouse gas emissions nationally and internationally. Although these facilities can be operated in both onshore and offshore areas, the latter infrastructures take advantage of wider usable space and larger quantities of wind intensity along with less adverse health effects (e.g., noise and light visual intrusion) on adjacent social communities [24]. Accordingly, offshore wind power farms (OWPFs) have drawn dramatic global attention in recent decades. It is observed that Europe is still proceeding with the highest investment, and nations such as the United Kingdom, Germany, and Scandinavia countries are well-equipped with OWPFs [25]. Meanwhile, academic research also has experienced increasing maturity and technological development since the 1990s. For instance, the International Energy Agency (IEA) reported that OWPFs had a 30% annually increasing rate in 2010 that will reach 15–20 MW in 2030 from 3 MW in 2010 the single-unit capacity. Moreover, the compounding turbine height grew to 230–250 m from 35 m [26]. Scientific research has focused on extending facilities in mainly four areas in this sector which can be categorized as sensors and instrumentation, regional potentials of wind energy, wind turbines design, characterization, and the last area related to the development and design of wind farms [27]. However, reaching a practical, efficient, and optimal decision in all areas faces serious challenges because numerous divergent factors must be considered to meet sustainable development continuously. Hence, decision-making in energy projects requires technical, economic, environmental, and social impacts and potential constraints that are often complicated and intertwined. This chapter aims at shedding some light on the importance and applications of MCDM approaches on the development and design of OWPFs. For instance, the optimal site selection of wind turbines entails considering many influencing factors to support decision-making. Kazak et al. [28] argued that three main factors, reflecting the sustainable progress principle, contribute to the decision-making process named social, environmental, and economic factors. The first cluster includes site characteristics directly or indirectly corresponding with the health effect of these farms. The rotor blades' sunlight reflections give rise to shadow flicker and stroboscopic effect phenomenon frequently reported by adjacent residents to onshore facilities [29,30]. Moreover, excessive noise hurts their quality through sleep disorders [31]. The second cluster contains elements that affect the facility's profits. For instance, natural features of the locations such as land roughness, the height of the terrain and elevation of the terrain, possible natural hazards, availability of communication infrastructures (energy transmission networks and the transportation system), and proximity of customers can play significantly in the profitability of the facilities. The last factors associated with the facilities' environment can be extended to many elements per people's awareness and understanding of environmental importance. Environmental noise level, collisions risk, intrusion into benthic and pelagic habitats, food chains, and ocean/sea pollution are important concerns related to OWPFs [32]. Moreover, safety, reliability, and resilience are associated with the environmental conditions and financial elements that should be determined through optimized trade-offs between maximizing safety and resilience and minimizing the installation, operation, and maintenance costs. Therefore, it is evident that without applying the MCDM models, it is difficult to reach the optimal solutions under uncertainty. In this sense, several academic efforts have been made using the MCDM models to optimize the development and design of OWPFs. Some researchers introduced spatial decision support systems to optimize the site selection process [28]. To illustrate the proposed model, Kazak et al. [28] analyzed the 13 location factors. They argued that applying the MCDM model assured a wider perspective in comparing different site selection scenarios, including conflicted interests groups, minimized subjective assessment, and defined priorities and weighted factors. Moreover, 12 studies that applied the Geographical Information System (GIS) based MCDA approach has been analyzed [33]. Furthermore, in a GIS-MCDM analysis for site selection of hybrid offshore wind and wave energy systems in Greece, AHP is used for eliminating the subjectivity significantly in judgments when criteria are related to economic, technical, and sociopolitical factors considering in the decision-making process and supported by developed GIS database [34]. Lee et al. [35] proposed a comprehensive evaluation model by incorporating interpretive structural modeling and fuzzy ANP to decide for suitable turbines when developing a wind farm in Taiwan. The criteria of machine characteristics, economic aspects, environmental issues, and technical levels along with 14 subcriteria and 4 alternatives were investigated in this study [35].

To optimal design of OWPFs, a trade-off between the capital investment, operational costs, efficiency and safety, and reliability should be developed by designers, which introduces the development and design as a multiobjective optimization challenge. Rodrigues S et al. (2016) applied evolutionary algorithms as the state of the art for solving multiobjective optimization problems and constraint-handling techniques for dealing with multiobjective wind farm layout optimization problems. They also analyzed the relation between problem dimensionality/complexity and the degrees of freedom offered by different turbine-placement grid resolutions [36]. In addition, it is indicated that offshore wind farm operations and maintenance costs currently total 6 million €/year or 25%–28% of total costs. In contrast, this cost is projected to be twice that of offshore wind for wave and tidal energy converters. Reliability-based design optimization in offshore renewable energy systems (i.e., wind, wave, and tidal) has been suggested as an effective way to reduce the operation and maintenance cost of this system [37]. Recently, Xue et al. [27] proposed a new fuzzy Bayesian MCDM approach for offshore wind turbine selection. They combined the principal component analysis and expert judgment findings into the model to deal with 11 contributing factors. The factors arise from the wind turbine parameters, wind turbine economy, wind turbine reliability, and navigation safety as the attributes, while fuzzy *IF-THEN* rule system and Bayesian network are used to quantifying the decision-making model [27].

3. Background of multiattribute decision-making tools

Multiattribute decision-making (MADM) tools are evaluation-based techniques to prioritize among different alternatives with conflicting attributes. MADM techniques are various and have a wide variety of applications in different management fields. Evaluating the preference of methods and other types of alternatives is the main task for MADM experts. In this section, we investigate the three well-known MADM techniques as follows.

3.1 VIKOR (VIseKriterijumska Optimizacija I Kompromisno Resenje)

The VIKOR (multicriteria optimization and compromise solution) technique is a well-known multicriteria optimization approach designed for complicated systems which concentrate on compromise ranking of alternatives despite conflicting criteria. In VIKOR, the optimal point computation is based on the value of “closeness” to the positive ideal solution. It is appropriate for those decision-making problems where the risk is less important than the profit [38]. The method was developed by Ref. [39] and extended during the past decades.

The technique is widely extended in years, and scholars developed modified versions to robust the solution procedure for decision-making problems. Opricovic and Tzeng [40] combined the VIKOR method with fuzzy triangular numbers in a postdisaster planning decision-making problem. Later, they [41] developed a new defuzzification model by integrating VIKOR and TOPSIS methods to determine right and left scores by fuzzy maximum and minimum, respectively, and a weighted average total score according to the membership function. They have extended the VIKOR in different researches and compared it with ELECTRE, PROMETHEE, and TOPSIS techniques [42]. Huang et al. [43] introduced a revised VIKOR model based on the regret theory where its levels of regret are evaluated by a fixed reference and some other alternatives and assessed the model’s two examples. Park et al. [44] extended VIKOR in a group decision-making problem. The interval-valued intuitionistic fuzzy hybrid geometric operator is used to aggregate all individual interval-valued intuitionistic fuzzy decision matrices. The score function is used to measure the score of each attribute value. Ju and Wang [45] proposed a new VIKOR method that criteria weights and values are linguistic information and evaluated the method using a 2-tuple linguistic TOPSIS technique. Kumar and Samuel [46] used the VIKOR method to choose the most suitable renewable energy source at a university campus in India. The results suggest that the wind turbine alternative is the best one. For detailed information about the extensions and their applications up to 2016, the interested readers could refer to Ref. [47].

The VIKOR method is widely applied in recently published articles. According to the Web of Science directory, more than 200 articles have been published during the past 3 years by including the word VIKOR in the title. The critical point is the recent applications of this method and its variants. Selecting among the construction methods for an energy-efficient building scheme was studied using the VIKOR method to provide practical decision-making guidance [48]. Kutlu Gündoğdu and Kahraman [49] extended the VIKOR method to spherical fuzzy VIKOR for a warehouse selection problem. Hospitality brand management is another subject that integrates the best-worst method and VIKOR to provide a decision-making guide for the managers [50].

Moreover, the VIKOR method [108] was implemented in the technical piston material selection challenge to prioritize the manufactured composites. Bakioglu and Atahan [52] developed an integrated version of the TOPSIS and VIKOR method to evaluate the risk associated with self-driving cars to provide helpful insight to the top decision-makers in this industry. According to the literature, there are many applications in industry and society using the VIKOR method and the extensions, especially the fuzzy extensions, and the scholars are very interested in utilizing it in their investigations.

3.2 PROMETHEE (Preference Ranking Organization Method for Enrichment Evaluation)

PROMETHEE and GAIA methods introduced and extended at the beginning of the 1980s [53,54] are widely used for group MCDM problems with various criteria. Lots of extensions have been developed during the past decades. Due to the simplicity of the ranking method in practice and concept, the number of scholars using PROMETHEE in their studies has increased in the past decade. PROMETHEE has been applied in various areas of applications such as management, business, chemistry, etc. Brans et al. [55] indicate that in PROMETHEE II, obtaining a complete preorder on a finite set of alternatives is proposed, unlike a partial preorder in PROMETHEE I. PROMETHEE II contains six steps as [56]:

- Evaluation matrix construction
- Performance difference determination
- Preference function construction
- Aggregated preference indices calculation

- Outranking flows calculation
- Net outranking flows calculation

Practitioners extended the PROMETHEE I and II methods to utilize in various types of applications in recent years. Rakotoarivelo et al. [57] applied PROMETHEE to assess the risk analysis related to microfinance and their operations to improve the management activities of these organizations. Metzner [60] investigated the applicability of the PROMETHEE method to solve the complicated real-state decision-making problems. He concluded that the method was consistent and efficient to cope with such real-state industry issues. Medical research studies [105] analyzed the breast cancer treatment procedures, including radiation therapy, chemotherapy, hormone therapy, and surgery, using the fuzzy-PROMETHEE approach to evaluate the side effects, costs, treatment duration, and survival rate factors. Butowski [60] developed an evaluation structure based on an integrated version of AHP and PROMETHEE to assess the European offshore and coastal sailing tourism sales for sailing tourism. A group of trained weights evaluated the attractiveness of 10 coastal areas. A heterogeneous multiattribute group decision-making problem [61] used five types of information formats, including real numbers, intervals, trapezoidal intuitionistic fuzzy numbers, triangular intuitionistic fuzzy numbers, and intuitionistic fuzzy sets. They investigate the alternatives in a PROMETHEE-FLP method, where FLP stands for fuzzy linear programming used to derive the group priorities. Chen [62] represented a novel outranking approach by implementing Pythagorean fuzzy sets in the PROMETHEE method for MCDM analysis. They provided the integrated version with PROMETHEE I and II and tested the technique on a real-world problem of bridged construction method selection. The results indicate the method's capability in terms of stability and reasonability. Military airport location problem is another application [63] that applied an integrated version of AHP, PROMETHEE, and VIKOR methods and compared the results with other well-known techniques to quantify the performance of the designed approach. Choosing a profitable startup corporation among various available firms is an issue in the venture capitalists area Tian et al. [64] constructed a structure by PROMETHEE with hesitant fuzzy linguistic information to support the decision-making procedure. Sidhu and Singh [65] used an improved version of PROMETHEE to select the dedicated cloud database servers, and their results approve the method's effectiveness. He et al. [66] developed a simulation-aided based PROMETHEE-TOPSIS method for the groundwater remediation selection strategy. They evaluated their approach on a real case study in Anhui, China.

However, in recent years, several extensions of the PROMETHEE method and different integrations of it with well-known methods such as regret theory [67], goal programming [68], best-worst method [69], TODIM [70], etc., and their applications have been investigated. Authors believe there is enormous potential to develop robust and practical methods based on the PROMETHEE I and II methods using linear programming tools. Moreover, there is a lack of a systematic review study for the variants and applications of PROMETHEE-based approaches.

3.3 ELECTRE (ELimination Et Choice Translating REality)

ELECTRE method was introduced in 1965 by Bernard Roy, a professor at Université Paris Dauphine, and later published [71] in a French journal. It was later modified to ELECTRE I, II, III, IV, IS, TRI versions, and many other variants. However, the different types of ELECTRE methods differ operationally due to the type of problem they can be applied for. For more detailed information, readers could refer to Ref. [72]. These ELECTRE-based methods have been widely used in operations research management science, information systems, business economics, environmental science, computer science, and energy sectors.

Yu et al. [73] studied the prioritized MCDM problems by the ELECTRE method, which, unlike traditional MCDM problems, the interdependencies within criteria are considered. Rouyendegh [74] developed an intuitionistic fuzzy-based ELECTRE model to cope with more complex situations and vagueness in option value assignment. Liang et al. [75] used picture fuzzy numbers to describe the experts' opinions in a fuzzy environment. By integrating the ELECTRE method with evaluation based on distance from average solution, they evaluated the cleaner production level of gold mines. Yadav et al. [76] designed a hybrid best-worst method based on ELECTRE framework to assist the decision-makers in offshore outsourcing adoption to achieve higher sustainable development in their business. Linguistic variables are another proper tool when both qualitative and quantitative criteria are included. Jasemi and Ahmadi [77] took advantage of linguistic variables in a novel fuzzy ELECTRE approach to a personnel selection problem. Fernández et al. [78] developed a novel evolutionary-based ELECTRE TRI-nB model using a genetic algorithm. The results showed the higher values of restore capacities and improved capacity of consistent assignments of new actions. Erdin and Ozkaya [79] studied the 2023 energy strategies of Turkey in a site selection ELECTRE-based methodology. In a plastic recycling problem, Geetha et al. [84] introduced a novel hesitant Pythagorean fuzzy ELECTRE III method to find the best recycling method. In a study related to the wind farm industry [85], the authors used ELECTRE method to choose the optimal version of wind turbine for installation in a wind farm in western Romania. The criteria considered for the turbine selection problem were time requirements, environmental considerations, and operation conditions. An old review of the ELECTRE method methodologies and applications is provided by Ref. [72] for interested readers.

4. Background of multiobjective problems in offshore and wind farms

Determining offshore wind farms and related designs and economic and scheduling decisions constantly confront contradictory objectives. A multi-objective decision-making (MODM) problem as a variant of MCDM is a decision-making problem of finding an optimal solution with more than one conflicting objective. Since all the objectives cannot be optimized simultaneously, a list of nondominated optimal solutions is obtained. Each solution among the list is an optimal solution by a particular objective value that is not dominated by any other solution. In simple words, all the members of the nondominated solutions list are optimal results, and decision-makers must select one based on their preferences according to the objectives. However, the preferences could be applied in the initial step in some cases, and one solution will be obtained. MODM is applied to various problems in the economy, management, safety, engineering, etc.

Vilfredo Pareto and Francis Ysidro are recognized as the founders of the MODM concept during the years 1845–1926. Optimization of an MODM problem in the current form was founded and developed in the 1940s [106], and [107] presented the earliest definition for a nondominated optimal set, also called Pareto optimal solution economics. During the past five decades, the concept of MODM has been applied in too many real-life problems, and the methods were applicable in engineering, business, safety, etc. Interested readers could refer to Refs. [80,82,83]

4.1 Practical studies

Several studies investigated multiobjective problems in the wind farm industry. Márquez et al. [88] used the actual data to generate the wind regime in a typical day by minimizing the standard deviation of energy produced in a day, minimizing unused installed power, and maximizing energy production. The study was conducted by different sets of objectives for different islands. Another study emphasizing reliability implications based on a hierarchical level [89] presented an MODM framework to appraise the combination of distant wind farms by transmission network development as an optimal power system plan. They used a fuzzy method to represent the intercommunication of objectives.

Transmission expansion planning (TEP) is a practical and essential problem in this subject [90]. The market competition viewpoint proposed a probabilistic MODM problem of TEP to determine the optimal connecting node in large-scale distant wind farms. The objectives consist of investment cost and congestion cost by considering security constraints. Abbasi et al. [91] addressed the uncertainty-related loads in TEP by developing a multiobjective problem of minimizing congestion cost, risk cost, and investment cost. The load correlation is modeled by the uncentered transformation method, and simulation-based methods compared the results. Taherkhani et al. [92] investigated TEP and generation expansion planning as a multiobjective problem by optimizing the capacity of wind farms, the number of expanded lines, and the number of wind farms lines. The problem is considered a probabilistic optimization problem due to the wind speed and system demand uncertainty. The results emphasize the main advantage of the wind generating system over the conventional generating systems. A similar study [93] proposed a multiobjective problem to minimize the investment cost of new offshore wind farm units, fuel cost, carbon emission generated, investment costs of transmission lines, and maximizing the incentives related to the new generating units. Their results indicate that the incentive support scheme must be improved and minimize the total cost to utilize offshore power generation effectively.

Another exciting field in operations research is the layout design problem. Thousands of studies investigated various layout problems with distinct objective functions and constraints in past decades. The wind farm layout problem is an expensive and challenging design that requires a wide range of information and considerations. Different types of objectives related to the waking effect, energy efficiency, implementation cost, etc., have been considered in the literature. Paul and Rather [94] studied a new biobjective wind farm layout problem to minimize the annual load payment related to cables and land use and maximize the yearly gross energy production. They used Jensen's wake model to calculate and consider the wake effect. They used a fuzzy approach to find the best trade-off and reliability among the Pareto optimal solutions. Variation in wind states is a very challenging issue in the initial design phase of the farms. Mittal and Mitra [95] developed a multi-objective robust optimization problem that optimizes the location and the number of wind turbines resistant to varying wind direction and speed. They adopted the minimum and maximum limit of variations in uncertain parameters by considering guaranteed service levels. They claim that the technique improves the power by 75% and decreases the cost by 1%–10% compared to other approaches. Furthermore, micrositing wind turbines gained more concerns recently. Huang et al. [96] proposed a three-dimensional wind turbine design as a multiobjective problem. They considered the new low-speed types of turbines, their operation and maintenance cost, and the wake effect calculations as the objectives. Their methodology is tested on a 100 MW wind farm to show the performance and flexibility of the approach. A two-level model is developed by Ref. [97], which investigates tuning the turbine locations and turbine heights. In the first level of their study, a set of block candidates are optimized, and then the whole wind farm layout is designed.

Due to the complexity of the structure and harsh operating environment of wind turbines, various types of failures are common in this industry, and it causes undesirable operation and maintenance costs. Maintenance scheduling is necessary to decrease unexpected costs and keep wind power alive according to its market competitiveness. Recently, this demand attracts researchers in the field of optimization. Zhang and Yang [98] developed a dynamic multiobjective scheme for the maintenance and resource scheduling of wind farms. Their model is designed for different types of wind farms with a periodically generating system. As a preventing maintenance scheduling research, Zhong et al. [99] presented a nonlinear biobjective model to minimize the maintenance cost and maximize the system reliability of an offshore wind farm. Ge et al. [100] developed a multiobjective framework to optimize the maintenance cost of offshore turbines considering the wake effect of arbitrary wind direction. The problem minimizes the maintenance cost while maximizing the amount of power generation and is tested in a short-term maintenance schedule of an offshore wind farm. Maintenance scheduling by applying optimization techniques has received significant attention in the past years and is expected to extend in the coming years.

4.2 Objectives and solution methods

As mentioned earlier, a multiobjective problem deals with contradicting objectives. These objectives shape the results and, generally, the goal of the research. Moreover, the solution methodology is the second significant part due to the complexity of the cases, especially in offshore wind farm working conditions. Table 6.1 shows some WoS-listed literature of offshore wind farm multiobjective problems since 2017 in terms of objectives and solution methodologies.

4.3 Future directions

The offshore wind farm industry is gaining more and more attention in recent studies. The need for clean energy with onshore limitations leads the decision-makers to investigate more on offshore wind farms with an optimized structure. Reviewing the related literature has identified that multiobjective topics will have great potential in cooperation with wind farm implementation and maintenance phases. Therefore, the following directions could be suggested:

- Stochastic programming due to the uncertainty of wind speed and direction conditions
- Maintenance scheduling and service operation vessel routing and scheduling
- Waste management optimization under the subjects of sustainable decommissioning
- Location analysis along with maintenance scheduling and vessel routing
- Three-dimensional optimized design by consideration of installation, maintenance costs, and more accurate wake analysis

TABLE 6.1 Objectives and solution procedures of multiobjective wind farm optimization problems.

Topic	Objectives	Solution methodology	Article
Installation scheduling in offshore wind farms	• Installation cost • The completion period of the installation	• Exact (CPLEX) • Variable neighborhood search	[101]
Multiobjective scheduling problem	• Total operating cost • Total power loss	• Benders' decomposition • NSGAII	[102]
Noise analysis of high tip speed wind turbine	• Tip speed • Cost • Noise	Gradient-based multiobjective evolutionary algorithm	[103]
Preventive maintenance scheduling	• System reliability maintenance related costs	NSGAII	[99]
Optimum offshore wind farm location problem	• Capital expenditure • Operational expenditure	• TOPSIS • NSGAII	[104]
Preventive maintenance scheduling	• Reliability • Manpower cost • Transportation cost	Fuzzy arithmetic and NSGAII	[105]
Maintenance scheduling	• Maintenance cost • Power generation	Exact	[100]
Optimal capacity of battery energy storage system	• Battery cost and lifetime • Availability of wind turbines • Expected energy not supplied • Loss of load hour	Heuristic	[106]
Designing the configuration of wind turbines and topology of the electrical collector system	• Wind turbines" daily profit rate • Daily average capacity factor • Power quality	• NSGAII • Binary PSO • Quadratic programming	[107]

- Considering potential development in numbers while designing wind turbines layout problems (interested readers could refer to Ref. [108] for wind farm layout problems)
- Cable routing problem by consideration of different cable capacities along with other mentioned objectives

References

[1] J.P. Huang, K.L. Poh, B.W. Ang, Decision analysis in energy and environmental modeling, Energy 20 (1995) 843–855, https://doi.org/10.1016/0360-5442(95)00036-G.

[2] P. Nijkamp, A. Volwahsen, New directions in integrated regional energy planning, Energy Pol. 18 (1990) 764–773.

[3] D.A. Haralambopoulos, H. Polatidis, Renewable energy projects: structuring a multi-criteria group decision-making framework, Renew. Energy 28 (2003) 961–973.

[4] E. Strantzali, K. Aravossis, Decision making in renewable energy investments: a review, Renew. Sustain. Energy Rev. 55 (2016) 885–898.

[5] G.-J. Jiang, H.-X. Chen, H.-H. Sun, M. Yazdi, A. Nedjati, K.A. Adesina, An improved multi-criteria emergency decision-making method in environmental disasters, Soft Comput (2021), https://doi.org/10.1007/s00500-021-05826-x.

[6] H. Li, J.-Y. Guo, M. Yazdi, A. Nedjati, K.A. Adesina, Supportive emergency decision-making model towards sustainable development with fuzzy expert system, Neural Comput. Appl. (2021), https://doi.org/10.1007/s00521-021-06183-4.

[7] M. Yazdi, Risk assessment based on novel intuitionistic fuzzy-hybrid-modified TOPSIS approach, Saf. Sci. 110 (2018) 438–448, https://doi.org/10.1016/j.ssci.2018.03.005.

[8] M. Yazdi, F. Khan, R. Abbassi, R. Rusli, Improved DEMATEL methodology for effective safety management decision-making, Saf. Sci. 127 (2020a) 104705, https://doi.org/10.1016/j.ssci.2020.104705.

[9] M. Yazdi, A. Nedjati, E. Zarei, R. Abbassi, A novel extension of DEMATEL approach for probabilistic safety analysis in process systems, Saf. Sci. 121 (2020b) 119–136, https://doi.org/10.1016/j.ssci.2019.09.006.

[10] M. Yazdi, A. Nedjati, E. Zarei, R. Abbassi, A reliable risk analysis approach using an extension of best-worst method based on democratic-autocratic decision-making style, J. Clean. Prod. (2020c) 120418, https://doi.org/10.1016/j.jclepro.2020.120418.

[11] M. Yazdi, A. Nedjati, R. Abbassi, Fuzzy dynamic risk-based maintenance investment optimization for offshore process facilities, J. Loss Prev. Process. Ind. (2019b) 194–207, https://doi.org/10.1016/j.jlp.2018.11.014.

[12] S. Daneshvar, M. Yazdi, K.A. Adesina, Fuzzy smart failure modes and effects analysis to improve safety performance of system : case study of an aircraft landing system, Qual. Reliab. Eng. Int. (2020) 1–20, https://doi.org/10.1002/qre.2607.

[13] Ö. Uygun, H. Kaçamak, Ü.A. Kahraman, An integrated DEMATEL and Fuzzy ANP techniques for evaluation and selection of outsourcing provider for a telecommunication company, Comput. Ind. Eng. 86 (2014) 137–146, https://doi.org/10.1016/j.cie.2014.09.014.

[14] C.-N. Wang, T.-T. Tsai, Y.-F. Huang, A model for optimizing location selection for biomass energy power plants, Processes (2019), https://doi.org/10.3390/pr7060353.

[15] M. Yazdi, A. Nedjati, E. Zarei, S. Adumene, R. Abbassi, F. Khan, Chapter Thirteen - Domino effect risk management: Decision making methods, in: F. Khan, V. Cozzani, Reniers, G.B.T.-M. in C.P.S. (Eds.), Domino Effect: Its Prediction and Prevention, Elsevier, 2021, pp. 421–460, https://doi.org/10.1016/bs.mcps.2021.05.013.

[16] M.C. Yu, M.H. Su, Using fuzzy DEA for green suppliers selection considering carbon footprints, Sustainability 9 (2017), https://doi.org/10.3390/su9040495.

[17] B.F. Hobbs, P.M. Meier, Multicriteria methods for resource planning: an experimental comparison, IEEE Trans. Power Syst. 9 (1994) 1811–1817.

[18] M. Yazdi, O. Korhan, S. Daneshvar, Application of fuzzy fault tree analysis based on modified fuzzy AHP and fuzzy TOPSIS for fire and explosion in the process industry, Int. J. Occup. Saf. Ergon. 26 (2020) 319–335.

[19] S. Kabir, M. Yazdi, J.I. Aizpurua, Y. Papadopoulos, Uncertainty-aware dynamic reliability analysis framework for complex systems, IEEE Access 6 (2018) 29499–29515, https://doi.org/10.1109/ACCESS.2018.2843166.

[20] H. Wang, X. Lu, Y. Du, C. Zhang, R. Sadiq, Y. Deng, Fault tree analysis based on TOPSIS and triangular fuzzy number, Int. J. Syst. Assur. Eng. Manag. 8 (2017) 2064–2070, https://doi.org/10.1007/s13198-014-0323-5.

[21] M. Yazdi, P. Hafezi, R. Abbassi, A methodology for enhancing the reliability of expert system applications in probabilistic risk assessment, J. Loss Prev. Process. Ind. (2019a) 51–59, https://doi.org/10.1016/j.jlp.2019.02.001.

[22] E. Zarei, M. Yazdi, R. Abbassi, F. Khan, A hybrid model for human factor analysis in process accidents: FBN-HFACS, J. Loss Prev. Process. Ind 57 (2019) 142–155, https://doi.org/10.1016/j.jlp.2018.11.015.

[23] E. Zarei, F. Khan, M. Yazdi, A dynamic risk model to analyze hydrogen infrastructure, Int. J. Hydrog. Energy 46 (2021) 4626–4643, https://doi.org/10.1016/j.ijhydene.2020.10.191.

[24] X. Sun, D. Huang, G. Wu, The current state of offshore wind energy technology development, Energy 41 (2012) 298–312.

[25] by fuel type-Exajoules, C., Emissions, C.D., BP Statistical Review of World Energy June 2020, 2006. https://www.bp.com/content/dam/bp/business-sites/en/global/corporate/pdfs/energy-economics/statistical-review/bp-stats-review-2020-full-report.pdf.

[26] IEA, Offshore Wind Outlook 2019: World Energy Outlook Special Report, 2019. Https://www.iea.org/reports/offshore-wind-outlook-2019. (Accessed 30 March 2021).

[27] J. Xue, T.L. Yip, B. Wu, C. Wu, P.H.A.J.M. van Gelder, A novel fuzzy Bayesian network-based MADM model for offshore wind turbine selection in busy waterways: an application to a case in China, Renew. Energy (2021), https://doi.org/10.1016/j.renene.2021.03.084.

[28] J. Kazak, J. Van Hoof, S. Szewranski, Challenges in the wind turbines location process in Central Europe–The use of spatial decision support systems, Renew. Sustain. Energy Rev. 76 (2017) 425–433.

[29] F. Girbau-Llistuella, A. Sumper, F. Díaz-González, S. Galceran-Arellano, Flicker mitigation by reactive power control in wind farm with doubly fed induction generators, Int. J. Electr. Power Energy Syst. 55 (2014) 285–296.

[30] C. Vilar, H. Amarís, J. Usaola, Assessment of flicker limits compliance for wind energy conversion system in the frequency domain, Renew. Energy 31 (2006) 1089–1106.

[31] I.J. Onakpoya, J. O'Sullivan, M.J. Thompson, C.J. Heneghan, The effect of wind turbine noise on sleep and quality of life: a systematic review and meta-analysis of observational studies, Environ. Int. 82 (2015) 1–9.

[32] H. Bailey, K.L. Brookes, P.M. Thompson, Assessing environmental impacts of offshore wind farms: lessons learned and recommendations for the future, Aquat. Biosyst. 10 (2014) 1–13.

[33] M. Mahdy, A.S. Bahaj, Multi criteria decision analysis for offshore wind energy potential in Egypt, Renew. Energy 118 (2018) 278–289, https://doi.org/10.1016/j.renene.2017.11.021.

[34] M. Vasileiou, E. Loukogeorgaki, D.G. Vagiona, GIS-based multi-criteria decision analysis for site selection of hybrid offshore wind and wave energy systems in Greece, Renew. Sustain. Energy Rev. 73 (2017) 745–757.

[35] A.H.I. Lee, M.-C. Hung, H.-Y. Kang, W.L. Pearn, A wind turbine evaluation model under a multi-criteria decision making environment, Energy Convers. Manag. 64 (2012) 289–300.

[36] S. Rodrigues, P. Bauer, P.A.N. Bosman, Multi-objective optimization of wind farm layouts – complexity, constraint handling and scalability, Renew. Sustain. Energy Rev. 65 (2016) 587–609, https://doi.org/10.1016/j.rser.2016.07.021.

[37] C.E. Clark, B. DuPont, Reliability-based design optimization in offshore renewable energy systems, Renew. Sustain. Energy Rev. 97 (2018) 390–400.

[38] M.K. Sayadi, M. Heydari, K. Shahanaghi, Extension of VIKOR method for decision making problem with interval numbers, Appl. Math. Model. 33 (2009) 2257–2262, https://doi.org/10.1016/j.apm.2008.06.002.

[39] L. Duckstein, S. Opricovic, Multiobjective optimization in river basin development, Water Resour. Res. 16 (1980) 14–20, https://doi.org/10.1029/WR016i001p00014.

[40] S. Opricovic, G.H. Tzeng, Multicriteria planning of post-earthquake sustainable reconstruction, Comput. Civ. Infrastruct. Eng. 17 (2002) 211–220, https://doi.org/10.1111/1467-8667.00269.

[41] S. Opricovic, G.H. Tzeng, Defuzzification within a multicriteria decision model, Int. J. Uncertain. Fuzziness Knowl. Based Syst. 11 (2003) 635–652, https://doi.org/10.1142/S0218488503002387.

[42] S. Opricovic, G.H. Tzeng, Extended VIKOR method in comparison with outranking methods, Eur. J. Oper. Res. 178 (2007) 514–529, https://doi.org/10.1016/j.ejor.2006.01.020.

[43] J.J. Huang, G.H. Tzeng, H.H. Liu, A revised VIKOR model for multiple criteria decision making - the perspective of regret theory, in: Communications in Computer and Information Science. Springer, Berlin, Heidelberg, 2009, pp. 761–768, https://doi.org/10.1007/978-3-642-02298-2_112.

[44] J.H. Park, H.J. Cho, Y.C. Kwun, Extension of the VIKOR method for group decision making with interval-valued intuitionistic fuzzy information, Fuzzy Optim. Decis. Making 10 (2011) 233–253, https://doi.org/10.1007/s10700-011-9102-9.

[45] Y. Ju, A. Wang, Extension of VIKOR method for multi-criteria group decision making problem with linguistic information, Appl. Math. Model. 37 (2013) 3112–3125, https://doi.org/10.1016/j.apm.2012.07.035.

[46] M. Kumar, C. Samuel, Selection of best renewable energy source by using VIKOR method, Technol. Econ. Smart Grids Sustain. Energy 2 (2017), https://doi.org/10.1007/s40866-017-0024-7.

[47] A. Mardani, E.K. Zavadskas, K. Govindan, A. Amat Senin, A. Jusoh, VIKOR Technique: A Systematic Review of the State of the Art Literature on Methodologies and Applications, Sustain 8 (2016), https://doi.org/10.3390/su8010037.

[48] F. Zhang, A study on energy-efficient building scheme selection by heterogeneous VIKOR, in: 2018 4th International Conference on Systems, Computing, and Big Data (ICSCBD 2018). Francis Acad Press, Guangzhou, 2019, pp. 6–9, https://doi.org/10.25236/icscbd.2018.002.

[49] F. Kutlu Gündoğdu, C. Kahraman, A novel VIKOR method using spherical fuzzy sets and its application to warehouse site selection, J. Intell. Fuzzy Syst. 37 (2019) 1197–1211, https://doi.org/10.3233/JIFS-182651.

[50] X. Mi, J. Li, H. Liao, E.K. Zavadskas, A. Al-Barakati, A. Barnawi, O. Taylan, E. Herrera-Viedma, Hospitality brand management by a score-based q-rung ortho pair fuzzy V.I.K.O.R. method integrated with the best worst method, Econ. Res. Istraz. 32 (2019) 3266–3295, https://doi.org/10.1080/1331677X.2019.1658533.

[51] S. Dev, A. Aherwar, A. Patnaik, Material Selection for Automotive Piston Component Using Entropy-VIKOR Method, Silicon 12 (2020) 155–169, https://doi.org/10.1007/s12633-019-00110-y.

[52] G. Bakioglu, A.O. Atahan, AHP integrated TOPSIS and VIKOR methods with Pythagorean fuzzy sets to prioritize risks in self-driving vehicles, Appl. Soft Comput. 99 (2021), https://doi.org/10.1016/j.asoc.2020.106948.

[53] J.-P. Brans, L'ingénierie de la décision: élaboration d'instruments d'aide à la décision. La méthode PROMETHEE. Press, l'Université Laval, 1982.

[54] J.P. Brans, P. Vincke, A preference ranking organisation method: (the PROMETHEE method for multiple criteria decision-making), Manag. Sci. 31 (1985) 647–656.

[55] J.P. Brans, P. Vincke, B. Mareschal, How to select and how to rank projects: the Promethee method, Eur. J. Oper. Res. 24 (1986) 228–238, https://doi.org/10.1016/0377-2217(86)90044-5.

[56] M. Abedi, S. Ali Torabi, G.H. Norouzi, M. Hamzeh, G.R. Elyasi, PROMETHEE II: a knowledge-driven method for copper exploration, Comput. Geosci. 46 (2012) 255–263, https://doi.org/10.1016/j.cageo.2011.12.012.

[57] J.B. Rakotoarivelo, P. Zaraté, D.M. Kilgour, Future risk analysis for bank investments using PROMETHEE, Estud. Econ. Apl. 36 (2019) 207, https://doi.org/10.25115/eea.v36i1.2525.

[58] S. Metzner, Transferring outranking models to real estate management: the assessment of potential investment markets using PROMETHEE, J. Property Invest. Finance 36 (2018) 135–157, https://doi.org/10.1108/JPIF-01-2017-0009.

[59] N.A. Isa, D. Uzun Ozsahin, I. Ozsahin, in: I. Ozsahin, D.U. Ozsahin, B.B.T.-A. of M.-C.D.-M.T. in H, B.E. Uzun (Eds.), Chapter 19 - Top cancer treatment destinations: a comparative analysis using fuzzy PROMETHEE, Academic Press, 2021, pp. 277–308, https://doi.org/10.1016/B978-0-12-824086-1.00019-0.

[60] L. Butowski, An integrated AHP and PROMETHEE approach to the evaluation of the attractiveness of European maritime areas for sailing tourism, Morav. Geogr. Rep. 26 (2018) 135–148, https://doi.org/10.2478/mgr-2018-0011.

[61] J. Dong, S. Wan, A PROMETHEE-FLP method for heterogeneous multi-attributes group decision making, IEEE Access 6 (2018) 46656–46667, https://doi.org/10.1109/ACCESS.2018.2865773.

[62] T.Y. Chen, A novel PROMETHEE-based outranking approach for multiple criteria decision analysis with Pythagorean fuzzy information, IEEE Access 6 (2018) 54495–54506, https://doi.org/10.1109/ACCESS.2018.2869137.

[63] B. Sennaroglu, G. Varlik Celebi, A military airport location selection by AHP integrated PROMETHEE and VIKOR methods, Transport. Res. Transport Environ. 59 (2018) 160−173, https://doi.org/10.1016/j.trd.2017.12.022.

[64] X. Tian, Z. Xu, J. Gu, Group decision-making models for venture capitalists: the promethee with hesitant fuzzy linguistic information, Technol. Econ. Dev. Econ. 25 (2019) 743−773, https://doi.org/10.3846/tede.2019.8741.

[65] J. Sidhu, S. Singh, Using the improved PROMETHEE for selection of trustworthy cloud database servers, Int. Arab J. Inf. Technol. (2019).

[66] L. He, F. Shao, L. Ren, Identifying optimal groundwater remediation strategies through a simulation-based PROMETHEE-TOPSIS approach: an application to a naphthalene-contaminated site, Hum. Ecol. Risk Assess. 26 (2020) 1550−1568, https://doi.org/10.1080/10807039.2019.1591267.

[67] X. Jia, X. Wang, A PROMETHEE II method based on regret theory under the probabilistic linguistic environment, IEEE Access 8 (2020) 228255−228263, https://doi.org/10.1109/ACCESS.2020.3042668.

[68] C.N. Roukounis, G. Aretoulis, T. Karambas, A combination of PROMETHEE and goal programming methods for the evaluation of water airport connections, Int. J. Decis. Support Syst. Technol. 12 (2020) 50−66, https://doi.org/10.4018/ijdsst.2020040103.

[69] A. Ishizaka, G. Resce, Best-Worst PROMETHEE method for evaluating school performance in the OECD's PISA project, SocioEcon. Plan. Sci. 73 (2021), https://doi.org/10.1016/j.seps.2020.100799.

[70] X. kang Wang, H. yu Zhang, J. qiang Wang, J. bo Li, L. Li, Extended TODIM-PROMETHEE II method with hesitant probabilistic information for solving potential risk evaluation problems of water resource carrying capacity, Expet Syst. (2021), https://doi.org/10.1111/exsy.12681.

[71] B. Roy, Classement et choix en présence de points de vue multiples (la méthode ELECTRE), La Rev. d'Informatique Rech. Opérationelle 8 (1968) 57−75.

[72] K. Govindan, M.B. Jepsen, ELECTRE: a comprehensive literature review on methodologies and applications, Eur. J. Oper. Res. 250 (2016) 1−29, https://doi.org/10.1016/j.ejor.2015.07.019.

[73] X. Yu, S. Zhang, X. Liao, X. Qi, ELECTRE methods in prioritized MCDM environment, Inf. Sci. 424 (2018) 301−316, https://doi.org/10.1016/j.ins.2017.09.061.

[74] B.D. Rouyendegh, The intuitionistic fuzzy ELECTRE model, Int. J. Manag. Sci. Eng. Manag. 13 (2018) 139−145, https://doi.org/10.1080/17509653.2017.1349625.

[75] W.Z. Liang, G.Y. Zhao, S.Z. Luo, An integrated EDAS-ELECTRE method with picture fuzzy information for cleaner production evaluation in gold mines, IEEE Access 6 (2018) 65747−65759, https://doi.org/10.1109/ACCESS.2018.2878747.

[76] G. Yadav, S.K. Mangla, S. Luthra, S. Jakhar, Hybrid BWM-ELECTRE-based decision framework for effective offshore outsourcing adoption: a case study, Int. J. Prod. Res. 56 (2018) 6259−6278, https://doi.org/10.1080/00207543.2018.1472406.

[77] M. Jasemi, E. Ahmadi, A new fuzzy ELECTRE-based multiple criteria method for personnel selection, Sci. Iran. 25 (2018) 943−953, https://doi.org/10.24200/sci.2017.4435.

[78] E. Fernández, J.R. Figueira, J. Navarro, An indirect elicitation method for the parameters of the ELECTRE TRI-nB model using genetic algorithms, Appl. Soft Comput. J. 77 (2019) 723−733, https://doi.org/10.1016/j.asoc.2019.01.050.

[79] C. Erdin, G. Ozkaya, Turkey's 2023 energy strategies and investment opportunities for renewable energy sources: site selection based on ELECTRE, Sustainability 11 (2019), https://doi.org/10.3390/su11072136.

[80] S. Geetha, S. Narayanamoorthy, J.V. Kureethara, D. Baleanu, D. Kang, The hesitant Pythagorean fuzzy ELECTRE III: an adaptable recycling method for plastic materials, J. Clean. Prod. 291 (2021), https://doi.org/10.1016/j.jclepro.2020.125281.

[81] Deleted in Review.

[82] M. Ehrgott, C.M. Fonseca, X. Gandibleux, J.-K. Hao, M. Sevaux, Evolutionary Multi-Criterion Optimization, Springer Berlin / Heidelberg, 2009, p. 586.

[83] M.D. Little, A Critique of Welfare Economics Oxford, Clarendon Press, 1950.

[84] S.A.N. Alexandropoulos, C.K. Aridas, S.B. Kotsiantis, M.N. Vrahatis, Multi-objective evolutionary optimization algorithms for machine learning: a recent survey, in: Springer Optimization and its Applications, Springer International Publishing, 2019, pp. 35–55, https://doi.org/10.1007/978-3-030-12767-1_4.

[85] A. Nedjati, M. Yazdi, R. Abbassi, A sustainable perspective of optimal site selection of giant air-purifiers in large metropolitan areas, Environ Dev Sustain (2021), https://doi.org/10.1007/s10668-021-01807-0.

[86] B.Y. Qu, Y.S. Zhu, Y.C. Jiao, M.Y. Wu, P.N. Suganthan, J.J. Liang, A survey on multi-objective evolutionary algorithms for the solution of the environmental/economic dispatch problems, Swarm Evol. Comput. 38 (2018) 1–11, https://doi.org/10.1016/j.swevo.2017.06.002.

[83] M. Yazdi, Introducing a heuristic approach to enhance the reliability of system safety assessment, Qual. Reliab. Eng. Int. 35 (2019), https://doi.org/10.1002/qre.2545.

[88] A.L. Márquez, C. Gil, R. Baños, M.G. Montoya, F.G. Montoya, F. Manzano-Agugliaro, A cooperative multi-objective island parallel model for wind farm planning, in: Civil-Comp Proceedings, Civil-Comp Press, 2011, https://doi.org/10.4203/ccp.95.49.

[89] M. Sadegh Javadi, M. Saniei, H. Rajabi Mashhadi, G. Gutiérrez-Alcaraz, Multi-objective expansion planning approach: distant wind farms and limited energy resources integration, IET Renew. Power Gener. 7 (2013) 652–668, https://doi.org/10.1049/iet-rpg.2012.0218.

[90] A. Hajebrahimi, M. Rashidinejad, A. Abdollahi, A fuzzy Analysis to connect the large-scale distant wind farm to grid in probabilistic multi objective transmission expansion planning, in: 22nd Iranian Conference on Electrical Engineering, ICEE 2014. Institute of Electrical and Electronics Engineers Inc., 2014, pp. 589–595, https://doi.org/10.1109/IranianCEE.2014.6999611.

[91] S. Abbasi, H. Abdi, S. Bruno, M. La Scala, Transmission network expansion planning considering load correlation using unscented transformation, Int. J. Electr. Power Energy Syst. 103 (2018) 12–20, https://doi.org/10.1016/j.ijepes.2018.05.024.

[92] M. Taherkhani, S.H. Hosseini, M.S. Javadi, J.P.S. Catalão, Scenario-based probabilistic multi-stage optimization for transmission expansion planning incorporating wind generation integration, Elec. Power Syst. Res. 189 (2020), https://doi.org/10.1016/j.epsr.2020.106601.

[93] S.L. Gbadamosi, N.I. Nwulu, Y. Sun, Multi-objective optimisation for composite generation and transmission expansion planning considering offshore wind power and feed-in tariffs, IET Renew. Power Gener. 12 (2018) 1687–1697, https://doi.org/10.1049/iet-rpg.2018.5531.

[94] S. Paul, Z.H. Rather, A new Bi-level planning approach to find economic and reliable layout for large-scale wind farm, IEEE Syst. J. 13 (2019) 3080–3090, https://doi.org/10.1109/JSYST.2019.2891996.

[95] P. Mittal, K. Mitra, In search of flexible and robust wind farm layouts considering wind state uncertainty, J. Clean. Prod. 248 (2020), https://doi.org/10.1016/j.jclepro.2019.119195.
[96] L. Huang, H. Tang, K. Zhang, Y. Fu, Y. Liu, 3-D layout optimization of wind turbines considering fatigue distribution, IEEE Trans. Sustain. Energy 11 (2020) 126–135, https://doi.org/10.1109/TSTE.2018.2885946.
[97] M. Song, K. Chen, J. Wang, A two-level approach for three-dimensional micro-siting optimization of large-scale wind farms, Energy 190 (2020), https://doi.org/10.1016/j.energy.2019.116340.
[98] C. Zhang, T. Yang, Optimal maintenance planning and resource allocation for wind farms based on non-dominated sorting genetic algorithm-II, Renew. Energy 164 (2021) 1540–1549, https://doi.org/10.1016/j.renene.2020.10.125.
[99] S. Zhong, A.A. Pantelous, M. Beer, J. Zhou, Constrained non-linear multi-objective optimisation of preventive maintenance scheduling for offshore wind farms, Mech. Syst. Signal Process. 104 (2018) 347–369, https://doi.org/10.1016/j.ymssp.2017.10.035.
[100] X. Ge, Q. Chen, Y. Fu, C.Y. Chung, Y. Mi, Optimization of maintenance scheduling for offshore wind turbines considering the wake effect of arbitrary wind direction, Elec. Power Syst. Res. 184 (2020), https://doi.org/10.1016/j.epsr.2020.106298.
[101] C.A. Irawan, D. Jones, D. Ouelhadj, Bi-objective optimisation model for installation scheduling in offshore wind farms, Comput. Oper. Res. 78 (2017) 393–407, https://doi.org/10.1016/j.cor.2015.09.010.
[102] H.Y. Kim, M.K. Kim, S. Kim, Multi-objective scheduling optimization based on a modified non-dominated sorting genetic algorithm-II in voltage source converter-multi-terminal high voltage DC grid-connected offshore wind farms with battery energy storage systems, Energies 10 (2017), https://doi.org/10.3390/en10070986.
[103] L. Wang, G. Chen, T. Wang, J. Cao, Numerical optimization and noise analysis of high-tip-speed wind turbine, Adv. Appl. Math. Mech. 9 (2017) 1461–1484, https://doi.org/10.4208/aamm.OA-2016-0171.
[104] V. Mytilinou, E. Lozano-Minguez, A. Kolios, A framework for the selection of optimum offshore wind farm locations for deployment, Energies 11 (2018), https://doi.org/10.3390/en11071855.
[105] S. Zhong, A.A. Pantelous, M. Goh, J. Zhou, A reliability-and-cost-based fuzzy approach to optimize preventive maintenance scheduling for offshore wind farms, Mech. Syst. Signal Process. 124 (2019) 643–663, https://doi.org/10.1016/j.ymssp.2019.02.012.
[106] S. Paul, A.P. Nath, Z.H. Rather, A multi-objective planning framework for coordinated generation from offshore wind farm and battery energy storage system, IEEE Trans. Sustain. Energy 11 (2020) 2087–2097, https://doi.org/10.1109/TSTE.2019.2950310.
[107] S. Tao, Q. Xu, A. Feijoo, G. Zheng, Joint optimization of wind turbine micrositing and cabling in an offshore wind farm, IEEE Trans. Smart Grid 12 (2021) 834–844, https://doi.org/10.1109/TSG.2020.3022378.
[108] S. Tao, Q. Xu, A. Feijóo, G. Zheng, J. Zhou, Nonuniform wind farm layout optimization: a state-of-the-art review, Energy (2020), https://doi.org/10.1016/j.energy.2020.118339.

Chapter 7

Recent advances of image processing techniques in agriculture

Helia Farhood, Ivan Bakhshayeshi, Matineh Pooshideh, Nabi Rezvani, Amin Beheshti
School of Computing, Macquarie University, Sydney, NSW, Australia

1. Introduction

Before harvesting, it is vital to keep track of agricultural productions to detect various diseases and defections on them since some might be infectious. If they are not recognized in the early stages, all agricultural productions can be infected. Performing these processes manually is time-consuming, expensive, and associated with human error and needs human resources. All the plants and fruits detection, classification, grading, and defect detection processes are done based on their visual features and appearance. Therefore, image processing and machine vision are widely used in agricultural plants and vegetable analysis to solve the mentioned issues with manual approaches. Image processing can be applied to all mentioned processes. By object detection and image segmentation techniques, fruits and plants are detected from the images, and by classification techniques, they are classified into different types. These methods can be used as a preprocessing step for grading to remove the background or defect detection to find defected areas. They can also be used for a robot to divide the crops based on their types or to recognize different crops [1,2].

Texture analysis, color analysis, and morphological feature analysis can be used for the grading and sorting phase. Image processing is used to detect different sizes, shapes, colors, and textures information from the crops and extracts adequate feature vectors. It then feeds the vector to different multi-class classifiers as an input to classify the crops into different quality levels or grades. The same image processing techniques, such as color analysis of the grading phase, can be used for defect detection. Besides this, monochrome, hyperspectral, and multispectral imaging techniques are also used in defect

Artificial Intelligence and Data Science in Environmental Sensing
https://doi.org/10.1016/B978-0-323-90508-4.00007-1

detection. The reason is that some of the defects cannot be captured by RGB cameras and need other hardware and reflected wavelengths [2,3].

Automatic visual recognition is a novel technology, and its history could be limited to few decades. After passing the era of manual face recognition, its progress has rapidly surged to the extent that it becomes one of the most active subjects in computer science. Since the recent development of artificial intelligence (AI) methods, academia and industry leaders have introduced various face recognition models with competitive accuracy rates. Visual identification technology has not been restricted only to human identification but also is appalled to animals, plants, and objects identification. Object detection refers to finding object examples in a picture from a specific class [4] and facial recognition refers to identifying individuals' faces according to their biological systems [5]. While this could be done by using eyes as natural visual sensors, in a similar process, artificial sensors such as video cameras and infrared cameras can capture the visual data for computer processing. To perform a face recognition process in humans or animals, the target spots should initially be identified from the image by the face detection system.

Applying machine learning approaches and particularly AI could revolutionize the method of image processing [6,7]. More complex features such as histogram of oriented gradient (HOG), aggregated channel features, scale-invariant feature transform (SIFT), and speeded-up robust features (SURF) required more devoted effort to raise the precision scores. Moreover, the robustness of detectors was enhanced by training a combination of detectors separately. Advanced methods of Al such as deep learning and particularly convolutional neural networks (CNNs) could inspire computer vision technology [6,8]. Face recognition applications, which enjoy AI techniques, can potentially replace the human vision to identify individuals from their faces [10]. In computer processing, AI probably increases the accuracy rate in target identification and enhances the capacity for work with data. This evolution of image processing could contribute to many other industries such as agriculture, energy, and transportation.

2. Application in plants detection

In this part, we focus on novel studies that discuss image processing approaches utilized for the extraction and recognition of plants, plant disease recognition in different situations such as underexposure or overexposure, and various colors of the plants and background three-dimensional (3D) monitoring techniques to study plant growth and their performance.

2.1 Plant segmentation and extraction in the field

Plant segmentation is one of the significant challenges in smart agriculture because it can be applied to several applications, such as plant species recognition and determination of the growing phase. Image-based segmentation methods often include two major steps: preprocessing and pixel classification. Preprocessing requires some significant primary processing on the main data from the camera, for instance, improving the contrast and eliminating noise. Image improvement is one of the main steps in AI and image processing because it plays a vital role in different usages, such as plant recognition and remote sensing. To enhance an image, the contrast of the obtained image should be improved and modified to address the problems associated with the changeability of luminance, such as shadow and sunlight [11]. Color conversion can be used to solve the problems of lighting in an image. The normalized difference vegetation index is one solution to decrease the illumination reaction and separate plants from the image's background [12]. Filtering is another way to improve an image. Homomorphic filtering is a kind of filtering technology used to minimize illumination problems and can be applied to different images under several environmental conditions [13]. The main aim of image processing techniques for weed and plant detection and recognition is to identify the pixels in an image to segment plants from the background. The background can involve soil and remains, which can be weeds and crops. Removing the background is one of the main stages in this process and must be done in a suitable way to avoid misclassification. Some techniques have been proposed for segmenting plant images.

Three popular methods for segmentations include (1) color index−based segmentation, (2) threshold-based segmentation, and (3) learning-based segmentation. In addition to the index-based methods, the image processing technique with color spaces is one of the main popular approaches to separate crops from a background in computer vision. Various color space models can be used for this purpose, such as the hue saturation value (HSV) color space or CIELAB (l: lightness, a: green−red, b: blue−yellow) color space. Several research works have indicated that separating plants from soil based on color can be challenging because of shading, overexposure, and underexposure, which causes problems in color spaces. One solution is to describe the color of a pixel singly of the brightness in different channels. Another solution can be a setting threshold manually on suitable color channels, although changing the RGB amounts into grayscale does not work properly in some segmentation methods because, for instance, plant and soil background pixels could have close grayscale values. As a result, to illustrate segmentation properly, the RGB space is usually converted to alternative spaces of color. Other techniques for the segmentation of plants include recognitions via machine learning methods such as decision trees or CNNs. Such methods have found significant applications in sensing systems for agriculture applications [14−17]. Recently, deep learning has been used in different applications. However, these approaches need huge labeled training databases and a significant amount of computing capability. In addition, the input of tagged data is required for machine learning systems and might not be suitable for various

TABLE 7.1 Some agricultural product classification and segmentation techniques.

Reference	Process	Approach	Classifier	Accuracy
[19]	Classification	Color layout descriptor segmentation, edge histogram descriptor	Support vector machine	100%
[20]	Classification	Scale-invariant feature transform	Random forest	96.97%
[21]	Segmentation and classification	Edge detection, color, texture, shape, and size feature extraction	Neural network	96.55%
[22]	Segmentation and classification	Color, texture, and shape feature extraction	PCA+BBO-FNN	89.11%
[23]	Recognition	Histogram, color, and texture features	Nearest neighbor	>95%

FNN, feedforward neural network; *PCA*, principal component analysis.

types of plants and crops if this was not investigated in the training dataset [18]. Table 7.1 indicates the accuracy of some existing methods in this area.

Sabzi et al. [24] presented a machine vision prototype for online detection and classification of the Marfona potato plant based on video processing and metaheuristic classifiers. Different footages were taken from two Marfona potato farms covering a total area of 6 hectares in order to adequately train their machine vision system. They used the grayscale level, color characteristics, and shape to extract features.

2.2 Plant diseases recognition

This part discusses recent studies that have addressed plant disease identification and recognition. We will touch on factors that help recognize these diseases through images and then review conventional and deep learning methods, including supervised and unsupervised approaches.

2.2.1 Datasets and problem setting

The datasets available for the task of plant disease detection are normally a number of images (mainly leaf of a plant) each labeled with a pair of plant disease (or plant-no disease). The aim of the predictor is to detect the label given the image pixels automatically. An example of these datasets is Plant-Village [25]. The images in this dataset are color images but some researchers have experimented with the grayscale version [26].

2.2.2 Conventional machine learning methods

As explained in Ref. [27], the process for detecting disease in plans through image analysis and machine learning is comprised of the following steps:

- Image acquisition and labeling: Taking high-quality photos and then labeling images to be ready for a supervised learning task.
- Preprocessing: This step encompasses applying standard image processing techniques like segmentation to highlight areas of interest in the image which leads to extracting leaf, stem, or root sections in the image.
- Feature extraction: In this step, more specific features like color, shape, and texture are extracted and will be fed to the classifier.
- Model training and classification: Obviously, in this step, a machine learning model is trained to perform prediction.

Although these methods follow a conventional and rather inefficient approach for feature extraction, they seem to have yielded acceptable results. Support vector machines (SVMs), artificial neural networks, K-nearest neighbors (KNNs), and fuzzy classifiers are compared as part of the survey and SVMs have reported the best results.

Deep learning methods are also mentioned in Ref. [27] as effective methods that have outperformed conventional methods. We will talk about those in the next part.

2.2.3 Deep learning methods

As mentioned earlier, adoption of the deep learning methods and especially deep convolution neural networks has been quite prevalent among more recent researches for detecting plant diseases. The main difference between deep learning methods and conventional ones is the elimination of the heavy feature extraction process which is carried out as part of the training of CNNs. Mohanty et al. [26] have surveyed these methods and suggested the following common steps among all of them:

- Model architecture: The choice of CNN model architecture may profoundly impact the result of prediction. In Ref. [26], AlexNet [28] and GoogleNet [29] have been experimented with (also other similar architectures like AlexNetOWTBn [30], Overfeat [31], VGG [32], etc.), but more recent researches have also presented their results with more modern CNN architectures [33] (like ResNet and Inception V4 [34]). The transfer learning setting that has been followed in different researches also varies which will be discussed in the next section.

- Training approach: There are mainly two main approaches when it comes to model training, either training the model from scratch or fine-tuning a pretrained model using transfer learning.
- Dataset type: As covered before, the input data that are used for the plant disease detection are different in that they contain either color or grayscale images and also whether the leaf segmentation is already applied on the image or not.

The result of experimenting with different aforementioned settings indicates that more recent models (e.g., ResNet and VGG) outperform their older counterparts. Fine-tuning of pretrained models yields better results training from scratch based on the results reported in Ref. [26]. However, utilizing more modern architectures and training the model from scratch is reported to result in impressive classifiers [35]. Moreover, prediction results for color images are proven to surpass other dataset types.

2.2.4 Transfer learning methods

As discussed earlier, transfer learning allows transferring the knowledge of the general-purpose learner model trained on a massive range of images to a specific task like plant disease detection. This is especially beneficial when dealing with the limited number of training samples in the dataset which is normally the case for plant disease detection datasets [25]. The following aspects of transfer learning have been taken into account in the context of plant disease identification:

- Freezing weights of initial layers: It is a standard practice to freeze initial weights of the pretrained model, but in some works, the whole network is trained and only initial weights are derived from the pretrained model [26].
- Gradual reduction of learning rate: This is common practice that has been employed by most researchers for plant disease detection [26].
- Batch normalization: This is a technique that helps in keeping the learning rate high and speeding up the training process by normalizing inputs of each layer by adjusting the mean and variance of the input over a mini batch [36]. This has been shown to be effective in the detection of plant diseases as well [37].

2.2.5 Analysis of effectiveness

It is worthwhile to have a look at the results derived from some research works to analyze the effectiveness of the methods mentioned above and also go through possible challenges. Barbedo [38] has elaborated on these challenges, highlights of which are as follows:

- Amount of data: The datasets in hand normally do not contain enough data for training a robust model from scratch. Using transfer learning and data augmentation techniques can mitigate that issue.

- Image quality: Images are normally taken in the controlled conditions which makes it harder for the trained model to be able to generalize for future disease recognition.
- Image background: Images normally contain backgrounds with unrelated textures and colors which might deceive the model during training. Getting users to select their area of interest can be a potential fix to this problem.
- Disorders with similar symptoms: Because there are a lot of factors that translate to symptoms for detecting disease, visual cues might not be sufficient to distinguish between similar classes. In these cases, leveraging some external information like weather, historical disease data, etc., can be helpful.

2.3 Three-dimensional monitoring for plant growth

This part presents various approaches to 3D growth plant monitoring and is concerned with the tracking of plant leaves. The significance of plant development and structure in calculating agricultural yields and controlling weather patterns is widely acknowledged, but computing the structure of plant characteristics is difficult, time-consuming, and demanding [39]. Research is being conducted on plant breeding to enhance and develop food production by improving crop performance attend to relationships between genotype (DNA) and phenotype (visual characteristics). While genotyping procedures are quickly improving, the majority of current phenotyping methods are laborious, intrusive, and sometimes harmful [40]. Phenotyping has lately been linked with automation and image processing to develop its constructs in order to bridge the gap between genetic and phenotype information.

Recently, several research works have focused on the automated monitoring of plant growth. Throughout the last few years, approaches based on 2D imaging have been developed for several applications, such as plant growth analysis and assessment of yield. However, when compared to 3D imaging, 2D imaging has numerous disadvantages.

In recent years, studies pointed to the use of 3D images as a means to automatically extract visual attributes of a plant rather than using 2D images. The most significant drawbacks of 2D photos relate to the occlusion of leaves of plants or stems because plants have complicated constructions. In addition, overlapping between leaves and the distance from the camera to the plant could make the use of 2D images problematic [41]. Both active and passive 3D imaging systems are being employed in precision agriculture and plant monitoring. LiDAR and other active technologies can generate a 3D matrix of objects in the surroundings and recover signals from the canopy. LiDAR, on the other hand, is costly and weather-dependent.

In contrast, passive methods can be utilized in applications that need high resolution and low cost, such as monitoring plant growth and structure, such as leaves, stems, flowers, and fruits. Plant structure is captured using passive 3D imaging technologies, which do not introduce new energy (such as light) into the environment. The main approaches and technologies which use a passive method include multiview stereo, shape from focus/shading/texture, structure from motion (SfM), light field (plenoptic) cameras, and space carving methods.

Passive techniques that use more than two sensors can have trouble detecting and aligning the same spots in successive photos, which can lead to erroneous 3D shape reconstruction. Plant leaves and canopies are particularly difficult to work with because they typically depict large, homogenous expanses with little distinct texture [42]. Two cameras are used to monitor the same thing in stereovision techniques. A baseline separates the two cameras, and it is assumed that the distance between them is known precisely. The two cameras simultaneously capture two images, which are then analyzed to see what the differences are. Features such as corners can be easily located in one image, and the other image can be searched to find the same feature. Conversely, the differences between the images can be found to assists to get the depth map which enables the creation of 3D. Stereovision creates high-resolution depth data with multiview compared with the active approaches but has some limitations such as self-occlusion and texture of the plants [43].

In forestry, the application of SfM photogrammetry is revolutionizing the way 3D remote sensing data is acquired. SfM photogrammetry provides 3D modeling at a low cost and with little technical knowledge. Plant characteristics may be extracted from SfM point clouds and derived 2D data in area-based techniques, according to the research. Furthermore, taking into account the increased spatial and spectral resolution of available sensors for SfM photogrammetry allows for 3D plant growth tracking [44]. Approaches based on shape from focus/shading/texture use one camera for estimating the 3D model; then these kinds of approaches are simple and do not suffer from the complex stereo correspondence. However, compared with SfM approaches, the result will not be very accurate because of the occlusion problem and the data which is missed by using a single image [39].

According to the research, only light field cameras have been successful in capturing 3D plant growth in a single shot [23,45]. Using a light field camera to generate 3D point clouds from a single photo is a simple approach to add variety to remote sensing monitoring and modeling applications [46]. Fig. 7.1 indicates a sample of real light field image with different focus and depth map, where the main goal is to create high-quality 3D point clouds with a single snapshot from real objects with the aim of having 3D models of plants or any other real objects for photogrammetry [45]. However, light field technology relies on costly camera technology to capture high-resolution data [42].

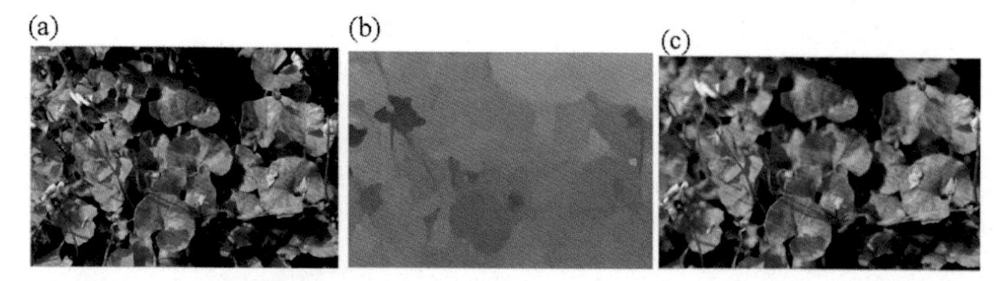

FIGURE 7.1 Real light field image captured by Lytro camera (A) focuses on the foreground of light field image, (B) the depth map of the image, which has been created by Lytro camera, and (C) the light field image with background focus.

3. Application in livestock recognition

This section first presents the history of visual recognition technology and discusses the livestock recognition approaches, particularly in the case of cattle identification.

The first reports of using face reconnection technology date back to 1871, which was a compression among face photographs in a British court [47]. In 1950, Alan Turing published a paper named "Computing Machinery and Intelligence," which can be deemed as a turning point of AI [48]. Even though there were many efforts for human face recognition, such as manual face measurement by Bledsoe in the 1960s, this technology achieved first remarkable accuracy improvement in 1970 [49]. The emergence of AI contributed imperatively to the development of facial recognition technology. Goldstein, Harman, and Lesk proposed a face recognition method based on 22 face points and criteria, including hair color and lip thickness. Their system still followed the manual computing of Bledsoe [49]. It took almost 20 years when Pentland and Trak reviled their method to detect the face from the pictures in 1991 [50]. They endeavored to identify the face of humans in two dimensions by employing technical and also environmental factors. Feature extraction and training traditional machine learning models were sophisticated and suboptimal tasks. DARA and NIST released the FIRST program which strikingly attracted the commercial market in the early years of the 21 century [51]. After a brief period, law enforcement officially utilized face recognition technology in order to prevent crimes in society. Viola and Jones put forward a practical method that could impact this technology in 2004. In their approach, Haar features and AdaBoost were employed in the training phase of face detector models, which led to a noticeable advancement from the previous milestone [52].

3.1 Livestock detection

AI-powered image processing technology can increase animal welfare. Visual identifications based on animal biometrics have received great attention both from the academic and industry. Automatic detecting visual features and analyzing livestock behaviors based on biometric characterizes could open up broad research opportunities [53]. From the marketing perspective, the market value of introducing a proper tool for the farming sector could be substantial given the scale of this industry. While traditional identification methods such as hot iron branding, tattooing, and freeze branding are not in line with animal welfare strategies, recent methodologies such as ear tags and rumen bolus are vulnerable in the case of loss and fraud [54]. So far, AI recognition methods indicated high performance in this regard and can be considered as an alternative approach.

3.2 Cattle recognition

Research in visual animal biometrics worldwide has presented a broad spectrum of applications and methods in recognition or tracking systems. As an illustration, in the case of cattle, they employed these methods for animal insurance claim identification by matching the registered livestock's image with the claimed image. Another instance of an automatic animal recognition application is related to recognizing the livestock as soon as they pass the farm gate. This is an important development in the sense that the traditional animal recognitions system based on the livestock framework could not realize the animal's ownership. Obtaining a practical animal detection tool based on visual animal biometrics would require developing and integrating advanced and innovative algorithms [55].

Intelligence animal recognitions are derived from two common steps of detection systems. In the first step, the data should be acquired, and the second step is animal detection. When the sensors, such as cameras, handle capturing the animal images, the noises and artifices are removed by image prepossessing techniques. Subsequently, animals' biometric factors are analyzed by applying statistical-based quality calculation to store the unique features of the case. This database plays a significant role in individual animal recognition and breed classification. Regarding the high level of resemblance among the same breed of animal, many notable visible features cannot support animal identification [54].

3.2.1 Muzzle point image pattern

Muzzle point image pattern (nose pattern) is a unique characteristic among the cattle, same as fingerprints in human identification. It includes a rich dense texture feature that can be utilized in the retrieval of cattle recognition systems. The limitation of this approach is that taking a muzzle picture of each cow automatically is difficult [56]. Kumar et al. proposed a recognition approach based on cattle muzzle point images (see Fig. 7.2) [57]. They applied various deep learning architectures, including the CNN approach with a combined of

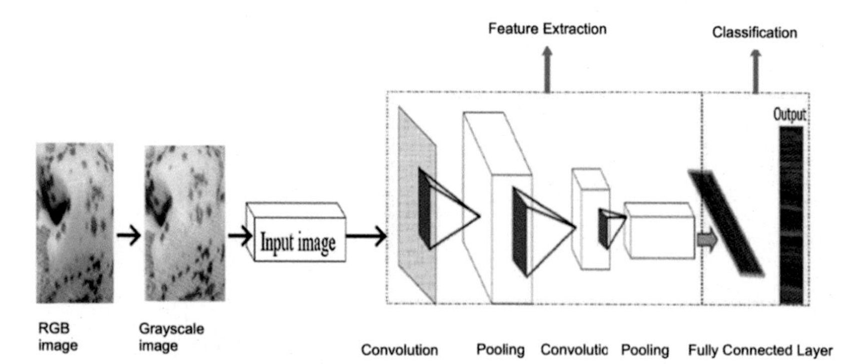

FIGURE 7.2 Muzzle point recognition diagram. *From Kumar, S., et al., 2018. Deep learning framework for recognition of cattle using muzzle point image pattern. Measurement 116, 1–17.*

deep belief network (DBN), restricted Boltzmann machines (RBMs), and stacked denoising auto-encoder (SDAE). After capturing the muzzle point images, in the prepossessing process, appropriate images were transformed to grayscale to alleviate the noises from the dataset and also speed up the training process. Then the image dataset was fed to the convolutional layers. SDAE technique is utilized for encoding the extracted features, whereas DBN with the RBM model was employed to learn the extracted features and their classification. This approach yielded an accuracy rate of 98.99% for cattle identification.

3.2.2 Coat pattern

In this system, cattle are recognized based on both color and body pattern. After sensors such as cameras which are capturing pictures of cows, the body of the animal is detected. Afterward, unique features of the detected images through a deep learning method are extracted. As a limitation, despite the superior performance of this strategy, it could hardly apply to some cattle breeds with a unicolor body. In a recent study, after 22 days of video recording 45 case studies, Zin et al. reached a 97.01% accuracy score by developing a deep convolutional neural network (DCNN) [56]. This study included two core components; firstly, the cattle's body regions are automatically detected and cropped by the histogram base and the interframe differencing approaches. Secondly, the DCNN is trained to recognize the cattle based on their RGB color images pattern. Fig. 7.3 demonstrates the workflow of livestock identification based on animals' body patterns.

3.2.3 Rear view

One of the challenges for visual animal recognition is the high level of resemblance in the same generation of livestock. While many researchers focused on developing AI tools for dairy cattle identification, beef cattle have

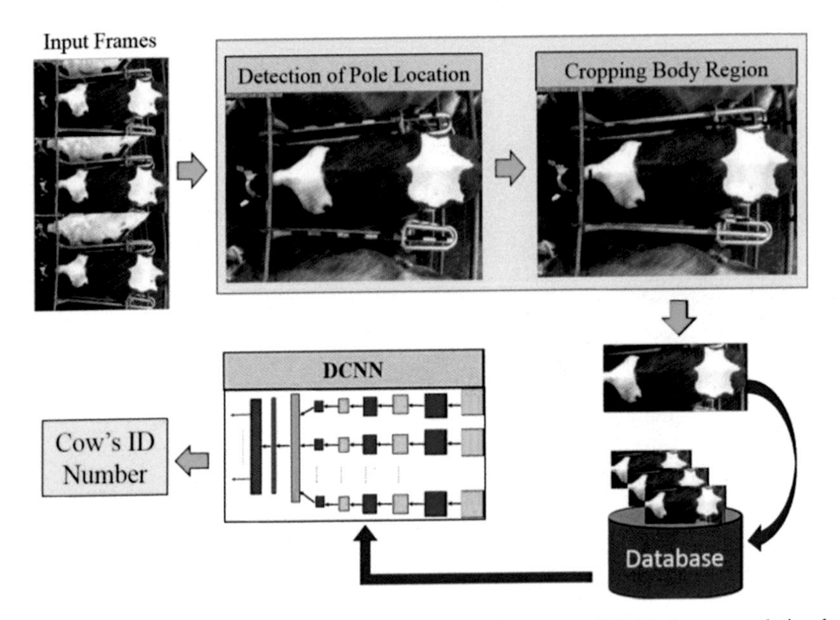

FIGURE 7.3 Cattle identification system based of body pattern. *DCNN*, deep convolutional neural network. *From Zin, T.T., Phyo, C.N., Tin, P., Hama, H., Kobayashi, I., 2018. Image technology based cow identification system using deep learning. In Proceedings of the International MultiConference of Engineers and Computer Scientists, 2018, vol. 1, pp. 236–247.*

received less attention since most breeds have similar coat colors misleading the visual recognition. In order to address this issue and improve the accuracy in beef cattle recognition, the shape of the animal body can be accurately deemed. Qiao et al. conducted a study on a sample of Australian beef cattle to identify individual cattle from the rear-view images [58]. The proposed model combined DCNN with long short-term memory (LSTM) techniques. In this approach, a video was recorded from the cow's backside and then broken into several image frames. Cattle features are extracted by DCNN model Inception-V3 based on the pretrained weights, whereas the LSTM network is used for identification. Eventually, an outstanding accuracy of 98.97% was reported in this study. The process of cattle identification through this method is shown in Fig. 7.4.

3.2.4 Multiviews face

Livestock recognition based on the animal face features has seen significant progress over the last few years. Substantial progress in human face recognition technologies and AI algorithms could be the main reason for this advancement. Although human face recognition methods could be applied to animals if extra adjustments and modifications are applied. The need for these algorithmic changes is driven by specific animal face shapes and high

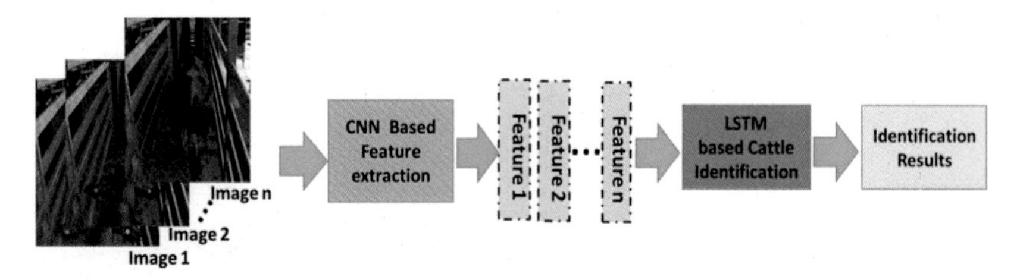

FIGURE 7.4 Cattle identification in combination of convolutional neural network (CNN) and long short-term memory (LSTM). *From Qiao, Y., Su, D., Kong, H., Sukkarieh, S., Lomax, S., Clark, C., 2019. Individual cattle identification using a deep learning based framework. IFAC-PapersOnLine 52(30,) 318–323.*

uncertainty in the data collection stage. For instance, Kumar et al. received a 92.5% identification accuracy rate by utilizing local binary patterns (LBPs) and SURF on 1200 images of cattle [59]. Besides, Berganmini et al. highlighted the poor link between the human recognition methods with their proposed approach for cattle reidentification. In this study, a couple of pictures were captured from both cattle profiles. Next, the pictures from the same cow were fed into an embedding DCNN. After that, separated convolutional branches carried the feature of two images subsequently, and both were connected to a single-output feature vector. Finally, the nearest neighbors between the dataset and input image could reidentify the cattle [60]. The reviewed cattle recognition approaches are summarized in Table 7.2.

4. Application in fruits and vegetables recognition

Fruit and vegetable analysis is highly essential in the agricultural industry. The quality of fruits affects their pricing. Sorting and grading fruits after harvesting and before exporting to the market is important to price different quality levels of fruits. In the grading stage, fruits are classified into grades such as A, B, C, and D to determine the quality levels determining a range of extra fancy to reject.

In this section, applications of image processing and machine vision are discussed in different processes.

4.1 Fruits and vegetables identification and classification

Fruits and vegetables identification and classification is the task of detecting fruits in images with backgrounds and classifying them into different types. This process can be used in the harvesting phase for harvesting robots or after harvesting to categories the fruits automatically.

Fangyuan Liu et al. [61] review fruits' identification methods. One study uses the fusion of four feature extractors to detect fruits: global histogram, Gabor-LBP, LBP, and HOG. SIFT model is also used to detect the shape of

TABLE 7.2 Cattle recognition studies with novel approaches.

References	Applications	Approach	Dataset	Models	Limitations	Performance
[57]	Cattle recognition	Muzzle pattern	500 images	DCNN, DBN, RBMS, and SDAE	Difficulty in the data collection	98.99%
[56]	Cattle identification	Coat pattern	22 days recording of 45 cows	DCNN	Limited to patterned bodies cattle	97.01%
[58]	Individual cattle identification	Rear view	10,320 images extracted from 526 videos	DCNN and LSTM	Time-consuming data collection	98.97%
[59]	Cattle face recognition	Face	1200 images	SURF and LBP	Not in real time	92.5%
[60]	Cattle reidentification	Multiviews (face)	17,802 images	DCNN	Computationally expensive	89%

DBN, deep belief network; *DCNN*, deep convolutional neural network; *LBP*, local binary pattern; *LSTM*, long short-term memory; *SDAE*, stacked denoising auto-encoder; *SURF*, speeded-up robust features.

fruits. Another study applies the split-and-merge technique to identify fruits. Multiple studies have also used the fusion of different colors and textures and morphological feature descriptors or different neural networks to find and detect fruits in images [61].

Khurram Hameed et al. [62] divide the classification process into four parts of data acquisition, preprocessing, feature extraction, and classification. Multiple visual sensors could be used for taking images of fruits. Black and white, RGB color, spectral, and thermal cameras are some of those sensors. The data preprocessing step is related to use image processing techniques to remove background, detect the region of interest (ROI), remove noise from images, etc. One of the essential techniques in preprocessing step is segmentation. Segmentation approaches are used for detecting the ROI and subtracting the background. Edge detection, texture and color analysis, thresholding, and brightness detection are some of the methods used for segmentation. Features are mainly divided into three steps of shape and size, texture, and color in the feature extraction part. Some shape feature descriptors are as follows: detecting convex Hull, bounding box, area and perimeter, contour curvature, shape moments, etc. Hameed et al. address various texture analysis techniques such as SIFT, HOG, LBP, Gabor, basic image feature, Fisher vector, Weber local descriptor, and different CNNs. Shiv Ram Dubey et al. [1] also discuss some feature descriptors used for the feature extraction part. Global color histogram, color coherence vector, border/interior classification, completed LBP, Unser's feature, and improved sum and difference histogram are some of the feature descriptors they address.

Various classifiers are used to classify the fruits into different types based on the features extracted in the previous phase. Some of the classifiers are as follows: KNN, linear discriminant analysis (LDA), SVM, feedforward neural network, K-means, principal component analysis (PCA), backpropagation neural networks, logistic regression, quadratic discriminant analysis, self-organizing map, decision trees, and fuzzy C-means [62].

José Naranjo-Torres et al. [63] review CNN applications in fruits' classification and identifications. Various studies using new network structures and pretrained networks are addressed. GoogleNet, AlexNet, and VGG are used with transfer learning technology to classify fruits, and modified VGG, ResNet, AlexNet, Inception, GoogleNet, and some other new structures are used for fruit detection purposes. In the fruit detection process, many Mask neural networks such as Mask R-CNN are addressed as well since the goal of identification is to find the ROI [63].

4.2 Fruits and vegetables grading and sorting

One of the essential steps before sending fruits and vegetables to market is sorting or grading them. Grading a fruit means categorizing it into different qualities based on its size, shape, color, texture, weight, etc. In the agriculture market, it is of utmost importance to grade the fruits for better pricing. Also, doing the grading fast and hygienic is exceptionally essential. Since grading fruits is mainly based on fruits' morphological features, photographs can carry those features, and hence image processing is vastly used in this domain.

Many image processing algorithms extract color features, textural features, shapes, and sizes from objects in the image. These algorithms are used in different studies to grade different fruits. The grading systems mainly have three steps. Firstly, the background is removed. Secondly, the features are extracted, and finally, a classier is used to classify the features into different grading levels. Rashmi Pandey et al. [64] review different studies done in image processing applications in automatic fruit grading. Some studies focus on grading unique fruits for the grading systems, and some others focus on general algorithms of grading that can be applied to all types of fruits. Pandey et al. focus on both general algorithms and specific fruit grading systems. Apple, orange, lemon, strawberry, mango, and date are some of the fruits mentioned in different studies.

Different methods are addressed for background removal, such as eliminating particular hue components, image segmentation, thresholding, etc. [64]. Pandey et al. mainly focus on studies that have used color features to grade different fruits. Some studies extract the fruit's color's mean value in RGB channels as the feature vector to grade the fruits. For example, this method is used to grade San-Fuji apples in one study [65], and another study uses this method as a general grading system [66]. Many studies use the color features extracted from different or multiple color spaces such as HSI, HSV, La*b*, Lu*v*, and RGB. Direct color mapping is another method used by different studies. Most of the mentioned methods extract a proper feature vector that can be given as the input to different classifiers to grade the fruits into different quality classes.

Different classifiers are then used to classify the extracted feature vectors. Pandey et al. address some of them, such as SVMs, fuzzy logic, neural networks, KNN, LDA, etc. [64].

Anup Vibhute et al. [67] review different grading systems. K-means, thresholding, segmentation, and filtering (such as Bayer filtering) techniques are some background elimination methods addressed in Ref. [67]. Vibhute et al. describe different feature extraction techniques used for grading purposes in their paper. Image segmentation, shape description and analysis, textural information extraction, pattern recognition, morphology, and color histogram extraction are some of those techniques. In some studies, extra hardware is used for grading purposes. X-rays images, monochrome images, and images taken under halogen lights are images taken from specific hardware and sometimes include more information than RGB photos [67].

Anuja Bhargava et al. [2] review the fruit quality detection methods in five steps. The first step is image acquisition, in which different hardware used to detect images from fruits are discussed. The second step is preprocessing of the images. Since the images always contain different noises, some noise elimination methods are addressed in this step. Some studies use filtration to avoid different noises. Some of the filters are median filter (used for eliminating high-frequency noises), simple filter, and modified unsharp filter. Another preprocessing that can be done is color space transformation, in which the image is transformed to other color spaces to be able to get more information out of the image. The third step is background removal. Segmentation methods are used in different studies to divide background and fruits. Thresholding and clustering are the segmentation methods used for background removal. Otsu method and simple grayscale thresholding are some examples of thresholding methods. K-means, C-means of fuzzy logic, rule-based clustering, hierarchical clustering, and minimum spanning tree are examples of clustering techniques used in different studies for this purpose.

The fourth step that Bhargava et al. address is the feature extraction step. They divide the feature extractions into three types of color extraction, texture feature extraction, and morphological feature extraction. In color extraction approaches, two actions are taken. The first action is defining the color space. Some studies use RGB color space, some use HSI, and some use CIELab (La*b*) color space. The problem with RGB could be its sensitivity to the camera; in other words, if a pixel is captured from different sensors, the RGB values of the captured images might not be the same. Hence HSI could be a better solution. However, for the grading purpose, the fruits' color changes are needed to be captured even if they are negligible. Hence HSI and RGB cannot be an excellent solution to detect tiny changes in the colors, and therefore CIELab color space can solve this issue. The second action in color extraction methods is extracting color features from selected color space in the previous action. Color histograms, coherence vectors, correlograms, and moments are different color feature extractors that different studies use [2].

Texture analysis is one of the other types of features that are used for fruit grading. Texture analysis is vital to detect the features of fruits' surfaces, such as contrast, roughness, line-likeness, energy, orientation, direction, etc. Some methods used in different studies are Gabor feature and PCA kernel, spectral Fourier transform, multilayer perceptron, and LBP. Morphological features are the other types of features that are used for grading purposes. They are related to the size and shape of fruits. Calculating the area, perimeter, length, and width of fruits are some techniques of extract features about the size of fruits. For shape detection, methods such as roundness, compactness, and aspect ratio are used.

Finally, in the fifth step, Bhargava et al. study different classifiers used to classify the features extracted in step four. Neural networks, LDA, AdaBoost, SVM, PCA, fuzzy logic, KNN, the minimum distance classifier, etc., are some of those classifiers [2].

4.3 Fruits and vegetables disease and defect detection

Detecting defects and fruits' diseases is one of the other essential steps that should be taken on. Dividing fresh and high-quality fruits and fruits with defects manually is time-consuming and too expensive and consists of human error. Also, it is crucial to detect diseases in the earliest stages since some kinds of them can be spread alongside all the trees.

Defects can be divided into different types, such as bruises, contamination, spots, pathogen infections, and chilling injury [3]. For example, apples can have scab, rot, blotch, fungi attack, frost damage, scar tissue, and insect attack defects [1]. Hence, detecting all these types of defects and diseases is important and should be processed fast, and therefore image processing and machine vision are widely used for this purpose.

Defect detection could also be interpreted as the last level of grading. Grading fruits is to classify them into different qualities, and infected fruits can be classified into the rejected or bad-quality class. However, defect detection is more challenging than grading. For example, stem and calyxes have the same color attributes as different defects such as bruises and spots, or the images of the fresh fruits can contain noises similar to defects, or RGB images might not capture all defects accurately. Hence, besides some studies that use the traditional machine vision approaches for defect detection using RGB images, many studies focus on monochrome, hyperspectral, and multispectral imaging techniques. Fruit defect detection and classification in image processing can be done in three steps. The first step is to segment the defective parts of the infected fruit. The second part is the feature extraction, and the third part is the classification [1].

In traditional image processing, different color feature extraction techniques are used to detect defects. Some studies use color channels such as hue (H) for feature extraction or color spaces such as YCbCr and RGB [68,69]. Some studies use PCA (in a study, second and third PCA components were used to detect bruises [70]). A study uses rotation to capture multiple images from one product to make sure the darker areas are not shadows and stem and calyxes [71].

Shiv Ram Dubey et al. [1] reviewed different studies focusing on fruits defect detection. Some image processing techniques addressed by them are as follows. The color co-occurrence matrix is used in some studies to detect defective regions. A study applies texture analysis of the ROI selected by the color co-occurrence matrix of HSI color space. ROI selection is made in another study using K-means clustering, and then MCSVM is applied to ROI for feature extraction and classification purposes. Pattern recognition is used to detect stem/calyx and defects. Another study creates a model based on the standard range of color of the fruit and compares each pixel to that model to recognize whether the pixel is healthy or damaged. Bayesian classification of fruits' color features is also shown to be an effective segmentation technique that can detect various defects. Sobel gradient is also used to segment the defective regions of fruit peel. Another defect segmentation approach is extracting a saliency map of the fruit. A study extracts that by applying an attention-based model to the Fourier spectra of the image.

J. B. Li et al. [3] also review the fruit defect detection field studies. The Gaussian model is applied to the fruit's color to segment the damaged regions in a study. Different region-based segmentation methods are also used in various studies to detect defective areas of fruits' skin.

As mentioned previously, not all defects are sensitive to RGB cameras. In order to detect all types of defects, monochromatic CCD cameras are used in various studies [3]. These cameras capture the image with a nonvisible light. Near-infrared is one of the most used lights in different studies of defect detection. Capturing images with monochromatic cameras with different band-pass and long-pass filters and then applying a classification model will lead to accurate defect detection. Many studies try filters at various wavelengths such as 700 nm, 740 nm, 840 nm, 950 nm, etc. The results show that different defects are sensitive to some of these wavebands. Besides all these studies, many studies combine monochromatic and color cameras and use color analysis as well to make the system more robust [3].

Some studies extract 3D information of the fruits by special hardware. The reason is that almost all of the 2D information of stem/calyx parts of fruits are similar to that of the defects. Hence it is hard to detect whether a segmented part is an actual defect or is stem/calyx. Nevertheless, the nature of stem/calyx is three dimensional, whereas the nature of defects is two dimensional. Hence using 3D information of a segmented part is helpful to distinguish between the two [3]. Other studies focus on multispectral or hyperspectral imaging to detect defects. These techniques would extract both spatial and spectral features from the image and lead to better defect detection. For these techniques, reflecting approach with the light waveband ranging (400−1000 nm or 900−1700 nm) is used in various studies to detect different types of defects and diseases [3]. Table 7.3 shows the accuracy of some existing approaches in the state-of-the-art fruit image processing techniques.

5. Conclusion

In this chapter, different applications of image processing in agriculture were discussed. Many agricultural processes are based on the farming products' visual features and properties; accordingly, image processing techniques are vastly used to automate these processes. Crops and plants detection, segmentation, and classification techniques were discussed. Most of these techniques have used RGB color space image data. Next, for the quality control processes such as sorting and grading, different visual feature extraction techniques were presented. These features are related to the texture, color, and morphological (shape and size) attributes of the products. Different texture

TABLE 7.3 Some state-of-the-art fruit image processing techniques.

Reference	Fruit	Process	Method	Classifier	Accuracy
[65]	Apple	Grading	Neural network + color	-	>95%
[72]	Fresh-cut star fruit	Grading	HSI color space, Fourier transform	LDA, MLP	100%
[73]	Mango fruits	Grading	Fractal dimension, L*a*b* color space	LS-SVM	100%
[74]	Grapefruit	Defect detection	39 color texture features, color co-occurrence, HSI color space	Discriminant analysis	96.7%
[75]	Apple	Classification	HSV, average red and green color components	SVM	100%
[76]	Lemon	Defect detection	Curvelet transform	SVM, PNN	96%
[77]	All	Recognition	Convolutional neural network	-	100%

LDA, linear discriminant analysis; *MLP*, multilayer perceptron; *SVM*, support vector machine.

descriptors, color descriptors, mathematical morphology feature extractors, pattern recognition, and different classifiers were introduced in this section. In most of the studies, a fusion of various features is used as the final feature vector. This chapter also suggested a combination of AI technology and can replace the conventional human vision in object recognition. Individual identification technology has become mature over the past few years. Automatic AI visual recognition would replace the traditional livestock identifications soon due to the animal welfare reasons and reliability. Different face characteristics, high similarity in the same breed, and the uncertainty in the environment caused difficulty in identifying livestock by AI methods. Deep learning and CNN are the typical approaches in the most accurate livestock recognition, no matter from which viewpoints they are seen. Furthermore, AI technologies and advanced sensors in the near future would be able to detect animal behaviors and emotions from multidimensions to help with tracking animal behavior patterns [78]. In the last part, the detection of disease and defects of agricultural products are discussed. Multiple types of infections and defects are described in this stage, and different color extraction methods are discussed to extract defect features. Also, using additional hardware with various light reflections is addressed since not all infections are recognized by visual light spectral, and to recognize all defects, some special wavebands are needed.

References

[1] S.R. Dubey, A.S. Jalal, Application of image processing in fruit and vegetable analysis: a review, J. Intell. Syst. 24 (4) (2015) 405–424.

[2] A. Bhargava, A. Bansal, Fruits and vegetables quality evaluation using computer vision: a review, J. King Saud Univ. Comput. Inf. Sci. 33 (3) (2021) 243–257.

[3] J. Li, W. Huang, C. Zhao, Machine vision technology for detecting the external defects of fruits—a review, Imag. Sci. J. 63 (5) (2015) 241–251.

[4] Y. Amit, P.F. Felzenszwalb, in: Object Detection, University of Chicago, 2014. Brown.edu. website.

[5] A.M. Martinez, Face recognition, overview, in: S.Z. Li, A. Jain (Eds.), Encyclopedia of Biometrics, Springer US, Boston, MA, 2009, pp. 355–359.

[6] M. Asadnia, L.H. Chua, X. Qin, A. Talei, Improved particle swarm optimization–based artificial neural network for rainfall-runoff modeling, J. Hydrol. Eng. 19 (7) (2014) 1320–1329.

[7] M. Asadnia, A.M. Khorasani, M.E. Warkiani, An accurate PSO-GA based neural network to model growth of carbon nanotubes, J. Nanomater. 2017 (2017).

[8] M. Asadnia, M.S. Yazdi, A.M. Khorasani, An improved particle swarm optimization based on neural network for surface roughness optimization in face milling of 6061-T6 aluminum, Int. J. Appl. Eng. Res. 5 (19) (2010) 3191–3201.

[9] Deleted in review.

[10] I. Adjabi, A. Ouahabi, A. Benzaoui, A. Taleb-Ahmed, Past, present, and future of face recognition: a review, Electronics 9 (8) (2020) 1188.

[11] E. Hamuda, M. Glavin, E. Jones, A survey of image processing techniques for plant extraction and segmentation in the field, Comput. Electron. Agric. 125 (2016) 184–199.

[12] A. Perez, F. Lopez, J. Benlloch, S. Christensen, Colour and shape analysis techniques for weed detection in cereal fields, Comput. Electron. Agric. 25 (3) (2000) 197–212.

[13] G. Pajares, J.J. Ruz, J.M. de la Cruz, Performance analysis of homomorphic systems for image change detection, in: Iberian Conference on Pattern Recognition and Image Analysis, Springer, 2005, pp. 563–570.

[14] M. Asadnia, et al., Mercury (II) selective sensors based on AlGaN/GaN transistors, Anal. Chim. Acta 943 (2016) 1–7.

[15] M.S. Syed, M. Rafeie, R. Henderson, D. Vandamme, M. Asadnia, M.E. J.L.o. a.C. Warkiani, A 3D-printed mini-hydrocyclone for high throughput particle separation: application to primary harvesting of microalgae, Lab Chip 17 (14) (2017) 2459–2469.

[16] F. Ejeian, et al., Biosensors for wastewater monitoring: a review, Biosens. Bioelectron. 118 (2018) 66–79.

[17] M. Mahmud, et al., Recent progress in sensing nitrate, nitrite, phosphate, and ammonium in aquatic environment, Chemosphere (2020) 127492.

[18] D. Riehle, D. Reiser, H.W. Griepentrog, Robust index-based semantic plant/background segmentation for RGB-images, Comput. Electron. Agric. 169 (2020) 105201.

[19] E. Rachmawati, I. Supriana, M.L. Khodra, Multiclass fruit classification of RGB-D images using color and texture feature, in: International Conference on Soft Computing, Intelligence Systems, and Information Technology, Springer, 2015, pp. 257−268.

[20] H.M. Zawbaa, M. Hazman, M. Abbass, A.E. Hassanien, Automatic fruit classification using random forest algorithm, in: 2014 14th International Conference on Hybrid Intelligent Systems, IEEE, 2014, pp. 164−168.

[21] M.T. Chowdhury, M.S. Alam, M.A. Hasan, M.I. Khan, Vegetables detection from the glossary shop for the blind, IOSR J. Electr. Electron. Eng. 8 (3) (2013) 43−53.

[22] Y. Zhang, P. Phillips, S. Wang, G. Ji, J. Yang, J. Wu, Fruit classification by biogeography-based optimization and feedforward neural network, Expet Syst. 33 (3) (2016) 239−253.

[23] R.M. Bolle, J.H. Connell, N. Haas, R. Mohan, G. Taubin, VeggieVision: a produce recognition system, in: Proceedings Third IEEE Workshop on Applications of Computer Vision. WACV'96, IEEE, 1996, pp. 244−251.

[24] S. Sabzi, Y. Abbaspour-Gilandeh, J.I. Arribas, An automatic visible-range video weed detection, segmentation and classification prototype in potato field, Heliyon 6 (5) (2020) e03685.

[25] D. Hughes, M. Salathé, An open access repository of images on plant health to enable the development of mobile disease diagnostics, arXiv preprint arXiv:1511.08060 (2015).

[26] S.P. Mohanty, D.P. Hughes, M. Salathé, Using deep learning for image-based plant disease detection, Front. Plant Sci. 7 (2016) 1419.

[27] U. Shruthi, V. Nagaveni, B. Raghavendra, A review on machine learning classification techniques for plant disease detection, in: 2019 5th International Conference on Advanced Computing & Communication Systems (ICACCS), IEEE, 2019, pp. 281−284.

[28] A. Krizhevsky, I. Sutskever, G.E. Hinton, Imagenet classification with deep convolutional neural networks, Adv. Neural Inf. Process. Syst. 25 (2012) 1097−1105.

[29] C. Szegedy, et al., Going deeper with convolutions, in: Proceedings of the IEEE Conference on Computer Vision and Pattern Recognition, 2015, pp. 1−9.

[30] H. Nazki, S. Yoon, A. Fuentes, D.S. Park, Unsupervised image translation using adversarial networks for improved plant disease recognition, Comput. Electron. Agric. 168 (2020) 105117.

[31] P. Sermanet, D. Eigen, X. Zhang, M. Mathieu, R. Fergus, Y. LeCun, Overfeat: integrated recognition, localization and detection using convolutional networks, arXiv preprint arXiv:1312.6229 (2013).

[32] K. Simonyan, A. Zisserman, Very deep convolutional networks for large-scale image recognition, arXiv preprint arXiv:1409.1556 (2014).

[33] J. Chen, J. Chen, D. Zhang, Y. Sun, Y.A. Nanehkaran, Using deep transfer learning for image-based plant disease identification, Comput. Electron. Agric. 173 (2020) 105393.

[34] C. Szegedy, S. Ioffe, V. Vanhoucke, A.A. Alemi, Inception-v4, inception-resnet and the impact of residual connections on learning, in: Thirty-first AAAI Conference on Artificial Intelligence, 2017.

[35] K.P. Ferentinos, Deep learning models for plant disease detection and diagnosis, Comput. Electron. Agric. 145 (2018) 311−318.

[36] S. Ioffe, C. Szegedy, Batch normalization: accelerating deep network training by reducing internal covariate shift, in: International Conference on Machine Learning, PMLR, 2015, pp. 448−456.

[37] E.C. Too, L. Yujian, S. Njuki, L. Yingchun, A comparative study of fine-tuning deep learning models for plant disease identification, Comput. Electron. Agric. 161 (2019) 272−279.

[38] J.G. Barbedo, Factors influencing the use of deep learning for plant disease recognition, Biosyst. Eng. 172 (2018) 84–91.

[39] Y. Zhang, P. Teng, M. Aono, Y. Shimizu, F. Hosoi, K. Omasa, 3D monitoring for plant growth parameters in field with a single camera by multi-view approach, J. Agric. Meteorol. 74 (4) (2018) 129–139.

[40] W. Gélard, A. Herbulot, M. Devy, P. Casadebaig, 3D leaf tracking for plant growth monitoring, in: 2018 25th IEEE International Conference on Image Processing (ICIP), IEEE, 2018, pp. 3663–3667.

[41] A. Paturkar, G.S. Gupta, D. Bailey, Non-destructive and cost-effective 3D plant growth monitoring system in outdoor conditions, Multimed. Tool. Appl. 79 (47) (2020) 34955–34971.

[42] G. Bernotas, et al., A photometric stereo-based 3D imaging system using computer vision and deep learning for tracking plant growth, GigaScience 8 (5) (2019) giz056.

[43] R. Kala, On-road Intelligent Vehicles: Motion Planning for Intelligent Transportation Systems, Butterworth-Heinemann, 2016.

[44] J. Iglhaut, C. Cabo, S. Puliti, L. Piermattei, J. O'Connor, J. Rosette, Structure from motion photogrammetry in forestry: a review, Curr. For. Rep. 5 (3) (2019) 155–168.

[45] H. Farhood, S. Perry, E. Cheng, J. Kim, 3D point cloud reconstruction from a single 4D light field image, in: Optics, Photonics and Digital Technologies for Imaging Applications VI, International Society for Optics and Photonics, vol. 11353, 2020, p. 1135313.

[46] H. Farhood, S. Perry, E. Cheng, J. Kim, Enhanced 3D point cloud from a light field image, Rem. Sens. 12 (7) (2020) 1125.

[47] A. Ouamane, A. Benakcha, M. Belahcene, A. Taleb-Ahmed, Multimodal depth and intensity face verification approach using LBP, SLF, BSIF, and LPQ local features fusion, Pattern Recogn. Image Anal. 25 (4) (2015) 603–620.

[48] A.M. Turing, Computing machinery and intelligence, in: Parsing the Turing Test, Springer, 2009, pp. 23–65.

[49] A.J. Goldstein, L.D. Harmon, A.B. Lesk, Identification of human faces, Proc. IEEE 59 (5) (1971) 748–760.

[50] M.A. Turk, A.P. Pentland, Face recognition using eigenfaces, in: Proceedings. 1991 IEEE Computer Society Conference on Computer Vision and Pattern Recognition, IEEE Computer Society, 1991, pp. 586, 587, 588, 589, 590,591–586, 587, 588, 589,590,591.

[51] P.J. Phillips, H. Wechsler, J. Huang, P.J. Rauss, The FERET database and evaluation procedure for face-recognition algorithms, Image Vis Comput. 16 (5) (1998) 295–306.

[52] P. Viola, M.J. Jones, Robust real-time face detection, Int. J. Comput. Vis. 57 (2) (2004) 137–154.

[53] J. Duyck, C. Finn, A. Hutcheon, P. Vera, J. Salas, S. Ravela, Sloop: a pattern retrieval engine for individual animal identification, Pattern Recogn. 48 (4) (2015) 1059–1073.

[54] S. Kumar, S.K. Singh, Cattle recognition: a new frontier in visual animal biometrics research, Proc. Natl. Acad. Sci. India Phys. Sci. 90 (4) (2020) 689–708.

[55] M. Lahiri, C. Tantipathananandh, R. Warungu, D.I. Rubenstein, T.Y. Berger-Wolf, Biometric animal databases from field photographs: identification of individual zebra in the wild, in: Proceedings of the 1st ACM International Conference on Multimedia Retrieval, 2011, pp. 1–8.

[56] T.T. Zin, C.N. Phyo, P. Tin, H. Hama, I. Kobayashi, Image technology based cow identification system using deep learning, Proc. Int. MultiConf. Eng. Comp. Sci. 1 (2018) 236–247.

[57] S. Kumar, et al., Deep learning framework for recognition of cattle using muzzle point image pattern, Measurement 116 (2018) 1–17.

[58] Y. Qiao, D. Su, H. Kong, S. Sukkarieh, S. Lomax, C. Clark, Individual cattle identification using a deep learning based framework, IFAC-PapersOnLine 52 (30) (2019) 318–323.

[59] S. Kumar, S. Tiwari, S.K. Singh, Face recognition for cattle, in: 2015 Third International Conference on Image Information Processing (ICIIP), IEEE, 2015, pp. 65–72.

[60] L. Bergamini, et al., Multi-views embedding for cattle re-identification, in: 2018 14th International Conference on Signal-Image Technology & Internet-Based Systems (SITIS), IEEE, 2018, pp. 184–191.

[61] F. Liu, L. Snetkov, D. Lima, Summary on fruit identification methods: a literature review, in: 2017 3rd International Conference on Economics, Social Science, Arts, Education and Management Engineering (ESSAEME 2017), Atlantis Press, 2017.

[62] K. Hameed, D. Chai, A. Rassau, A comprehensive review of fruit and vegetable classification techniques, Image Vis Comput. 80 (2018) 24–44.

[63] J. Naranjo-Torres, M. Mora, R. Hernández-García, R.J. Barrientos, C. Fredes, A. Valenzuela, A review of convolutional neural network applied to fruit image processing, Appl. Sci. 10 (10) (2020) 3443.

[64] R. Pandey, S. Naik, R. Marfatia, Image processing and machine learning for automated fruit grading system: a technical review, Int. J. Comput. Appl. 81 (16) (2013) 29–39.

[65] K. Nakano, Application of neural networks to the color grading of apples, Comput. Electron. Agric. 18 (2–3) (1997) 105–116.

[66] W.C. Seng, S.H. Mirisaee, A new method for fruits recognition system, in: 2009 International Conference on Electrical Engineering and Informatics vol. 1, IEEE, 2009, pp. 130–134.

[67] A. Vibhute, S.K. Bodhe, Applications of image processing in agriculture: a survey, Int. J. Comput. Appl. 52 (2) (2012).

[68] J.D. Pujari, R. Yakkundimath, A.S. Byadgi, Grading and classification of anthracnose fungal disease of fruits based on statistical texture features, Int. J. Adv. Sci. Technol. 52 (1) (2013) 121–132.

[69] J.D. Pujari, R. Yakkundimath, A. Byadgi, Reduced color and texture features based identification and classification of affected and normal fruits' images, Int. J. Agric. Food Sci. 3 (3) (2013) 119–127.

[70] J. Xing, C. Bravo, P.T. Jancsók, H. Ramon, J. De Baerdemaeker, Detecting bruises on 'Golden Delicious' apples using hyperspectral imaging with multiple wavebands, Biosyst. Eng. 90 (1) (2005) 27–36.

[71] B. Bennedsen, D. Peterson, A. Tabb, Identifying defects in images of rotating apples, Comput. Electron. Agric. 48 (2) (2005) 92–102.

[72] M.Z. Abdullah, J. Mohamad-Saleh, A.S. Fathinul-Syahir, B. Mohd-Azemi, Discrimination and classification of fresh-cut starfruits (*Averrhoa carambola* L.) using automated machine vision system, J. Food Eng. 76 (4) (2006) 506–523.

[73] H. Zheng, H. Lu, A least-squares support vector machine (LS-SVM) based on fractal analysis and CIELab parameters for the detection of browning degree on mango (*Mangifera indica* L.), Comput. Electron. Agric. 83 (2012) 47–51.

[74] D.G. Kim, T.F. Burks, J. Qin, D.M. Bulanon, Classification of grapefruit peel diseases using color texture feature analysis, Int. J. Agric. Biol. Eng. 2 (3) (2009) 41–50.

[75] M. Suresha, N. Shilpa, B. Soumya, Apples grading based on SVM classifier, Int. J. Comput. Appl. 975 (2012) 8878.

[76] S.A. Khoje, S. Bodhe, A. Adsul, Automated skin defect identification system for fruit grading based on discrete curvelet transform, Int. J. Eng. Technol. 5 (4) (2013) 3251–3256.

[77] S. Sakib, Z. Ashrafi, M. Siddique, A. Bakr, Implementation of fruits recognition classifier using convolutional neural network algorithm for observation of accuracies for various hidden layers, arXiv preprint arXiv:1904.00783 (2019).

[78] S. Neethirajan, Happy cow or thinking pig? WUR wolf–facial coding platform for measuring emotions in farm animals, bioRxiv (2021).

Chapter 8

Tuning swarm behavior for environmental sensing tasks represented as coverage problems

Shadi Abpeikar, Kathryn Kasmarik, Phi Vu Tran, Matthew Garratt, Sreenatha Anavatti, Md Mohiuddin Khan

School of Engineering and Information Technology, University of New South Wales, Canberra, ACT, Australia

1. Introduction

In recent years, industrial residues in the environment have increased as the consequence of increasing industrial activities, their products, and consumers of these products [1]. These industrial residues are a significant threat for natural ecosystems and their habitants [2]. One of the efficient ways to control and reduce the dispersion of these industrial residues is to predict and detect pollution [3]. Environmental sensing is an effective way to detect pollution [4–6]. However, complete sensing and accurate prediction needs more than just human monitoring [3]. LiDAR systems, ultraviolet, infrared, or radar systems, and image processing are some examples of remote environmental sensing tools and sensors [7,8]. These can reduce the risk of humans being involved with the harmful residues [3]. As another tool, artificial intelligence (AI) can provide the ability of residue detection and prediction, by the aid of artificial agents like robots, multiagent systems, and single-agent systems [3]. One example of such a system is a robot swarm. Swarm robots consist of many robots, which can communicate with each other in a decentralized manner, and monitor their surroundings with the aid of their sensors. Some of these sensors are gas sensors, temperature sensors, and humidity sensors [9] to name a few. Because robots are mobile, swarm robots can have an advantage over other tools such as fixed sensor networks, to sense wider areas. Due to these advantages, many marine, terrestrial, and airborne swarm robotic systems are being developed for remote environmental sensing [3]. To name a few, in Ref. [10], an experiment based on a red-green-blue (RGB) camera examined

Artificial Intelligence and Data Science in Environmental Sensing
https://doi.org/10.1016/B978-0-323-90508-4.00001-0

the possibility of spectral remote monitoring by a swarm of unmanned aerial vehicles. In Ref. [11], an odor localization task by autonomous mobile robots was examined. These experiments show that by increasing the communication between groups of agents, odor localization performance increases. In Ref. [12], semiautonomous robots are used to detect spills on ocean surfaces and remove them with the help of a magnetic conveyor and magnetic powder. In Ref. [13], a system of mobile robots was developed to detect mines and explosive residues and to execute landmine clearance tasks. In Ref. [14], a robot swarm was designed to quickly localize contaminant sources in an indoor environment. In Ref. [15], Reynolds' boid model [16] was applied to build a swarm of marine robots for chemical detection in underwater environments. The quick chemical detection done by Ref. [15] is very helpful, since in indoor environments late detection can result in a disaster, like inhaling toxic gas or an explosion. In Ref. [17], schools of robots are used for pollution and harmful contaminant detection in ports. These robots are inspired by the behavior of fish schools, and they can find pollution sources and notify other swarm members. Efficient exploration, fast detection, and effective communication are the consequences of many robots equipped with multiple sensors, maintaining their swarming behavior and using decentralized communications [3]. However, a key challenge of robot swarming is tuning the behavior of the swarm. In general, different robots have slightly different operating constraints, and thus, the weights controlling the swarm behavior must be set slightly different for each new group of robots. This step is often assumed. However, in practice it can be time-consuming and difficult to get right. The main contribution of our reinforcement learning for swarm behavior automatic tuning (RL-SBAT) in this chapter is an approach to behavior tuning of the swarm, and evaluation of tuned swarms in an architecture to do environmental sensing represented as a coverage problem. The other differences between our work and the works mentioned above are highlighted in Table 8.1.

The remainder of this chapter is organized as follows. Section 2 provides some preliminary discussions on related works: Reynolds' boid model, reinforcement learning (RL) and coverage problems. Section 3 presents our system for environmental sensing by swarm robots, and Section 4 discusses the performance of this system in a point-mass simulator and simulated robots. Finally, Section 5 concludes the chapter and discusses some areas for future work.

2. Preliminaries

In this section fundamental methods for environmental sensing by swarm robots are discussed. To this end, Section 2.1 discusses related work on swarm robots as artificially intelligent agents, which use machine learning for environmental sensing. In Section 2.2, Reynolds' boid model inspired by natural swarms will be briefly introduced. Then Sections 2.3 and 2.4 will introduce RL and coverage problem, as the two components of our system.

TABLE 8.1 Comparison between reinforcement learning for swarm behavior automatic tuning (RL-SBAT) and the other related works.

Main goal of swarms for environmental sensing	References	Comparison to RL-SBAT	
Navigation	[12,13,15,17]	The proposed approach uses a general, organic behavior rather than a navigation, formation, or target search behavior.	In addition, RL-SBAT has the ability to tune swarm behavior from a random behavior.
Formation control	[11]		
Target search (source localization)	[11,14]		
Leader–follower	[15]	In the proposed method all agents are equal (no leaders or followers).	
Image collection for image processing	[10]	The proposed approach focuses on feature-based sensing instead of image processing.	
Bio-inspired swarms	[15,17]	The proposed approach considers a diverse range of swarm behaviors.	

2.1 Related work

Environmental sensing aims to detect harmful changes in an ecosystem [18]. Traditional environmental sensing methods and tools had some limitations in terms of both time and space coverage [19,20]. The data gathered by these methods and tools were unable to match fast changes of the environment in a timely manner [21]. To cover these limitations, a faster and more reliable method was required, which could deal with real-time sensing. Also, due to threats of industrial residues on human health and safety, a new system was required to reduce the risk of human involvement in residue detection tasks. Robot systems can fulfill these requirements [3,22]. Also, for robot systems to be more efficient, some supportive methods of path planning [23], navigation [24,25], coverage [23,24], and mapping [25] have been developed. These methods work based on models of machine learning [26], data mining [27], and expert system design [28].

A multi-entity Bayesian network is used in Ref. [29], to provide a swarm of robots ability to understand and exchange information for chemical pollution monitoring. The pollution monitoring is done for underwater environments by a swarm of underwater robots. A particle swarm optimization (PSO) method is applied in Ref. [30], aiming to develop a system of robots for odor source localization. In PSO, each robot senses the environment and informs other robots about the existence or nonexistence of an odor source. Robots move in the direction that increases local and global strength of odor sensing. An artificial neural network is used in Ref. [31], to provide pollution source localization in a complex environment using a single-robot system. The fitness function of the neural network considers both energy saving and accurate positioning. A modified version of this artificial neural network system could successfully find the optimal path to pollution resources in indoor environments [32]. The optimal path is measured in terms of search time and energy saving of the robot. An environmental monitoring problem is considered in Ref. [33], where an RL approach can effectively cover the environment and predict future events by the aid of pattern recognition. A bio-inspired behavior of aggregation and pheromone tracking is introduced in Ref. [34], which permits robots to do pollution source localization. A fuzzy controller system for oil spill prediction by swarm robots is developed in Ref. [35]. Also the same hybrid fuzzy algorithm approach is used for oil spill tracking in Ref. [36]. An exploration method based on a genetic algorithm is developed in Ref. [37]. The genetic algorithm provides a search zone for hazardous gas leak detection by a swarm of robots. An offline optimization algorithm based on spiral surge is proposed in Ref. [38], in which a swarm of robots solves an optimal odor localization task.

Considering the ability and efficient performance of the mentioned related works, swarm robots can play a significant role in environmental sensing for industrial residue detection. The advantages of swarm robots in environmental sensing are as follows:

- Their autonomous movement reduces the need for human involvement in this task.
- The spatial and temporal changes of the environment can be better addressed by a swarm of robots than what can be done by human or other static sensors. This can be achieved by AI algorithms to optimize the environmental sensing, number of robots, and their distributed sensing.
- The sensing task can be done more accurately and with less energy consumption by the aid of coverage, mapping, and path planning solutions.
- Swarm robots can solve environmental sensing quickly, in cases where fast detection of contaminant resources is crucial.

In this chapter an RL approach is applied in a point-mass simulator to tune swarm behavior from random motions. The aim of swarm behavior is to keep robots in motion together in a way that they can communicate efficiently. However, the difficulty of making robots swarm lies in tuning parameters such as their vision angle, communication radius, and the weights of their preferences for different sub-behaviors. The work in this chapter addresses the question of how to tune robot behavior to achieve swarming. It further shows that such swarms can cover an environment more quickly than groups with untuned behavior. The RL method presented in this chapter is different from the works mentioned above, as it can automatically tune swarm behavior very quickly, when the best parameters of the system are initially unknown. The next section introduces the swarm model used in this chapter.

2.2 Reynolds' boid model

Swarm behavior refers to the way that schools of fish, herds of land animals, and flocks of birds move in nature. Inspired by these behaviors, Reynolds [16] created one of the early computer models of swarming. This computer-based swarm is known as the boids (bird android) model, which works based on three simple rules. These rules are:

1. Collision avoidance (repulsion): following this rule, each boid avoids collision with the other boids.
2. Velocity matching (alignment): based on this rule, each boid should attempt to match its velocity with the other boids moving in its neighborhood.
3. Flock centering (attraction): regarding this rule, the boids should attempt to move close to the other boids within their neighborhood.

Later extensions of the Reynolds' boid model were designed which could be applied in robots and unmanned aerial and ground vehicles. These extensions, called the Boid Guidance Algorithms (BGAs) [39], include further rules for speed and turn rate, to maintain the agent within the operating conditions of robot hardware. The swarm behaviors considered in this chapter are achieved from a BGA model. For more information on these behaviors, see, e.g., Ref. [40]. The main limitations of these approaches are that BGAs need to be retuned for every new robotics platform to which they are applied. The work in this chapter provides a way to do the tuning automatically.

2.3 Reinforcement learning

RL is a sequence of decisions in a trial-and-error approach, which can learn to solve complex real problems over time [42]. The learning procedure occurs in number of episodes, with the overall aim to maximize the cumulative reward value and to reach convergence of reward values to the maximum expected reward [43]. The cumulative reward based on Markov decision process of RL is computed by Eq. (8.1) [44].

$$G_t = \sum_{k=0}^{T} r_{t+k+1} \tag{8.1}$$

where r_k is the reward signal of each step, and it is computed by

$$r_k = \overline{\rho}(x_k, u_k) \tag{8.2}$$

where $\overline{\rho}$ is the policy, (x_k, u_k) are the state and action of current step k, respectively, and t is the time.

Also, to provide a more reliable reward, a discount factor is applied to the cumulative reward. This discount factor is a weight value $\gamma \in [0, 1)$ which affects the cumulative reward of Eq. (8.1) as provided in Eq. (8.2). The bigger the γ, the greater the preference of the agent for the long-term reward, while smaller values result in preference for short-term reward by the agent.

$$G_t = \sum_{k=0}^{\infty} \left(\gamma^k \times r_{t+k+1}\right) \tag{8.3}$$

Each learning episode starts from an initial state, executes an action, updates its current state, receives a reward or a penalty, and updates its policy mapping states and/or actions to cumulative reward. These rewards and penalties help the agent to solve a problem. The agent only knows about the constraints and rules and does not have any idea on how to solve the problem [43]. The constraints and rules contain information about the action and state space, when and in which condition reward or penalty will be obtained, and when to stop the training of an episode or the whole training process. For more information on RL algorithms, see, e.g., Ref. [43]. In this work we train an RL agent to be able to adjust the parameters of a group of randomly moving agents to make the agents swarm together and navigate as a group. We then use the tuned swarm to solve a coverage problem. Such problems are introduced in the next section.

2.4 Coverage problems

Coverage refers to a task in which one or more robots try to sense the entirety of a large environment. The robots may be static, or they may move and communicate with each other [45]. In coverage problems, robots disperse around the environment in a way that permits them to cover most of the environment [45]. By solving the coverage problem, every part of the environment will be visited at least once [46]. Coverage can be achieved either directly or indirectly. Direct coverage by a robot implies that the robot itself visits a location. Indirect coverage implies that a specific robot does not visit every location but receives communication from other robots indicating at least one other robot has visited every location. This means that not every robot needs to visit every location. Solving coverage problems is required in many applications like exploration, surveillance, patrolling, and multiple target searching [47]. Environmental sensing to detect industrial residue sources is also an example that can be framed as a coverage problem. In these applications, robots sense the environment, exchange information with their neighbors, and consequently "cover" the area [47].

One way to provide a system of robots a way to move while staying in the communication range, and without running into each other, is to maintain a swarm behavior while performing a coverage task. However, tuning the behavior of a group of robots by hand to make them swarm is a nontrivial task. The following section provides a solution for both swarm tuning and achieving environmental coverage.

3. System design: swarming for coverage tasks

As mentioned in Section 2.1, swarming robots can provide many advantages in environmental sensing. These advantages are due to the wide range of available robots' sensors and their efficient communication with each other. However, tuning the parameters of a swarm system is difficult. Thus, the system described in this chapter has two parts. The first tunes the behavior of a group of robots (for example, in a scenario where a new group of robots rolls off a truck). This first subsystem uses an RL approach that in turn uses an if-then rule engine reward function. Then, the tuned swarm solves a coverage problem in the second subsystem. The system is presented in Fig. 8.1. The two subsystems of autonomous swarm behavior tuning and the coverage algorithm will be discussed in what follows.

3.1 Autonomous tuning of swarm behavior by the reinforcement learning subsystem

Autonomous swarm behavior tuning is a very hard task [49,50]. RL is a solution for swarm behavior tuning, which has been investigated by many

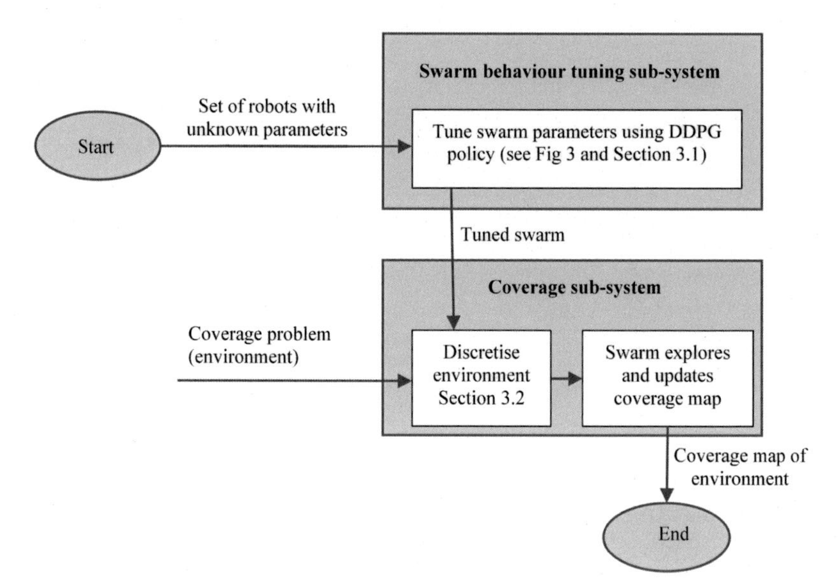

FIGURE 8.1 Full system architecture.

researchers [49,51]. However, the challenging part to design a reinforcement learner that can bootstrap swarm behavior in multiple scenarios is to design a general enough reward function [49]. In this section, a reinforcement learner to tune swarming behavior is proposed. The reward function works based on a knowledge-based system. This knowledge base generates a reward signal based on an if-then rule engine. The rule engine is extracted from a decision tree, which was trained from a human-labeled dataset. The labeled dataset is the aggregation of opinions from 90 participants engaged in a swarm behavior recognition task [52,53], which was conducted online [53]. The online survey contains 16 questions, each related to a short video clip capturing the motion of 200 boids in a point-mass simulation. The parameter setup for these motions includes eight structured movements and eight random movements [40]. Participants were asked to move slider ranges to label these videos as flocking or not flocking. Responses were averaged and the average decision used to label features extracted from the simulation underlying each video. Features included position, velocity, cohesion, alignment, separation forces, and number of neighbors, as outlined in Table 8.3 in the Appendix. For more information on the swarm behavior datasets, please refer to Refs. [40,52]. Also, it is downloadable at Ref. [54].

For the current work, a decision tree was trained with these labeled data. Results showed that it was able to achieve 99% accuracy in swarm behavior recognition [52]. However, this decision tree resulted in a very big if-then rule engine (1000 rules), which was unwieldy for the design of a knowledge-based system. We thus applied a pruning algorithm to the large tree. The pruned version of this decision tree was able to classify swarming with an accuracy of 96.59% using only 73 rules. Of these rules, 28 rules consequent in flocking, and the consequence of the others is not flocking. This rule-based engine provides the reward signal for swarm behavior tuning.

The components mentioned above to develop the knowledge-based reward signal for the RL subsystem are summarized in Fig. 8.2.

The next stage of our system is the reinforcement learner itself. In our RL agent, the state and action spaces are 14-dimensional continuous spaces. These 14 dimensions refers to the features (parameters) of the boid parameters in our point-mass simulator. Thus, the state space was defined as

$$S = \left[\mathrm{W_s},\ \mathrm{W_a},\ \mathrm{W_c},\ \mathrm{V_{max}},\ \mathrm{V_{min}}\ ,\ \mathrm{R_s},\ \mathrm{R_{c,a}}\ \theta, \mathrm{P_{sa}},\ \mathrm{P_{fullscan}},\ \mathrm{P_{rule}},\ \mathrm{P_s},\ \mathrm{P_a},\ \mathrm{P_c}\right] \tag{8.4}$$

The definitions of these parameters and their acceptable ranges are provided in Table 8.3 of the Appendix. The action space encompasses "increase," "decrease," and "do-not-change" actions for each of the state variables. It is a continuous space, and the changes occur in the range of $[-\omega, \omega]$, where

$$\omega = [0.5; 0.5; 0.5; 0.1; 0.1; 2; 10; 50; 0.1; 0.1; 0.1; 0.1; 0.1; 0.1].$$

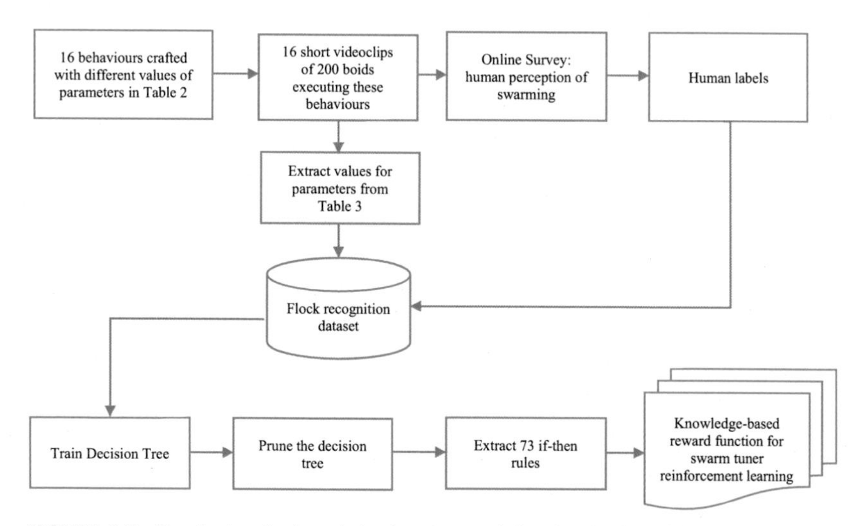

FIGURE 8.2 Developing the knowledge-based reward function in the reinforcement learning subsystem.

These ω values were selected after experimentation with the sensitivity of the learning algorithm, which indicated that some consideration was required to these values. This action will change the agent's state, while keeping the new state in its acceptable range.

Deep deterministic policy gradient (DDPG) RL [55] was used to train the learner in this 14-dimensional continuous action/state space. The MATLAB RL toolbox was used. This DDPG consists of 2000 episodes. Each episode could iterate for a maximum of 100 steps. The episode training stops in earlier steps when it reaches a flocking behavior (as determined by the reward signal of Eq. 8.5). The training procedure of each episode is presented in Fig. 8.3. In this flowchart, B is the boid behavioral parameters, as follows:

$$B=\left(\mathrm{V_x},\ \mathrm{V_y},\mathrm{A_x},\ \mathrm{A_y},\ S_x,\ S_y,\ C_x,C_y,\mathrm{n}_{(\mathrm{R_a},\mathrm{R_C})},\ \mathrm{n}_{(\mathrm{R_s})}\right)$$

B parameters are extracted from a point-mass boid simulator (implemented in Java) that simulates the swarm configuration with the given S parameters for 10 timesteps. The B parameters are defined in Table 8.4 of the Appendix. To compute the reward value, the if-then rules are applied for each of $i=1,\ldots,200$ boids in each $t=1,\ldots,10$ timesteps. α_t^i is the decision of the rule engine for each boid at each timestep. The mean value α is the average of α_t^i values for all boids and in all 10 timesteps. Thus, when most of the α_t^i decisions are flocking (the average α decides in flocking), the whole motion will be considered as flocking by the reward signal generator. The reward signal is computed by Eq. (8.5).

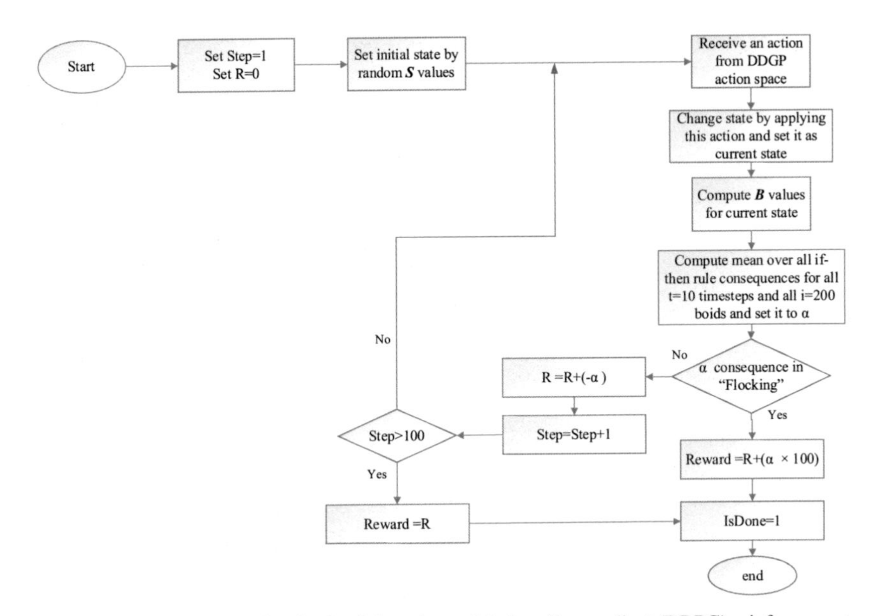

FIGURE 8.3 Training each episode of deep deterministic policy gradient (DDPG) reinforcement learning.

$$\text{Reward} = \begin{cases} \alpha \times 100 & \text{if rule engine results in 'flocking'} \\ -\alpha & \text{if rule engine results in 'not flocking'} \end{cases} \tag{8.5}$$

3.2 Coverage algorithm subsystem

In the previous section we discussed the RL subsystem and how it can be trained to know how to tune a swarm behavior. This section considers the coverage subsystem. The coverage subsystem has two parts. First, a group of agents with randomized parameters is tuned for 10 steps. Then this group of agents is applied to the coverage task. Tuning uses the trained DDPG policy described in the previous section, which is permitted to take 10 actions. The second part is the coverage process. First, the environment is discretized using a square grid. This grid-based representation of the environment is shared with all agents. The swarm is then permitted to explore the environment and the grid is updated to record when any agent enters a particular location. Coverage at any given time can be computed as a percentage of the total environment that has been visited by any agent. We use a centralized approach in this chapter. However, it can also be decentralized.

Algorithm 1. Computing the cumulative coverage.

1. Divide the environment into equal grid cells
2. For each timestep $t = 1, ..., T$
3. For each grid cell
4. If this cell is visited at least by one boid and has not been visited before
5. Count this cell as a covered-cell.
6. Compute the cumulative coverage rate by $\frac{\textit{number of cells marked as covered}}{\textit{total number of cells}}$

4. Experimental analysis

This section presents four experiments. The first experiment examines the performance of the DDPG RL (RL-SBAT) as it learns how to tune a swarm. The second experiment tunes a number of known swarms and then examines their performance at a coverage task. The third and fourth experiments consider the tuning and coverage ability of RL-SBAT in some unseen scenarios of both point-mass boids and simulated robots controlled by a BGA, respectively. The results of the latter three experiments have dual purpose: reconfirming that the tuning modifies the behavior of the swarm, as well as showing that the tuned behavior is more effective compared to the random behavior for solving the coverage problem.

4.1 Experiment 1: learning to tune a swarm

4.1.1 Experimental setup

The setup for this experiment follows the methodology described in Section 3.1. The learning rate for the DDPG is set to 0.008 and the discount factor is 0.95. In this experiment, reward is computed by Eq. (8.5). We found through sensitivity analysis (out of the scope of this work) that the value of α is somewhat important, and that it is important that the top branch of Eq. (8.5) in orders of magnitude is larger than the bottom branch. This is done because the RL could better train sparse reward values [57,58].

4.1.2 Results

The average results from 10 runs training the DDPG learner is presented in Fig. 8.4. As shown in this figure, the average reward increases as the number of episodes increases. Moreover, the long-term reward converges to this average reward in the latter episodes of the RL. This shows that DDPG RL trains efficiently to tune swarm behavior, by reaching swarming behavior in almost

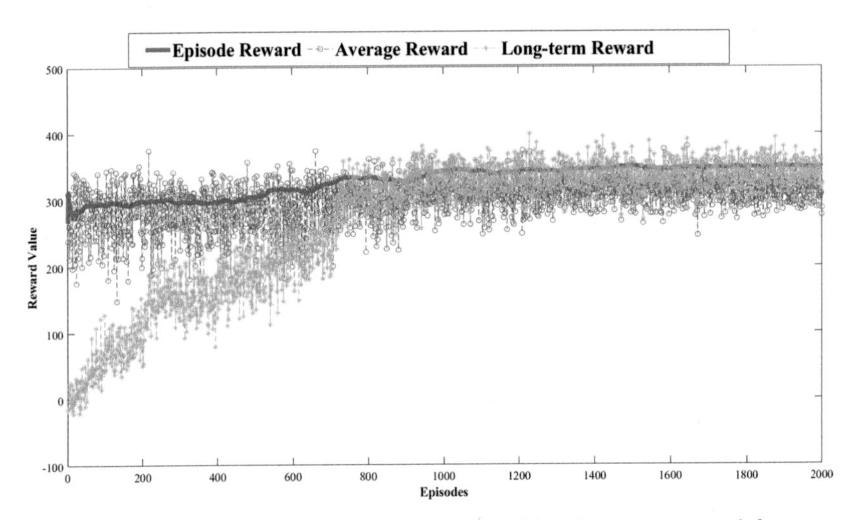

FIGURE 8.4 Average reward acquired from 10 runs training the swarm tuner reinforcement learner.

all the latter episodes. Moreover, we observed that the maximum number of actions required to tune a swarm behavior in the final episodes is less than 5. In Section 4.2, we will further validate these results by showing the actual ability of this trained network interacting with an untuned swarm.

4.2 Experiment 2: using a tuned swarm to solve a coverage problem

4.2.1 Experimental setup

The environment used in this experiment is a 2000×2000 pixel wraparound area. This area is discretized into 100 equal squares of 200×200 pixels. The cumulative coverage rate is computed over time for tuned behaviors. It is computed for 10 timesteps movement of 200 boids based on Algorithm 1. The environment and the corresponding 100 squares are presented in Fig. 8.5. In this figure, the red dot represents a boid; therefore, the green squares show a covered square at the current timestep.

To validate both subsystems of RL-SBAT, i.e., the tuning and the coverage subsystem, the eight random behaviors from our online survey [53] are fed to one of our trained reinforcement learners. These eight random behaviors are the results of different parameter setups, provided in Ref. [40]. The coverage percentage is computed over a period of 10 steps after tuning the swarm behavior. The aim of this evaluation is twofold. The first aim is to visualize how random motions change after being tuned by our RL-SBAT algorithm. The second aim is to compare the ability of the trained swarm to achieve coverage.

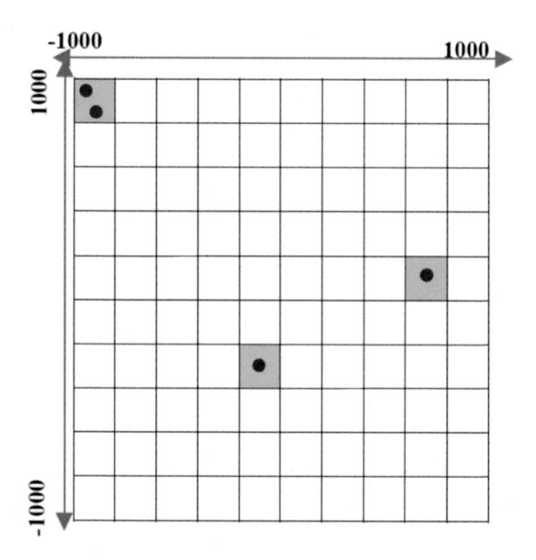

FIGURE 8.5 An example of covered squares within the test environment.

4.2.2 Results

The results of this experiment are presented in Fig. 8.6. The first column shows the initial random motions. Then these random motions change to a new tuned motion (swarm motion) presented in the middle column as the RL-SBAT algorithm modifies the swarm parameters. The blue arrows in columns one and two show the direction of boid motions, and the green dashed arrows show some embedded motions within these tuned motions. For example, in the tuned motion of part (a) of Fig. 8.6, all boids move to the right top corner of the arena, by circling around their path, which looks like a storm that moves to a specific direction. The tuning takes less than a second for each action in each case, and the maximum number of actions is 10.

Finally, the last column shows the cumulative coverage achieved by both the random and tuned motions over 10 timesteps after tuning. In Fig. 8.6, timestep $T = 1$ refers to the first timestep after the tuning procedure is finished, as presented in Fig. 8.7. The coverage rate at the firs timestep ($Timestep = 1$) is good, since the boids disperse in the area, so that they could cover most of the cells, in both random and tuned behaviors. The cumulative coverage rate of tuned motions improves over time, and it is higher than the cumulative coverage of the random motions. An example of how better coverage is achieved by tuned behaviors in this wraparound environment is shown in Fig. 8.8. It is clear that the coverage path is influenced by the wraparound environment here. However, in our experiments in simulated robots in walled environments (Section 4.4), we find that the structured coverage behaviors, although different, are still able to achieve coverage.

Returning to the experimental results of Fig. 8.6, it is clear that the individual tuned behaviors provide better coverage than the individual random behaviors. The average results obtained from each timestep of all random and tuned behaviors of Fig. 8.6 are further summarized in Fig. 8.9. This further analysis shows that, on average, the tuned behaviors outperform random behaviors.

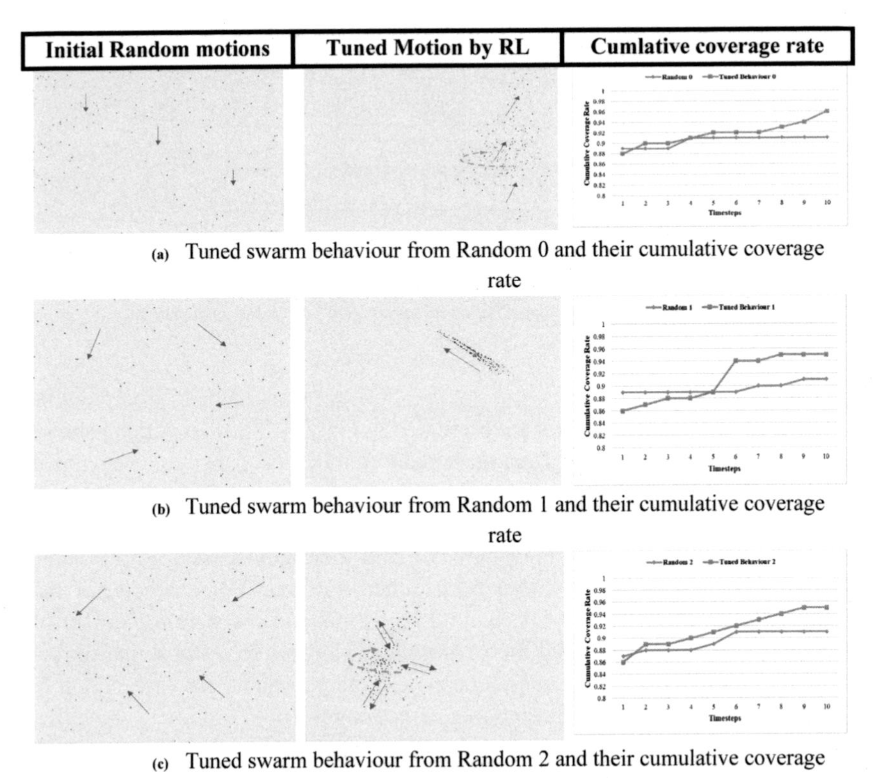

FIGURE 8.6 Evaluation of autonomous swarm behavior tuning by reinforcement learning (left and center) and their coverage ability (right). (A) Tuned swarm behavior from Random 0 and their cumulative coverage rate, (B) tuned swarm behavior from Random 1 and their cumulative coverage rate, (C) tuned swarm behavior from Random 2 and their cumulative coverage rate, (D) tuned swarm behavior from Random 3 and their cumulative coverage rate, (E) tuned swarm behavior from Random 4 and their cumulative coverage rate, (F) tuned swarm behavior from Random 5 and their cumulative coverage rate, (G) tuned swarm behavior from Random 6 and their cumulative coverage rate, and (H) tuned swarm behavior from Random 7 and their cumulative coverage rate.

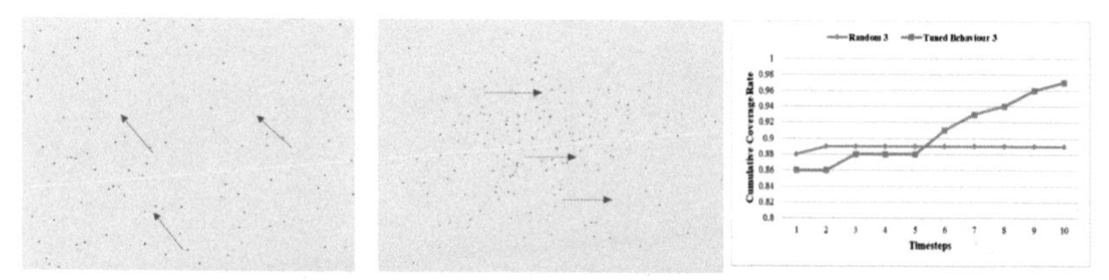

(d) Tuned swarm behaviour from Random 3 and their cumulative coverage rate

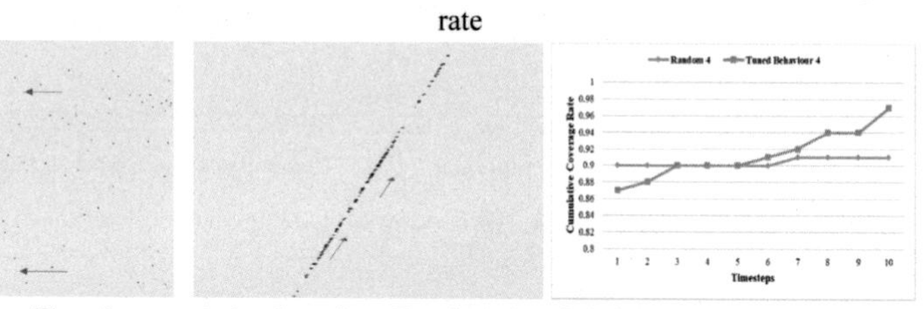

(e) Tuned swarm behaviour from Random 4 and their cumulative coverage rate

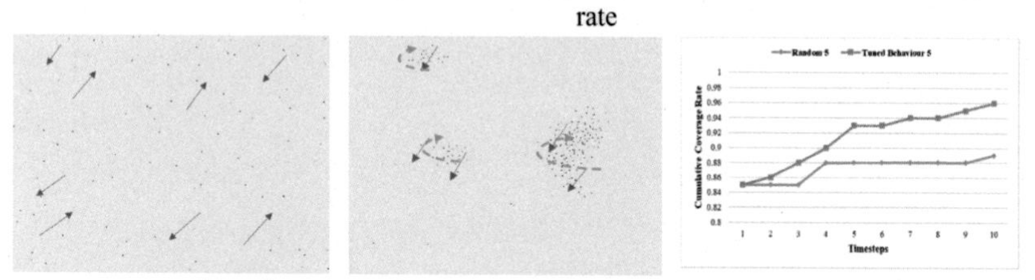

(f) Tuned swarm behaviour from Random 5 and their cumulative coverage rate

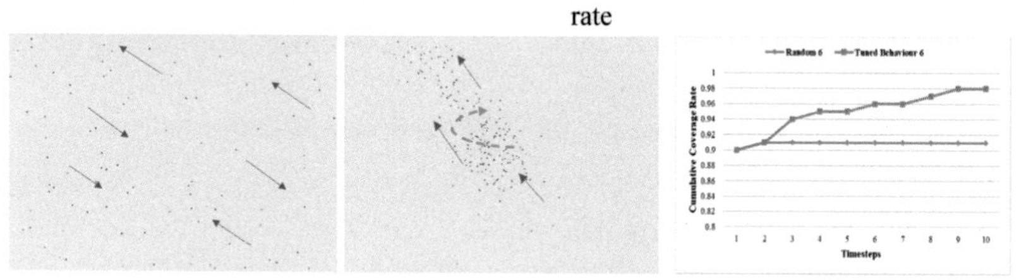

(g) Tuned swarm behaviour from Random 6 and their cumulative coverage rate

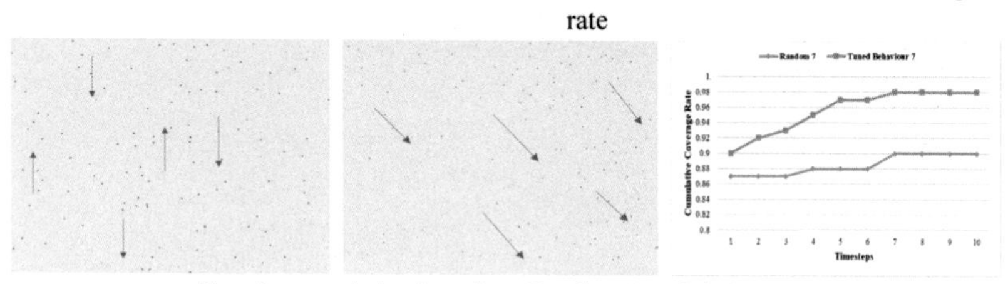

(h) Tuned swarm behaviour from Random 7 and their cumulative coverage rate

FIGURE 8.6 cont'd

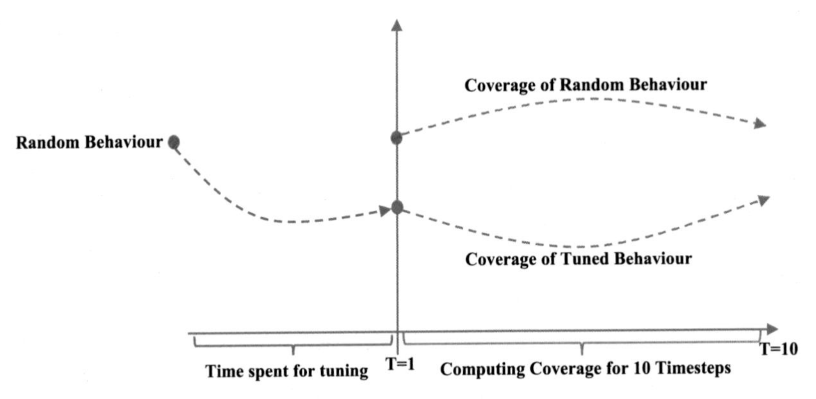

FIGURE 8.7 Time series to do tuning and coverage.

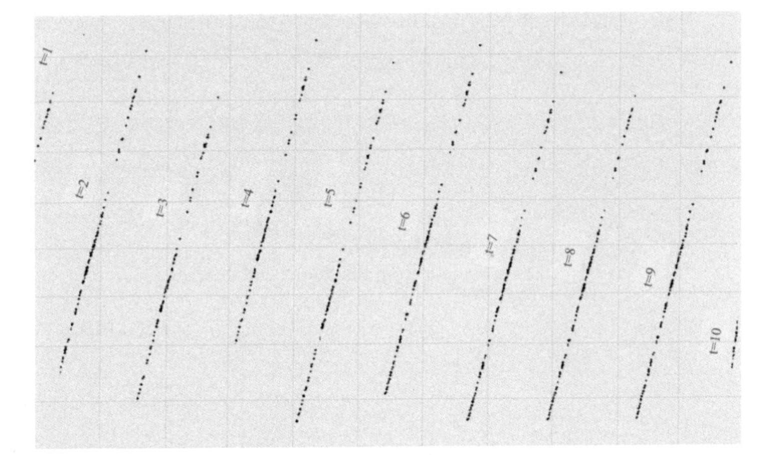

FIGURE 8.8 Example of coverage path achieved by one of the swarms (Fig. 8.6E) in the wraparound environment.

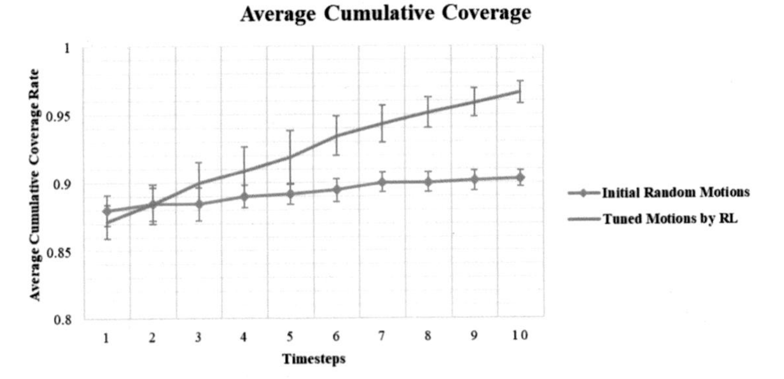

FIGURE 8.9 Average cumulative coverage of random and tuned behaviors over time. RL, reinforcement learning.

4.3 Evaluating the tuning and coverage ability of RL-SBAT on unseen random boids

4.3.1 Experimental setup

In Section 4.2 we used the eight random motions of Ref. [40] on which our knowledge-based system was tuned to examine its tuning performance. In this experiment we consider 100 unseen random motions. The aim of this experiment is to evaluate the performance of RL-SBAT algorithm in swarm behavior tuning and coverage, for random motion samples that it has not seen before. To do this experiment, 100 initial behaviors are randomly generated. Then, each of these 100 behaviors is fed to the RL-SBAT algorithm. In consequence, the RL-SBAT simulates a tuned behavior for each of the 100 random behaviors.

4.3.2 Results

The result of this experiment is presented in Fig. 8.10. Part (a) of this figure shows the reward values of both initial and tuned behaviors, and part (b) shows

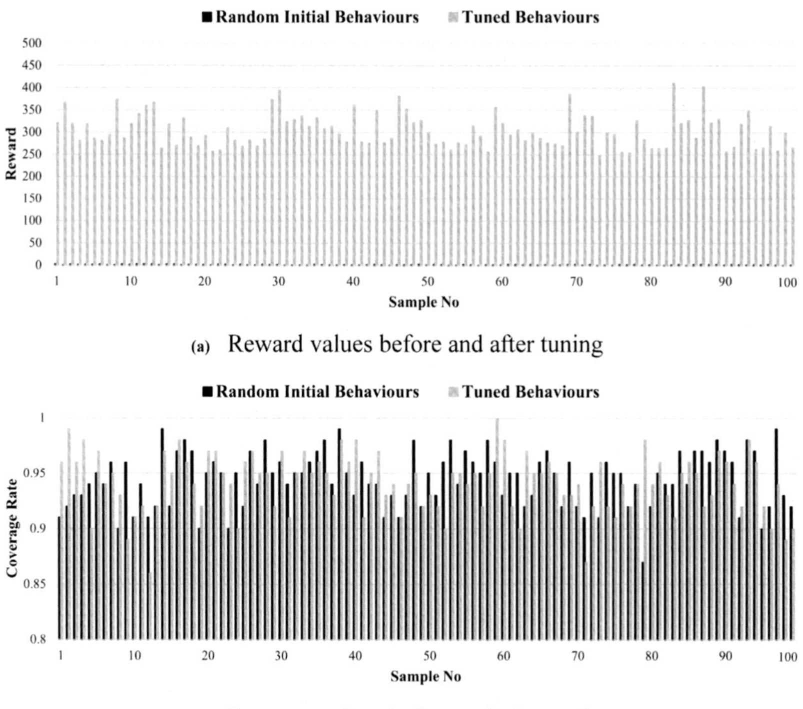

FIGURE 8.10 Evaluation of tuning and coverage ability of RL-SBAT on unseen random movement samples. (A) Reward values before and after tuning. (B) Coverage values before and after tuning.

the coverage they achieve in some steps (actions) after tuning. As shown in part (a) of Fig. 8.10, the reward value of tuned motions is increased compared to the initial behaviors of boid scenarios. The reward values of the tuned behaviors are well within the range shown in Fig. 8.4. As we saw in Experiment 1, the increased reward value of tuned motions is indicative that the group of agents is tuned.

The cumulative coverage rate for both random and tuned behaviors of boids is presented in part (b) of Fig. 8.10. As shown in this figure, the final coverage rate of tuned behaviors is higher than the initial random behaviors in most of the cases. This provides further evidence of successful tuning as Experiment 2 indicated that tuned swarms are more effective than random movement.

4.4 Evaluating the tuning and coverage ability of RL-SBAT on unseen random movement of robots

4.4.1 Experimental setup

In this experiment, three random parameter setups were applied to eight Pioneer3DX robots. The Pioneer3DX is a small, wheeled robot, equipped with 2 wheels, 2 motors, and 16 ultrasonic sensors. Each simulation on these eight robots is conducted in a 20×20 m walled environment as shown in Fig. 8.11. The robots have a reflective obstacle avoidance rule for walls, such that they will reflect off a wall at an angle equal to the angle of incidence.

We used the random scenarios 1, 2, and 5 from Ref. [40] as the basis for this experiment. These scenarios differ in the weight values of the cohesion, alignment, and separation forces as shown in Table 8.2. In this experiment we permit a trained RL-SBAT to tune these weights in each of these scenarios. Before tuning, the average reward attained in these scenarios was negative.

4.4.2 Results

Screenshots showing the tuned behaviors of our Pioneer3DX robots are presented in Fig. 8.11. After tuning, the average reward attained by the RL

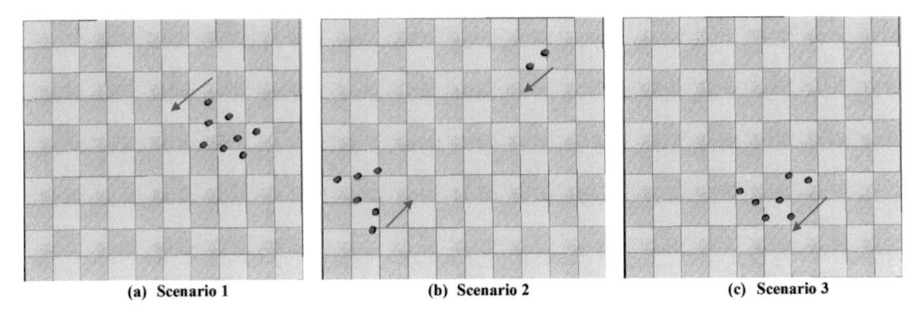

FIGURE 8.11 Tuned behaviors of Pionner3DX robots in three scenarios implemented in CoppeliaSim. (A) Scenario 1, (B) scenario 2, and (C) scenario 3.

TABLE 8.2 Simulation parameters with their values used to generate three scenarios of Pioneer3DX robot random motions.

Parameters	1	2	3
W_a (alignment weight)	1.0	1.2	1.0
W_c (cohesion weight)	0.0	0.1	0.05
W_s (separation weight)	1.2	1.2	1.2

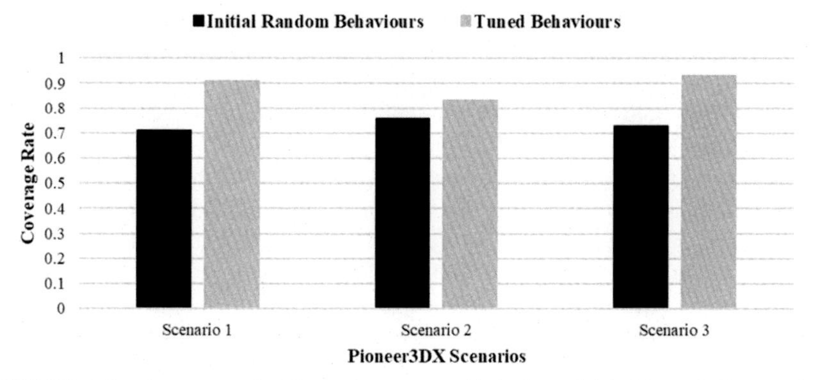

FIGURE 8.12 Coverage attained by random and tuned behaviors of eight Pioneer3DX robots in CoppeliaSim.

was 200–300, well within the range of reward values shown in Fig. 8.4. Visually, we can see from the screenshots that the robots have grouped themselves together. We observed that they were then able to move in the same direction and maintain this even after they encounter a wall.

The total coverage achieved by robots with random and tuned behaviors is provided in Fig. 8.12. This cumulative coverage is achieved from 20-min runtime of each tuned behavior in CoppeliaSim.[1] As this figure shows, the total coverage achieved by the random behaviors in this time is less than that achieved using the tuned behaviors.

5. Conclusions and future work

Recently, environmental sensing to reduce the spread of pollution from industrial residue has received attention due to increasing residue sources. AI has brought many advantages for environmental sensing, including

1. https://www.coppeliarobotics.com/.

accurate and real-time monitoring. Swarm robots are one of the approaches of AI in environmental sensing. They reduce the risks associated with human involvement, and they can sense the environment accurately and fast utilizing their distributed sensing and communication.

In this chapter, a system is designed to provide a swarm movement which can cover an area for environmental sensing purposes. The first subsystem is a machine learning approach using RL, which aims to rapidly tune a swarm behavior from any random behavior. The second subsystem conducts a coverage task using this tuned motion. These two subsystems are examined in this chapter for converting a number of random group movements to swarming. The results show that our trained reinforcement learner can tune swarm movement from these initial random motions. Moreover, the tuned movements of boids can cover more area over time, compared to the initial random motions. Also, the swarm tuning of the trained RL is done in less than a second. Therefore, both subsystems together could provide a tuned swarm behavior for boids in point-mass simulator, which can cover most of the environment.

The proposed system has demonstrated effective performance in our point-mass simulation environment, along with promising results in simulated robots performing coverage tasks. In future, we will examine the application of the system to real robots, including groups of robots that are not homogeneous. Correspondingly, we will also examine more realistic environments, including environments with obstacles and irregularly shaped environments.

Another possible direction for future work is to examine whether it is possible to include parameters of the coverage problem and/or the environment in the DDPG training phase. This would permit the DDPG policy to be optimized for (1) more rapid area coverage and (2) the particular type of environment in which the agents are located. This would permit the tuning architecture to tune specific swarm configurations adapted for achieving coverage in environments with different characteristics.

Appendix

TABLE 8.3 ***P*** **parameters boundaries.**

Parameter	Description	Range
W_s	Weight for separation rule	0–4
W_a	Weight for alignment rule	0–4
W_c	Weight for cohesion rule	0–4
V_{max}	Maximum speed: Maximum distance a boid will cover per tick	5–20
V_{min}	Minimum speed: Minimum distance a boid will cover per tick	2–10

Continued

TABLE 8.3 *P* **parameters boundaries.—cont'd**

Parameter	Description	Range
R_s	Separation radius: Boids under this distance qualify for separation rule	2–62
$R_{c,a}$	Cohesion/alignment radius: Boids under this distance qualify for cohesion rule	5–1005
θ_v	Vision angle: Agents within this angle relative to the agent's heading may be visible, if they are in range	17°–360°
P_{sa}	SA likelihood: Agents will update SA each tick with this probability	0–1
$P_{fullscan}$	Full scan likelihood: The probability with which agents will look all around and do a 360-degree scan	0–1
P_{rule}	Rule likelihood: Probability agent will update velocity each tick	0–1
P_s	Separation likelihood: Probability agent will apply separation rule each tick	0–1
P_a	Alignment likelihood: Probability agent will apply alignment rule each tick	0–1
P_c	Cohesion likelihood: Probability agent will apply cohesion rule each tick	0–1

From M.M. Khan, K. Kasmarik, M. Barlow, Autonomous detection of collective behaviours in swarms. Swarm Evol. Comput. 57 (2020) 100715.

TABLE 8.4 Boid parameters (*B*).

Parameter	Description
$\left(V_x^i t, V_y^i t\right)$	*x* and *y* components of the boid's velocity
$W_s \overrightarrow{\mathbf{s}}_t^i = \left(S_{x_t}^i, S_{y_t}^i\right)$	*x* and *y* components of separation vector multiplied by the separation weight
$W_a \overrightarrow{\mathbf{a}}_t^i = \left(A_{x_t}^i, A_{y_t}^i\right)$	*x* and *y* components of alignment vector multiplied by the alignment weight
$W_c \overrightarrow{\mathbf{c}}_t^i = \left(C_{x_t}^i, C_{y_t}^i\right)$	*x* and *y* components of cohesion vector multiplied by the cohesion weight
$n(R_a, R_s)$	The number of boids in the separation/alignment radii
$n(R_c)$	The number of boids in the cohesion radii

Table from M.M. Khan, K. Kasmarik, M. Barlow, Autonomous detection of collective behaviours in swarms. Swarm Evol. Comput. 57 (2020) 100715.

References

[1] L.M. Tufvesson, M. Lantz, P. Börjesson, Environmental performance of biogas produced from industrial residues including competition with animal feed–life-cycle calculations according to different methodologies and standards, J. Clean. Prod. 53 (2013) 214–223.

[2] J. Chen, M. Song, L. Xu, Evaluation of environmental efficiency in China using data envelopment analysis, Ecol. Indicat. 52 (2015) 577–583.

[3] M. Dunbabin, L. Marques, Robots for environmental monitoring: significant advancements and applications, IEEE Robot. Autom. Mag. 19 (1) (2012) 24–39.

[4] M. Asadnia, et al., Ca2+ detection utilising AlGaN/GaN transistors with ion-selective polymer membranes, Analytica Chimica Acta 987 (2017) 105–110.

[5] S. Foorginezhad, et al., Recent advances in sensing and assessment of corrosion in sewage pipelines, Process Saf. Environ. Prot. 147 (2021) 192–213, https://doi.org/10.1016/j.psep.2020.09.009.

[6] M. Mahmud, et al., Recent progress in sensing nitrate, nitrite, phosphate, and ammonium in aquatic environment, Chemosphere (2020) 127492.

[7] M.A. Dias, et al., An incongruence-based anomaly detection strategy for analyzing water pollution in images from remote sensing, Rem. Sens. 12 (1) (2020) 43.

[8] M.N. Jha, J. Levy, Y. Gao, Advances in remote sensing for oil spill disaster management: state-of-the-art sensors technology for oil spill surveillance, Sensors 8 (1) (2008) 236–255.

[9] M. Trincavelli, et al., Towards environmental monitoring with mobile robots, in: 2008 IEEE/RSJ International Conference on Intelligent Robots and Systems, IEEE, 2008.

[10] M. Dolia, et al., Information technology for remote evaluation of after effects of residues of herbicides on winter crop rape, in: 2019 3rd International Conference on Advanced Information and Communications Technologies (AICT), IEEE, 2019.

[11] A.T. Hayes, A. Martinoli, R.M. Goodman, Distributed odor source localization, IEEE Sensor. J. 2 (3) (2002) 260–271.

[12] S. Vigneshwaran, K. Yuvaraj, Fabrication and performance evaluation of semi-autonomous oil spills removing robot, Mater. Today Proc. 45 (2) (2021) 1305–1307, https://doi.org/10.1016/j.matpr.2020.05.285.

[13] D. Sudac, et al., Inspecting minefields and residual explosives by fast neutron activation method, IEEE Trans. Nucl. Sci. 59 (4) (2011) 1421–1425.

[14] Y. Chen, et al., Using multi-robot active olfaction method to locate time-varying contaminant source in indoor environment, Build. Environ. 118 (2017) 101–112.

[15] I. Berman, et al., Trustable environmental monitoring by means of sensors networks on swarming autonomous marine vessels and distributed ledger technology, Front. Robot. AI 7 (2020).

[16] C.W. Reynolds, Flocks, herds, and schools: a distributed behavioral model, Comput. Graph. 21 (4) (1987) 25–34.

[17] H. Hu, J. Oyekan, D. Gu, A school of robotic fish for pollution detection in port, in: Y. Liu, D. Sun (Eds.), Biologically Inspired Robotics, 2011, pp. 85–104.

[18] S.R. Carpenter, W.A. Brock, Rising variance: a leading indicator of ecological transition, Ecol. Lett. 9 (3) (2006) 311–318.

[19] F. Ejeian, et al., Design and applications of MEMS flow sensors, Review 295 (2019) 483–502.

[20] M.A. Parvez Mahmud, et al., Recent advances in nanogenerator-driven self-powered implantable biomedical devices, Adv. Energy Mater. 8 (2) (2017) 1701210, https://doi.org/10.1002/aenm.201701210.

[21] A. Ballesteros-Gómez, S. Rubio, Recent advances in environmental analysis, Anal. Chem. 83 (12) (2011) 4579–4613.
[22] B. Bayat, et al., Environmental monitoring using autonomous vehicles: a survey of recent searching techniques, Curr. Opin. Biotechnol. 45 (2017) 76–84.
[23] E. Horvath, C. Pozna, R.-E. Precup, Robot coverage path planning based on iterative structured orientation, Acta Polytech. Hung. 15 (2) (2018) 231–249.
[24] C.-F. Juang, Y.-C. Chang, Evolutionary-group-based particle-swarm-optimized fuzzy controller with application to mobile-robot navigation in unknown environments, IEEE Trans. Fuzzy Syst. 19 (2) (2011) 379–392.
[25] J.A. Rothermich, M.İ. Ecemiş, P. Gaudiano, Distributed localization and mapping with a robotic swarm, in: International Workshop on Swarm Robotics, Springer, 2004.
[26] B. Apolloni, et al., Machine Learning and Robot Perception, vol. 7, Springer Science & Business Media, 2005.
[27] N. Lu, Y. Gong, J. Pan, Path planning of mobile robot with path rule mining based on GA, in: 2016 Chinese Control and Decision Conference (Ccdc), IEEE, 2016.
[28] S. Lee, M.H. Kim, Learning expert systems for robot fine motion control, in: Proceedings 1988 IEEE International Symposium on Intelligent Control, IEEE Computer Society, 1988.
[29] X. Li, et al., SWARMs ontology: a common information model for the cooperation of underwater robots, Sensors 17 (3) (2017) 569.
[30] J. Zhang, D. Gong, Y. Zhang, A niching PSO-based multi-robot cooperation method for localizing odor sources, Neurocomputing 123 (2014) 308–317.
[31] D. Xiao, Y. Wang, Z. Cheng, Agent-based autonomous pollution source localization for complex environment, J. Ambient Intell. Humaniz. Comput. (2020) 1–9.
[32] D. Xiao, et al., Optimized neural network based path planning for searching indoor pollution source, J. Ambient Intell. Humaniz. Comput. (2021) 1–15.
[33] H. Zheng, D. Shi, A multi-agent system for environmental monitoring using boolean networks and reinforcement learning, J. Cybersecur. 2 (2) (2020) 85.
[34] A.S. Amjadi, et al., Cooperative pollution source localization and cleanup with a bio-inspired swarm robot aggregation, arXiv preprint (2019) arXiv:1907.09585.
[35] M. Pashna, R. Yusof, S. Yazdani, Analysis of prediction methods for swarm robotic—in the case of oil spill tracking, in: 2015 10th Asian Control Conference (ASCC), IEEE, 2015.
[36] M. Pashna, R. Yusof, R. Rahmani, Oil Spill trajectory tracking using swarm intelligence and hybrid fuzzy system, in: 2014 IEEE International Conference on Fuzzy Systems (FUZZ-IEEE), IEEE, 2014.
[37] J. Wang, et al., Locating hazardous gas leaks in the atmosphere via modified genetic, MCMC and particle swarm optimization algorithms, Atmos. Environ. 157 (2017) 27–37.
[38] A.T. Hayes, A. Martinoli, R.M. Goodman, Swarm robotic odor localization: off-line optimization and validation with real robots, Robotica 21 (2003) 427–441. ARTICLE.
[39] J.B. Clark, D.R. Jacques, Flight test results for UAVs using boid guidance algorithms, Conf. Syst. Eng. Res. 8 (2012) 232–238.
[40] M.M. Khan, K. Kasmarik, M. Barlow, Autonomous detection of collective behaviours in swarms, Swarm Evol. Comput. 57 (2020) 100715.
[41] Deleted in review.
[42] R.Q. Wang, Reinforcement learning: an introduction, in: Proceedings of 2006 International Conference on Artificial Intelligence, 2006, pp. 632–637.
[43] R.S. Sutton, A.G. Barto, Reinforcement Learning: An Introduction, second ed., 2018, pp. 1–526. Reinforcement Learning: An Introduction, 2nd Edition.

[44] L. Busoniu, R. Babuska, B. De Schutter, A comprehensive survey of multiagent reinforcement learning, IEEE Trans. Syst. Man Cybern. C Appl. Rev. 38 (2) (2008) 156–172.

[45] D. Spears, W. Kerr, W. Spears, Physics-based robot swarms for coverage problems, Int. J. Intell. Control Syst. 11 (3) (2006) 11–23.

[46] H.S. Jeon, et al., A practical robot coverage algorithm for unknown environments, in: Mexican International Conference on Artificial Intelligence, Springer, 2010.

[47] X. Liu, Y. Tan, Adaptive potential fields model for solving distributed area coverage problem in swarm robotics, in: International Conference on Swarm Intelligence, Springer, 2017.

[48] Deleted in review.

[49] M.B. Bezcioglu, B. Lennox, F. Arvin, Self-organised swarm flocking with deep reinforcement learning, in: 2021 7th International Conference on Automation, Robotics and Applications (ICARA), IEEE, 2021.

[50] V. Sperati, V. Trianni, S. Nolfi, Self-organised path formation in a swarm of robots, Swarm Intell. 5 (2) (2011) 97–119.

[51] T. Costa, et al., Automated discovery of local rules for desired collective-level behavior through reinforcement learning. Fundamentals and Applications of AI: An Interdisciplinary Perspective, Front. Phys. 8 (2020), https://doi.org/10.3389/fphy.2020.00200.

[52] K. Kasmarik, et al., Autonomous recognition of collective behaviour in robot swarms, in: Australasian Joint Conference on Artificial Intelligence, Springer, Australia, 2020, pp. 281–293.

[53] S. Abpeikar, et al., Human Perception of Swarming (Online Survey), 2019. Available from: https://unsw-swarm-survey.netlify.com/.

[54] S. Abpeikar, et al., Swarm Behaviour Dataset on UCI Data Repository, UCI Data Repository: UCI Data Repository, 2020.

[55] T.P. Lillicrap, et al., Continuous control with deep reinforcement learning, arXiv preprint (2015) arXiv:1509.02971.

[56] Deleted in review.

[57] R. Siraskar, Reinforcement learning for control of valves, Mach. Learn. 4 (2021) 100030.

[58] H. Ren, P. Ben-Tzvi, Advising reinforcement learning toward scaling agents in continuous control environments with sparse rewards, Eng. Appl. Artif. Intell. 90 (2020) 103515.

Chapter 9

Machine learning applications for developing sustainable construction materials

Hossein Adel, Majid Ilchi Ghazaan, Asghar Habibnejad Korayem
School of Civil Engineering, Iran University of Science and Technology, Tehran, Iran

1. Introduction

The environmental impact of building materials is negligible. Examples include the volume of carbon dioxide produced by the construction industry, natural aggregate resource usage, demolition wastes, and oil pollution from asphalt. For example, 5%–7% of the global carbon dioxide is produced by the cement industry, and the demand for cement is expected to reach 3.7–4.4 billion tons by 2050 [1]. These create the need to move toward materials with less environmental impact, which are more durable, consume less cement, use industry by-products and demolition wastes, and have the most optimal mixing design according to need.

In terms of environmentally friendly materials, rapid developments of novel construction materials types are undergone. This expansion can be divided into four approaches. The first approach is replacing or adding new components to the traditional mix design. Examples include partially replacing cement with supplementary cementitious materials (e.g., fly ash and silica fume), partially replacing natural aggregate with crumb rubber or recycled aggregate or reclaimed asphalt pavement aggregates, and switching asphalt to concrete pavement. The second is to move toward materials with more durability and strength. It is clear that with enhanced durability and strength of materials, the demand for retrofit and reconstruction diminishes, and the quantity of material consumption declines. Examples include the operation of fibers, superplasticizers, and nanoparticles that each with its own mechanism enhances durability and strength. Another approach is to detect and classify damages quickly so that the lifetime of materials can be maximized at a lower cost, which finally results in achieving optimal maintenance planning. Detection of cracks in pavements and their classification according to the type

Artificial Intelligence and Data Science in Environmental Sensing
https://doi.org/10.1016/B978-0-323-90508-4.00002-2

of damage is a famous case in this field. The last attitude is the optimal mixture design in which the components are together in optimal ratios.

For efficient operation of the above approaches, reliable models are needed that accurately predict the fresh properties, mechanical properties, and durability parameters of these novel binder systems. But the reactions and interactions of the mixture components are so complex and even unexplored that an exact chemical relationship cannot be obtained. The classical solution to this problem is to adopt statistical methods to develop empirical relationships for estimation. But with the advent of new binder systems, these types of models came into question. The time-consuming and high cost of samples besides the high nonlinearity of inputs and outputs made these models unable to answer the need. As a result, creating predictive models that can appropriately deal with complex materials to obtain accurate relationships between inputs and measure the effect of each parameter in the outputs became a significant task for civil engineers. In terms of damage detection, while visual detection and evaluation of defects are operated, this process is time-consuming, subjective, and based on experience, skill, and engineering judgment. So acquiring tools that automatically detect defects and categorize them based on damage type will be extremely advantageous. Furthermore, for both prediction and damage detection cases, models are required to efficiently deal with image inputs and perform feature extraction tasks appropriately.

Recently, the use of machine learning to predict the behavior of materials is increased [2,3]. Many studies have compared the performance of machine learning with empirical methods and have confirmed its better performance. Machine learning–based models can recognize complex and nonlinear relationships between variables, apply a large number of variables, extract features from images and implement them as input variables, detect and categorize damages from images, measure the importance of variables, generate multioutput models, distinguish influential variables from inconsequential, and solve multiobjective optimization problems. Machine learning has a wide range and includes many techniques. Among these techniques, artificial neural networks (ANNs), evolutionary algorithms (EAs), and support vector machines (SVMs) have been widely used in the field of construction materials.

The ANN's structure inspired by the human brain consists of an enormously parallel system composed of processing units [4]. These units, called neurons, are disposed of layers linked by weights. To map and approximate a function, the network should be learned through a learning method such as backpropagation or Levenberg–Marquardt. To boost the performance of ANN, the number of layers and neurons in each layer should be tuned via methods such as the trial-and-error method or optimization algorithms. In general, a well-tuned ANN can accurately approximate the complex relationship between inputs and outputs. SVM is applied for both regression and classification analysis. In general, using a nonlinear mapping function, SVM converts the nonlinear problem into a linear model. EAs such as gene expression programming (GEP) are based on population evolutionary theorem. The population in each generation is made up of individuals called chromosomes and can be encoded into a mathematical function or an expression tree. Each chromosome is made up of one or more genes that are linked to mathematical operators. By giving more chances to individuals with better performance, individuals of each generation form the next generation by means of mutation, transposition, and gene recombination. This process will continue until the desired error is reached. GEP performance depends on the number of genes in chromosomes, the mathematical functions of each gene, the size of the head and tail of each gene, mathematical operators linking genes, and the fitness function selected.

2. Prediction

2.1 Fresh properties

Fresh properties play an important role in characterizing the workability and filling ability of construction materials. Besides filling ability, transportation of freshly mixed materials is a particular concern for civil engineers designing construction methods because of blockage and material segregation. In addition, complex dynamic chemical processes occurring within the construction material's microstructure can cause significant time-dependent variations in all early-age properties (e.g., slump flow, t50, and J-ring tests). As a result, implying that minor differences in the time of measurement of early-age properties could result in significant variability, which causes double difficulty in early-age properties prediction.

In this context, Chandwani et al. [5] developed hybrid genetic algorithm—artificial neural network (GA-ANN) model to predict the slump of ready-mix concrete. To preventing ANN from getting trapped in local minimums and accelerating convergence, they used GA to apply the initial optimal weights and biases of the ANN and reached promising results. Furthermore, Moayedi et al. [6] applied three optimization algorithm (i.e., ant lion optimization [ALO], biogeography-based optimization, and grasshopper optimization algorithm [GOA]) for fine-tuning of a neural network (NN) and compared their applicability and found that ALO-NN is more accurate. Additionally, Nguyen et al. [7] implemented history-based adaptive differential evolution with linear population size reduction (L-SHADE) to optimize support vector regression hyperparameters (L-SHADE-SVR) to estimate the plastic viscosity of fresh concrete as a crucial parameter in inspecting the pumpability.

Self-compacting concrete (SCC), as a high-performance type of concrete, has little resistance to flow without any segregation and moderate viscosity. SCC should flow easily into complex shapes and through reinforcement bars with its weight and, without applying vibration, should settle into the mold. Developing a predictive model for fresh properties (slump flow, T50, T60, V-funnel, Orimet, and blocking ratio) of SCC as critical parameters to evaluate its filling ability is necessary. Due to highly time dependency of fresh properties and incorporating several types of fillers and admixtures (e.g., ground granulated blast furnace slag [GGBFS], fly ash, limestone powder, superplasticizer, increasing the sand-to-aggregate ratio, stabilizing agent), conventional methods cannot reach high accuracy. To address this problem, Sonebi et al. [8] applied radial basis function (RBF) and polynomial kernel-based SVMs to predict slump flow, T50, T60, V-funnel time, Orimet time, and blocking ratio (L-box) of SCC and operated a sensitivity analysis. Results showed that RBF kernel-based SVM was more accurate. Meanwhile, Kaloop et al. [9] used several models to predict SCC's slump flow, including the group method of data handling, minimax probability machine regression, emotional neural network (ENN), and hybrid artificial neural network-particle swarm optimization (ANN-PSO). Their results indicated that ENN outperformed other models and was the best with a smaller dataset for modeling the slump. Kaveh et al. [10] utilized two decision tree algorithms to predict several properties (i.e., compressive strength, slump flow, the L-box ratio, and the V-funnel) of the SCC containing fly ash. They operated M5′ for developing new practical equations and multivariate adaptive regression splines (MRAS) to determine the most important parameters besides its high predictive ability.

In fiber-reinforced SCC, in addition to SCC variables, we face several additional variables such as fiber type, fiber distribution, mineral admixtures, and viscosity-enhancing chemical admixtures than normal concrete. As the number of our variables increased, the empirical models lose their efficiency. To consider this issue, Kina et al. [11] implemented models based on extreme learning machine and long short-term memory (LSTM) to estimate fresh properties (slump flow, t50, and J-ring tests) of hybrid fiber-reinforced self-compacting concrete (HR-SCC) mixtures with different types and combinations of fibers. Additionally, Altay et al. [12] applied weighted K-nearest neighbors (W-KNN) and quadratic support vector machine (Q-SVM) models to predict fresh performance (slump flow and V-funnel time) of fiber-reinforced SCC.

High-performance concrete (HPC) can play an active role in improving sustainability and reliability. Due to the usage of SCMs, HPC is more environmentally friendly and exhibits superior engineering performances. However, the multiplicity and nonlinearity of variables make it almost impossible to predict their fresh properties by empirical relationships. To consider this problem, Chen et al. [13] presented a parallel hypercubic gene expression programming (GEP) model to estimate the slump flow of HPC and evaluate it with basic GEP, two types of regression models backpropagation neural network (BPNN). Results showed that the new model was more accurate than the basic GEP and two types of regression model and was preferable to BPNN because it provided formulas with measurable parameters. In addition, Kaveh et al. [7] operated support vector regression, M5P trees, random forest (RF), and multilayer perceptron regression (MLPReg) to predict slump flow. Their results showed MLPReg is the most accurate model.

Portland cement (PC) concrete is the most well-known and most widely demanded material for constructing infrastructure. However, the production of PC is considerable energy consumption and has enormous environmental impacts. Alkali-activated concrete (AAC), as a sustainable alternative to PC, is prepared by mixing amorphous aluminosilicate materials or industrial by-products (fly ash and blast furnace slag). For conscious use, an efficient model is required to predict its behavior. Gomaa et al. [14] presented a classification-and-regression tree–based RF model to predict two fly ash–based AACs properties (slump flow and compressive strength) in relation to their composition and relevant process parameters (the sequence of addition of various components of AAC and the duration of mixing at each step of the sequence) and distinguish the influential compositional parameters from the inconsequential ones. Also, used grid-search method and 10-fold cross-validation method to optimize two hyper-parameters of the RF model (i.e., number of trees in the forest; and number of leaves per tree) about the nature and volume of the databases. By using the outcomes of this work, we can achieve a pathway to optimize AAC's properties.

Using waste and industry's by-product materials instead of cement or natural resources is a firm stem to produce eco-friendly and economic materials. Studies showed that the ratio of waste foundry sand to produced metal in the casting process was approximately one [15]. To efficiently use these by-products, Tavana Amlashi et al. [16] applied ANN for predicting several parameters, including slump, and employed various training algorithms, including backpropagation and Levenberg–Marquardt. Finally, they performed a sensitivity analysis.

Solid waste is the main mining industry's by-product that causes loss of human lives, the devastation of agricultural and forestry lands, and financial losses for the mining companies. Occupation of land resources and the catastrophic tailings ponds failure causes conventional disposal into tailings ponds cannot be widely utilized in the modern era. Cemented paste backfill (CPB) has been considered as a cleaner and more sustainable problematic tailings management strategy in which tailings are placed back into underground mines in many leading countries in mining such as Australia [17]. Generally, CPB has enough strength to remain stable, flowable enough for efficient pumping, and be economical because it typically represents almost 20% of all mining expenditure. CPB mixtures are prepared on the ground and through pipes transformed underground by pumping or gravity-flow so that fresh properties (i.e., pressure drops, slump) of CPB are vital to prevent pipe blockage. Due to complex mapping from tailings characteristics, cement-tailings ratio, inlet velocity, and solids content, the Prediction of CPB fresh properties is high-dimension and challenging. To address this problem, Qi et al. [18] presented a decision tree regression (DTR) model coupled with PSO and 5-fold cross-validation for tuning hyperparameters and two ensemble learning techniques (i.e., RF and gradient boosting regression tree [GBRT]) to predict pressure drops of fresh CPB. This research considered 13 preditors for characterizing chemical and physical characteristics and cement-tailings ratio, inlet velocity, and solids content. Results show that RF and GBRT outperformed DTR.

2.2 Mechanical properties

Mechanical Properties, primarily compressive strength, are essential input parameters in many existing design practices and give an overview of the quality of the additives impact and the durability of the materials. In general, laboratory tests can be expensive and time-consuming so we may use empirical-based models or linear regression—based models. Due to the high input dimension and sophisticated relationship among inputs of new construction materials, conventional methods maybe are unsuitable. For example, unlike conventional concrete, other types such as HPC or fiber concrete have much more parameters like several SCMs and chemical admixtures. So that more sophisticated techniques such as machine learning—based techniques should be implemented to develop more accurate and reliable models. Overall, an accurate and reliable estimation can save time and cost and reduce the need for trial samples.

The cement industry harms the environment by contributing approximately one-20th of the CO_2 emissions, which is accepted as one of the factors of global warming. On the other hand, it is facing year by year demand increase. Benhelal et al. [1] predicted that world annual cement production would reach 3.7—4.4 billion tons by 2050. To consider the problem, SCMs (e.g., silica fume) have been used to partially replace cement in concrete mixtures and at the same time improve mechanical properties.

Due to the efficient use of steel fiber-reinforced concrete for enhancing the durability and frost resistance of concrete, it is necessary to have a model to predict the mechanical properties. To address this problem, Kang et al. [19] developed several models (i.e., linear regressor, lasso regressor, ridge regressor, KNN, decision tree, RF, AdaBoost regressor, gradient boost regressor, and extreme gradient boosting (XGBoost)) to predict compressive and flexural strength of steel fiber-reinforced concrete. To prevent overfitting, which creates unreliable models, they applied K-fold validation. Results showed that XGBoost regressor and Gradient Boosting regressor are the best models to predict compressive and strength, respectively.

Furthermore, another fiber that is widely utilized in construction materials is carbon fibers (CF). Due to its high electrical conductivity holds the prospect for the future and can be utilized in ice—snow melting pavements and self-monitoring structures. Besides plurality of inputs, accurate insight into fibers' distribution is vital to the prediction of model performance. To consider this problem, Tong et al. [20] implemented a fully convolutional network (FCN) to identify fibers in slices of X-ray images. To analyze the CF morphology, via using the well-trained FCN, different components in the X-ray images were segmented and labeled and were employed for three-dimensional reconstruction and volume calculation. Finally, the resistivity and bending strength of specimens were predicted by using an RBF-based network. Meanwhile, in a similar approach, Yuan et al. [21] used a fully convolutional network (FCN) to identify carbon fiber components' distribution from the SEM images of carbon fiber—reinforced cement—based composites (CFRC). To reach adequate samples for network training, they cropped 960 SEM images into near 20,000 smaller images. In the next step, the electrical conductivity of the CFRC specimens was predicted by using an RBF-based network. Finally, the network's weights were employed to evaluate the effects of CF distribution on electrical conductivity.

Like fiber-reinforced concrete, many parameters affect HPC's mechanical properties, including mix proportions, material characteristics, curing, and environmental conditions. While using chemical admixtures and additional SCMs that are necessary for HPC's performance, it causes a highly complex cementitious matrix which ambiguity exists with its mix design. As a result, we need models that can find the complex relationship between many variables with a higher degree of nonlinearity, unlike empirical and linear regression models. To solve this problem, Farooq et al. [22] implemented several ensembles and individual models (i.e., ANN, decision tree, RF, multiple linear regressions, and SVM). Ensemble approach adopted to enhance reliability and reduce over-fitting by combining several weak models. Ensemble models based on boosting (i.e., Adaboost and XGboost), bagging, and RF algorithms were applied. The performance comparison results showed that by deploying bagging and boosting, models' efficiency was enhanced and outperformed individual models. Furthermore, to solve the multiplicity of input variables, Han et al. [23] developed an RF-based model that selects proper variables among independent and nonindependent variables and predicts the compressive strength of HPC that shows a strong generalization capacity.

Another issue that contributes to the sustainability of materials is the use of recycled aggregates. By using recycled aggregates, in addition to compensating for the limitations of natural resources, we can also help bury garbage. The mass production of concrete has resulted in the gradual extinction of natural resources (such as river sand, gravel, or crushed rock) makes the use of alternative materials such as recycled concrete or crumb rubber inevitable. By using recycled aggregate, we can improve resource sustainability in the construction industry besides eliminating construction waste environmental impacts. As mentioned above, to move toward widespread use of recycled materials, we need efficient models that can accurately predict properties of those. In this regard, Xu et al. [24] used a mathematical approach, namely gray system theory (GST), to identify key parameters via parametric sensitivity analysis. In the next stage, they employed multiple nonlinear regression (MNR) and ANNs. Results illustrated that ANN-based models for all properties (i.e., compressive strength, flexural strength, splitting tensile strength, and elastic modulus) outperformed MNR-based models. Finally, they compared the ANN-based model with empirical and experimental models from the literature. Statistical results demonstrated that the ANN approach had the slightest error. In another study, Deng et al. [25] established a convolution neural network model to predict the compressive strength of recycled concrete. Results showed that this model exhibits higher precision, efficiency, and generalization than traditional neural networks. Furthermore, Kandiri et al. [26] developed three models by hybridizing ANN and optimization algorithms (i.e., GA, salp swarm algorithm (SSA), and GOA) to predict recycled aggregate concrete's (RAC) compressive strength. Results showed that the ANN that was hybridized with SSA had better performance among all models.

Table 9.1 summarizes different statistical metrics employed by researchers. Due to the multiplicity of studies performed to predict the mechanical properties of cementitious materials by machine learning methods, Table 9.2 summarizes some other studies.

Asphalt and concrete pavement are among widely used construction. Using machine learning to predict its mechanical properties along with reducing time and cost and increasing speed can greatly help its sustainability. Hitherto, many studies have implemented machine learning in this field, some of which are discussed below.

An important parameter that is a crucial input in pavement design guides is the dynamic modulus of asphalt mixture. Accurate measurement of dynamic modulus is achieved by laboratory-based tests at various simulated temperatures and loading frequency for several samples under various traffic and climate conditions. This method is time-consuming and expensive, so it is necessary to develop models that can predict this parameter accurately. Although several predictive models such as G*-based Witczak and viscosity-based Witczak in Levels 3 and 2 of the MEPDG and modified Hirsch [34,35] are utilized, their accuracy has been questioned in several previous studies [36,37]. As a result, several studies tried to develop machine learning–based models. In this regard, Behnood and Golafshani [38] adopted biogeography-based programming (BBP) to develop two models with four and eight input variables. In the next stage, they compared this model with the Witczak model, Hirsch model, and ANN model. Finally, they applied a parametric study and a sensitivity analysis to figure out the most influential factors. Their results showed that BBP-based models outperformed all of those.

Due to the multiplicity of studies performed to predict the dynamic modulus of asphalt by machine learning methods, Table 9.3 summarizes some other studies.

TABLE 9.1 Statistical metrics.

Statistical index	Formula
Error	$e_i = (y'_i - y_i)$
Mean square error (MSE)	$MSE = \frac{\sum_{i=1}^{n} e_i}{n}$
Root mean square error(RMSE)	$RMSE = \sqrt{\frac{\sum_{i=1}^{n} e_i^2}{n}}$
Mean absolute error (MAE)	$MAE = \frac{1}{n}\Sigma_{i=1}^{n}\lvert e_i\rvert$
Average absolute error (AAE)	$AAE = \frac{1}{n}\Sigma_{i=1}^{n}\lvert e_i\rvert$
Mean absolute percentage error (MAPE)	$MAPE = \frac{1}{n}\Sigma_{i=1}^{n}\left\lvert\frac{e_i}{y_i}\right\rvert \times 100$
Correlation coefficient (R)	$R = \frac{n\cdot\Sigma_{i=1}^{n} y'_i y_i - (\sum_{i=1}^{n} y'_i)\cdot(\sum_{i=1}^{n} y_i)}{\sqrt{\left[n\cdot(\sum_{i=1}^{n} y_i^2) - (\sum_{i=1}^{n} y_i)^2\right]\cdot\left[n\cdot(\sum_{i=1}^{n} (y'_i)^2) - (\sum_{i=1}^{n} y'_i)^2\right]}}$
Coefficient of determination (R^2)	$R^2 = 1 - \frac{\sum_{i=1}^{n} e_i^2}{\sum_{i=1}^{n} (y_i - \bar{y})}$
Mean (μ)	$\mu = \frac{1}{n}\Sigma_{i=1}^{n}\left(\frac{y_i}{y'_i}\right)$
Standard deviation (σ)	$\sigma = \sqrt{\frac{1}{n}\Sigma_{i=1}^{n}\left(\frac{y_i}{y'_i} - \mu\right)^2}$
Scatter index (SI)	$SI = \frac{RMSE}{\frac{\Sigma_{i=1}^{n} y_i}{n}}$
Absolute fitness	$f_i = \sum_{i=1}^{n}(M - \lvert e_i\rvert)$
Relative fitness	$f_i = \sum_{i=1}^{n}\left(M - \left\lvert\frac{e_i}{y_i}\right\rvert \times 100\right)$

Note: M, range of selection; n, number of samples; y, actual value; y', predicted value.

Another critical parameter in designing flexible pavements is phase angle which proper knowledge about it can ensure the maximum service life of pavements. In this regard, Hussain et al. [43] applied a convolutional neural network (CNN) to predict the phase angle behavior of asphalt concrete mixtures from nonimage inputs. They compared this model with an ANN, linear regression, extreme gradient boosting (XGBoost), and a variant of CNN. Results showed that their presented model outperformed all of these models. Ghorbani et al. [44] utilized ANN-based models to predict the thermal

TABLE 9.2 Further examples of the use of machine learning models in the modeling of mechanical properties.

References	Dataset size	Train set percent	Validation set percent	Test set percent	Material type	Methods used	Input variables	Output	Statistical index
[27]	196	70	15	15	Cementitious composite with fly ash	ANN and PSO-ANFIS and GA-ANFIS	$SiO_2 + Al_2O_3 + Fe_2O_3$ content, age, and fly ash replacement ratios	Compressive strength	MSE, RMSE, R^2, and MAPE
[28]	1030	70		30	High-performance concrete	Gradient tree boosting machine	Cement (C), fly ash, blast furnace slag (BS), water (W), superplasticizer (SP), coarse and fine aggregate (CA and FA, respectively) and age	Compressive strength	R, normalized percentage of RMSE (NRMSE%), MAE, ratio of RMSE to σ, coefficient of persistence (*cp*), and degree of index (*d*)
[29]	2817	70	15	15	Normal and high-performance concretes	GWO-ANN and GWO-ANFIS	Coarse aggregate (CA), fine aggregate or sand (S), water (W), cement (C), blast furnace slag (BFS), fly ash, age, and superplasticizer (SP)	Compressive strength	RMSE, SI, MAE, R^2

[30]	175 and 132 for each output respectively	75	25		Fiber-reinforced concrete incorporating nanosilica	MARS, M5P tree, LS-SVM, MLP-NN, and MLR	Volumetric percentage of fibers (Vf), amount of limestone powder and nanosilica, coarse to fine aggregate ratio (CA/FA), water to binder ratio (W/B), superplasticizer to binder ratio (SP/B), and age	Compressive strength and ultrasonic pulse velocity	R, RMSE, MAPE, performance index, AAE, σ, and μ
[31]	131	70		30	Lightweight self-compacting concrete	Beetle antennae search (BAS) algorithm based random forest (RF)	W/b, macrosynthetic polypropylene (PP) fibers, steel fiber, Scoria, natural fine aggregate, natural coarse aggregate, and crumb rubber	Uniaxial compressive strength	R and RMSE
[32]	26				Polypropylene fiber-reinforced self-compacting composites incorporating nano-CuO	Wavelet weighted least square SVM (WWLSSVM), least square SVM (LSSVM) and standard support vector machine (SVM)	Cement, polycarboxylic ether based high range water reducer, nano CuO, water, polypropylene fibers, sand and gravel	Compressive strength	R^2, the relative root mean squared error (RRMSE) and MAPE
[33]	412	80		20	Geopolymer self-compacting concrete	Genetic programming and ANNs	Fly ash, ground granulated blast furnace slag (GGBFS), silica fume, slump flow, T50, L-box, V-funnel, J-ring, age	Compressive strength, split-tensile strength, flexural strength	MSE, RMSE, R^2

TABLE 9.3 Further examples of the use of machine learning models in the modeling of dynamic modulus.

References	Dataset size	Train set percent	Validation set percent	Test set precept	Material type	Methods used	Input variables	Output	Statistical index
[38]	4022	90 (10-fold cross-validation)		10	Asphalt	Biogeography-based programming	Temperature, frequency, low-temperature PG of binder, voids in mineral aggregate	Dynamic modulus	RMSE, MAPE, MAE, R^2, SI
[38]	90	90 (10-fold cross-validation)		10	Asphalt	Biogeography-based programming	Temperature, frequency, HPG, LPG, voids in mineral aggregate, V_{bef}, recycled asphalt pavement content	Dynamic modulus	RMSE, MAPE, MAE, R^2, SI
[39]	7400	90.5	2.7	6.7	Asphalt	Gradient decision tree boosting	ρ_{200}, ρ_4, $\rho_{3/8}$, $\rho_{3/4}$, V_a, Effective asphalt content by volume of asphalt mixture, temperature, phase angle of asphalt binder, complex modulus of asphalt binder, binder viscosity, frequency	Dynamic modulus	R^2, MAPE, σ

[40]	5058	80		20	Hot mix asphalt	Deep convolution neural network	ρ_{200}, ρ_4, $\rho_{3/8}$, $\rho_{3/4}$, effective asphalt content by volume of asphalt mixture, V_a, binder viscosity, loading frequency of asphalt concrete, phase angle of asphalt binder, complex modulus of asphalt binder	Dynamic modulus	R^2, σ
[40]	1701	20		80	Hot mix asphalt	Deep convolution neural network	ρ_{200}, ρ_4, $\rho_{3/8}$, $\rho_{3/4}$ effective asphalt content by volume of asphalt mixture, V_a, loading frequency of asphalt concrete, phase angle of asphalt binder, complex modulus of asphalt binder	Dynamic modulus	R^2, σ
[41]	7400	90.5	2.7	6.7	Hot mix asphalt	Artificial neural network	ρ_{200}, ρ_4, $\rho_{3/8}$, $\rho_{3/4}$, effective asphalt content by volume of asphalt mixture, V_a, binder viscosity, loading frequency of asphalt concrete, phase angle of asphalt binder, complex modulus of asphalt binder, temperature, constants for determining η	Dynamic modulus	R^2, σ, MAPE

Continued

TABLE 9.3 Further examples of the use of machine learning models in the modeling of dynamic modulus.—cont'd

References	Dataset size	Train set percent	Validation set percent	Test set precept	Material type	Methods used	Input variables	Output	Statistical index
[42]	4112	98		2	Asphalt concretes	M5P model tree algorithm	ρ_{200}, ρ_4, $\rho_{3/8}$, $\rho_{3/4}$, Nominal maximum aggregate size, V_a, V_{bef}, voids in mineral aggregate by volume, voids filled with asphalt by volume, Asphalt content by weight, HPG, LPG, RAP content, testing frequency, testing temperature	Dynamic modulus	R^2, σ, MAE, MSE, RMSE, discrepancy ratio (DR), Residuals

Note: *HPG*, High-temperature PG of binder; *LPG*, Low-temperature PG of binder; V_a, air voids in the mixture by volume; V_{bef}, Effective bitumen content by volume; ρ_{200}, retained on the No.200 (0.075-mm) sieve; $\rho_{3/4}$, retained on the 3/4-in. (19-mm) sieve; $\rho_{3/8}$, retained on the 3/8-in. (9-mm) sieve; ρ_4, retained on the No.4 (4.75-mm) sieve.

conductivity and deformation response under combined thermal and mechanical loads of construction and demolition waste materials containing recycled concrete aggregate, crushed brick, waste rock, and reclaimed asphalt pavement which can be used in geothermal pavements. Based on the good accuracy of ANN-based models and sensitivity analysis results, we can use this model as a reliable and helpful tool to predict the deformation behavior of recycled waste materials and geothermal pavement production.

Another approach for producing sustainable materials with less environmental pollution is using roller-compacted concrete pavement (RCCP) instead of asphalt. Due to effectively use the long-term durability of RCCP, implementing predictive models is essential. Ashrafian et al. [45] applied GEP based model to predict the compressive strength of roller compacted concrete pavement. Debbarma et al. [46] developed ANN-based model to estimate the compressive strength of RCCP containing reclaimed asphalt pavement (RAP) aggregates. Results showed that among three training methods, namely Levenberg–Marquadart (LM), Bayesian regularization (BR), and scaled conjugate gradient (SCG), after network architecture optimization, BR-ANN showed slightly better performance.

Furthermore, machine learning was used for presenting predictive models for fracture energy by GEP and hybrid ANN/simulated annealing [47], Marshall Test by intuitive k-NN estimator [48], and international roughness index by RF regression [49].

2.3 Durability

Other major tasks in the sustainability of construction materials are durability design and service life prediction. To efficiently address these tasks, we need consistent models which can account for several inputs and estimate outputs with high accuracy. Chemical reactions and physical interactions related to durability are complex, and direct mathematical equations cannot be easily obtained for them, so the existence of predictive models is a good tool for describing the deterioration mechanisms and input variables' impact measurement. We are faced with complex processes and reactions that cannot be predicted with high accuracy by empirical models, so in this section, we address studies on the application of methods based on machine learning as an efficient method.

Corrosion as an aggressive process significantly affects composite material's behavior and imposing much cost for reconstructing and repairing structures. The concrete carbonation, which can accelerate corrosion, consists of many complex mechanisms related to the cement paste hydration reactions and the CO_2 dissolution in the concrete matrix that cannot mathematically formulate easily. Though several pieces of research tried to model concrete carbonation depth via linear, nonlinear, and physical-chemical formulations, these formulas cannot present a reliable answer due to the degree of nonlinearity of variables. Felix et al. [50] established an ANN-based model to predict the carbonation depth of fly ash containing concrete. To tune hyper-parameters, they tested several topologies and selected the best model that performed better in both training and test data. To assess the importance of variables, data dispersion analysis was applied, and to verify the influence of each available parameter, Pearson's and Spearman's correlation coefficients were calculated. Results showed that cement consumption, fly ash content, CO_2 rate, and relative humidity were the most important parameters. In another research, Lee et al. [51] generated an ANN-based model to predict the carbonation rate coefficient and compare the model's performance with the Architectural Institute Japanese model and FEM analysis. Results showed that with optimal learning rate ANN-based model, the differences in carbonation rate coefficient between experimental data and the ANN-based model were minor, and this model is slightly better than FEM analysis.

Carbonation Prediction models also are presented in other researches. For instance, Akpinar et al. [52] presented an ANN-based model that used 18 variables as input and trained the network by employing an SCG function and compared its performance with the Levenberg–Marquardt function as a widely performed learning algorithm. In the next step, SCG trained model was selected to evaluate the individual influence selected input parameters. Taffese et al. [53] developed four machine learning models (NN, decision tree, bagged decision tree, and boosted decision tree). Their result demonstrated that while all models predict the carbonation depth with rationally low error, the NN method outperformed other models.

For structures in marine and coastal environments for durability and service life prediction, surface chloride concentration (CS) of concrete is an essential parameter. Chloride ions from seawater when reaching the concrete matrix surrounding the reinforcing steel, the passive film will be distracted, which leads to accelerating the corrosion. In this regard, Cai et al. [54] employed five standalone machine learning models, that is, linear regression (LR), Gaussian process regression (GPR), SVM, multilayer perceptron artificial neural network (MLP-ANN), and RF models, as well as a weighted voting-based ensemble machine learning model. Results illustrated that while RF, MLP, and SVM were the top three standalone models, the LR was the worst, and among all models, the ensemble ML model is the most accurate. For the next step, they compared conventional quantitative models which previously presented with the ensemble ML model. After testing and validating, it was turned out that the ensemble ML model outperformed conventional quantitative models and can be operated as an objective function in the optimization of concrete mixture design to enhance its durability.

Several papers presented machine learning models related to the impact of chloride ions on construction materials. Liu et al. [55] generated three ANN models to predict the chloride diffusion coefficient of concrete. First, they used raw inputs, then for the second model, they normalized input data, and finally, they excluded the datasets that deviated significantly from global trends. Results demonstrated that the model's accuracy increased step by step. Taffese and Sistonen [56] adopted decision trees based ensemble of machine learning algorithms to measure the importance of variable that controls the chloride ingress in concrete exposed to the de-icing environment. Hoang et al. [57] predicted the chloride diffusion in cement mortar by applying the multigene genetic programming (MGGP) and MARS and compared the results with models developed based on ANN and least squares support vector regression. Results demonstrated that while MARS had the best accuracy, ANN and MGGP ranked second and third, respectively.

Long-term creep prediction of concrete structures is needed to yield a reliable, safe, and economical design. While the subject of creep prediction of HPC containing SCMs has not been sufficiently addressed in design codes, some of the infrastructures that used high-strength concrete containing SCMs have shown unacceptable excessive deflection, cracking, and prestress loss attributed to the inaccurate predictions of long-term deformation [58]. To address this issue, Chen et al. [58] employed the PSO algorithm to modify ACI committee 318-84 formula's parameters which widely uses to predict concrete creep. Overall, due to consideration of SCM's effect in the utilized dataset, modified creep models can predict HPC's creep more accurately.

One of the harmful distress mechanisms in the concrete field that causes a reduction in durability, service life, and mechanical properties and causes an increase in the maintenance cost is the alkali-silica reaction (ASR). ASR is known as a complex process that cannot be accurately predicted by empirical and curve-fitting approaches. To achieve adequate accuracy, Yu et al. [59] developed five machine learning–based models, that is, SVM, ANN, adaptive neuro-fuzzy inference system (ANFIS), M5P model, and gene expression programming (GEP) to predict the elastic modulus of concrete which is more sensitive to ASR. To guarantee the high accuracy of models, they employed the trial-and-error method and PSO algorithm to achieve optimal hyperparameters of ANN and SVR, respectively. In the next step, to illustrate the superiorities of these models, they compared their performance with three empirical models. The result showed that machine learning–based models had better performance, and among these models, ANFIS offers the optimal capacity.

As mentioned above, ASR is known as one of the most deteriorating phenomena for concrete. Allahyari et al. [60] divided ASR-induced expansion into two subproblems: kinetics and final expansion of ASR. As a challenge for each subproblem, they found one variable which did not report in the dataset, so they considered those constant for all samples. To address this challenge, they modified the value of those variables until achieving the best fit model via implementing the trial-and-error process. This model was compared with a chemo-mechanical one (Gao's model), and results showed that the model had higher accuracy. Finally, to make a black-box model in-site useable model, they developed user-friendly charts.

As mentioned above, while SCMs can reduce CO_2 emission, they can influence concrete's mechanical and durability properties, which can be addressed by predictive models. To this end, Lau et al. [61] established an ANN-based model to predict fly ash–based geopolymer abrasion resistance, which can be appropriated to enhance the abrasion resistance of fly ash–based geopolymer in the mixture design procedure.

Although some durability parameters related to construction material were discussed, cracking, which is a crucial parameter that can reduce the load-carrying capacity and diminish the resistance of materials against other potentially aggressive species, such as de-icing salt, chlorides, sulfates, freezing water, CO_2 is remaining. In this regard, Mermerdaş et al. [62] implemented two models, namely GEP and multiple linear regression (MLR), to predict the drying shrinkage behavior of concretes incorporated with high reactivity commercial metakaolin (MK) and calcined kaolins (CKs). They considered mineral admixture characteristics, concrete composition, and drying period to have an accurate prediction. Their results indicated that the GEP model is a more reliable and accurate prediction tool than the MLR model. Although most existing models considered drying shrinkage of concrete, these models are not calibrated with concretes that incorporate additives. To address this issue, Kiani et al. [63] applied the PSO algorithm to modify one of the widely utilized models, that is the ACI-209R-08 shrinkage model, and enhance its performance for concrete containing pozzolans. They collected a comprehensive database of concrete containing three types of pozzolans, including silica fume, fly ash, and slag (SL), and employed PSO to modify the ACI-209R-08 shrinkage model. In another research, Liu et al. [64] operated the SVM technique to develop a predictive model of autogenous shrinkage of concretes containing silica fume, fly ash, and high-range water-reducing admixture. To verify the performance of SVM based model, they compared it with the ANN prediction model.

Evaluation of the fatigue performance of asphalt binder is also one of the challenging problems that cannot be modeled easily by traditional constitutive models due to the complexity of polymer compositions. To accurately predict the fatigue performance of asphalt under cyclic loads, Yang et al. [65] exploited a popular type of recurrent neural network called LSTM, a type of ANN that facilitates the processing of time-dependent sequences. Results illustrated that this model could have an excellent performance in predicting the fatigue level of asphalt under a wide range of cyclic loads. In another research, Gong et al. [66] established a gradient boosted model (GBM) to predict the alligator cracking (AC) and longitudinal cracking (LC), which for flexible pavements are critical load-related distresses. Since GBM is convenient as a tree-based machine learning model for measuring the importance of variables, they interpreted the model to determine variable Importance and identify the critical variables.

One of pavement distress, which can profoundly influence road safety and quality, is rutting. Various variables should be considered in rutting prediction, which makes the task hard for empirical methods. To address this issue, Majidifard et al. [67] developed a GEP to predict Hamburg Wheel Tracking Test (HWTT), which is a rutting test operated in the United States. To measure variable importance, they carry out a sensitivity analysis. In another research, Gong et al. [68] developed two ANN-based models for rutting prediction. They employed the exact data for training models that the rutting transfer function of the MEPDG used and compared results of ANN-based models with the rutting transfer function of the MEPDG. One of the networks employed the same variable that MEPDG utilized, and the second network adopted 20 variables. They applied RF to interpret the model, which is a convenient and accurate approach to model interpretation.

3. Damage segmentation and detection

Although visual detection and evaluation of defects are operated, this process is time-consuming, subjective, and based on experience, skill, and engineering judgment. As will be explained below, machine learning as an automated model can be advantageous.

Crack detection segmentation in concrete as a widely demanded material can be financial and sustainably advantageous. To address this issue, Das et al. [69] employed Gaussian mixture modeling (GMM) to clustering cracks and divided them into two categories (tensile cracks and shear cracks). In this approach, acoustic emission (AE) results as a passive structural health monitoring technique was operated. The waveform parameters (i.e., RA values (RA) and average frequency (AF)) were utilized as input parameters of GMM, and linear SVM models operated to create a hyperplane to separate clusters. In another research, Dong et al. [70] used deep convolutional neural network (DCNN) to segment microstructural cracks of concrete and distinguished voids and microcracks. Using the results of DCNN, they reconstructed the three-dimensional concrete microcrack patterns subjected to different cycles of freeze-thaw action. To train the network, a dataset of CT images which is a nondestructive imaging technology to obtain the morphology of voids and microcracks with manual segmentation results was applied. In another study, Szeląg [71] implemented the Trainable Weka Segmentation and Analyze Skeleton plugin for ImageJ software to perform a quantitative and qualitative analysis of the cracking patterns to calculate total crack area, crack density, and fractal dimension of fracture line of low-alkali cement matrix modified with micro silica subjected to loading with increased temperature in the two stages, i.e., at 350°C and then at 450°C.

Since fast and accurate identification and segmentation of the distresses, defects, and damages is an essential step in maintaining pavements as vital infrastructure, researches are done in several cases to address this issue. Liu et al. [72] divided approaches of deep learning–based models, which are widely employed for pavement inspection, into three groups, that is, patch-based models, region-based models, and pixel-wise semantic segmentation. In the first approach, digital pavement images are cropped into small patches and classified into one of the distress types. For instance, AlexNet and LeNet-5, which are typical CNNs, are in this category. In the second approach, region-based deep learning models describe distresses' size and position in the form of regions of interest. For instance, region-based CNN and its variants are in this group. In the last approach, a pixel-wise semantic segmentation model classifies each pixel in a digital image into one of the distress classes. For example, fully convolutional networks (FCNs) and their modified types are in this category.

Liu et al. [72] combined pixel-wise (FCNs) and region-based (Faster RCNN) deep learning methods to identify distress classes, locations, and geometric information. In this deeply integrated model, several convolutional and pooling layers were employed to feature extraction. Faster RCNN was adopted to propose a manageable number of potential distress regions. In the next step, these regions are utilized for classification and bounding-box by Faster RCNN. Finally, FCN was applied to perform pixel-wise semantic segmentation. Hou et al. [73] utilized an adaptive lightweight deep learning model as a faster model in computation than CNN to identify the pavement cracks. This model achieved this goal by reducing input image size, using group convolution, and using global average pooling. In addition, to obtain an adequate number of samples which is essential for the performance of deep learning–based models, they used mirror and rotation techniques. Finally, they compared the presented model and other CNN-based models, including AlexNet, VGG16, lightweight model MobileNet, and SqueezeNet. The results showed that the proposed model works well. Hoang et al. [74] employed a multiclass SVM assisted by an artificial bee colony (ABC) algorithm (mcSVM-ABC) to build a model to integrate with image processing tools. Image processing tools, including nonlocal means (NLMs) and steerable filter, were adopted to noise negation and highlight the crack patterns. Based on the established prominent map, projective integral was utilized to characterize different types of cracks, and a hybrid model of Otsu's thresholding algorithm and Min-Max Gray Level Discrimination (M2GD) was applied to analyze the properties of identified cracks. Finally, mcSVM-ABC was carried out to generalize the mapping function between influencing factors and the pavement crack conditions. In a similar approach, Hoang and Nguyen [75] implemented machine learning algorithms, including the SVM, the ANN, and the RF to asphalt pavement crack classification based on data extracted by image processing tools including, steerable filters, projective integral of image, and the min-max gray level discrimination. In addition, some studies have combined deep learning and machine learning with image processing. For example, Tran et al. [76] employed Mask RCNN to detect and identify linear and fatigue cracks, and for both longitudinal and transverse cracks, utilized image processing to determine the severity.

4. Mixture design

In practical engineering applications, simultaneous reverse prediction of multiple construction materials components is complicated but very important because it is necessary to simultaneously find an optimal mixture to fulfill all the desired quality characteristics. Reach an optimal mixture design that satisfies our environmental concerns such as resource depletion, energy consumption, and greenhouse gas emission and at the same time provides a good performance still is a challenge that needs much more research. Although challenging to solve than the forward prediction, accurate reverse prediction of mixture components based on strength, cost, stability, etc., can be crucial for saving workforce and financial resources. However, up to date, most research has focused on the forward prediction of mechanical properties and has used machine learning models to predict the properties of construction materials from ingredients; predicting the ingredients for the intended performance is still a challenge.

Due to diverse and numerous combination methods, cement-based composite framework design is not simple. These features prevent researchers from an available route. For example, in the mixture design of fiber-reinforced cement (FRC)−based composite, we have many parameters: fiber types and sizes, fiber distribution, interface interaction zones, temperatures and curing conditions, mixing methods, etc. Considering each parameter as a variable, one needs to probe the high-dimensional combination space constructed by these variables to find a set of the optimal mixture condition leading to the desirable cement-based composite formation. This approach's cost exponentially increases with only a unit increase in the number of variables. For instance, to search space of 10 variables that individually have three options, a million experiments are needed.

As mentioned above, fiber-reinforced concrete can play an active role in increasing concrete's strength and durability. As a result, it is necessary to have accurate models to design its proportions. In this field, Tong et al. [77] applied a fully convolutional network integrated with a Gaussian-conditional random field which received SEM images to extract the microstructure features, such as fiber distribution and interface interaction. Then, NNs, whose input data were the extracted features from SEM images and other direct design variables of cement-based composite, were operated as the relationship describers to build the relationship between the input data and macro-properties of cement-based composite. Later, a gradient-based high-throughput method was proposed to design a FRC-based composite.

Due to the increasing development of building materials, we have large amounts of tested data, but our information may have missing values, wrong measurements, or outliers. Outliers could occur during the evaluation of samples' measurements that might include human or system errors, and identifying these outliers can boost our machine learning model's accuracy. So

Alsini et al. [78] utilized an anomaly-based outlier detection algorithm (IFS-LOF) to quantify the anomalies and outliers during the design phase of concrete mixtures.

In most previous studies, compressive strength is considered the most important parameter to design an appropriate mixture and other criteria such as sustainability have not received enough attention in the concrete industry. Accordingly, Naseri et al. [79] developed a mixture design model with a bayesian machine learning approach that estimates optimal mixture according to six types of objective functions and presents the most sustainable, the most economical, the eco-friendliest, and the least material-consuming mixtures.

In terms of ordinary concrete, several studies have been conducted. For instance, Gou et al. [80] presented a variant SVM that used particle swarm optimization (PSO) to minimize the maximum relative error in a multiple-input multiple-output problem that predicts the components from concrete strength, slump, and flow. Additionally, Fan et al. [81] operated a fuzzy weighted relative error SVM for reverse prediction of concrete components. Furthermore, Ziolkowski and Niedostatkiewicz [82] developed an ANN-based model to predict compressive strength, transform ANN into an actual mathematical equation, and simplify it into one general equation that can be adopted as a rapid tool for a concrete mix design check. Jafari and Mahini [83] implemented GEP to obtain equations to predict the compressive strength of a specific mixture of lightweight concrete that can be employed to generate the different mixture designs.

SCMs are widely utilized in the concrete industry either in blended cement or added separately to enhance concrete mixtures' mechanical and durability properties and reduce the environmental pollution associated with the landfilling of industrial by-products. Most of them with low carbon emission can decrease concerns around CO_2 emission from the cement industry by producing much more environmentally friendly materials; furthermore, these elements also exhibit superior engineering performances, such as higher strength and better durability. To design optimal proportions, Golafshani and Behnood [84] adopted a hybrid model based on BBP and constrained biogeography-based optimization to develop a simple formula for predicting silica fume concrete's compressive strength and optimal mix design.

HPC can play an active role in improving sustainability and reliability. However, due to SCMs, HPC is more environmentally friendly and exhibits superior engineering performances, multiplicity variables and nonlinear relationships between them make us use the machine learning method in mixture design. In this regard, Ke and Duan [85] proposed a performance-based design method that combines the Gaussian processes emulator with the bayesian inference method to infer a list of potential HPC formulae of a targeted performance.

Classic single-objective methods are not proficient at reaching optimal mixtures for complex materials such as concrete pavements. Generally, several performances, including flexural strength, abrasion strength, slump, drying shrinkage, and freezing-thawing resistance, and its unit cost should come together to achieve a suitable solution. So Shirzadi Javid et al. [86] operated GA, PSO, differential evolution, and interior−point optimizer to develop models for optimal mixture design of concrete pavements, considering abrasion strength, drying shrinkage, freeze and thaw resistance, flexural strength, and cost as objectives.

5. Multiobjective optimization

In real-world problems, we are constantly faced with more than one objective function that must be optimized simultaneously. Although single-objective optimization problems have been reviewed, we are usually faced with more than one objective function in construction materials design problems. For example, in the concrete mixture design, we should deal with the strength-workability-cost paradox. In this section application of machine learning tools in multiobjective optimization problems will be reviewed. Figs. 9.1 and 9.2 demonstrate general steps of multiobjective optimization and the general form of a Pareto Front graph, respectively.

To move toward more sustainable construction, the use of GGBFS, a by-product of manufacturing pig iron, is a partial replacement for cement in concrete mixtures that provide many technical and economic benefits and also decreases energy consumption and reduces greenhouse gas emissions. In many cases, the complexity of NN prediction models is important, along with the model's accuracy. The correct choice of the number of layers and the number of neurons in each layer dramatically improves the model's performance. For example, Kandiri et al. [87] used the multiobjective salp swarm algorithm (MOSSA) to obtain ANN models with the most acceptable error and complexity, and they reached a Pareto front, including 19 ANN models with various structures and precisions, which allows choosing among them based on the required

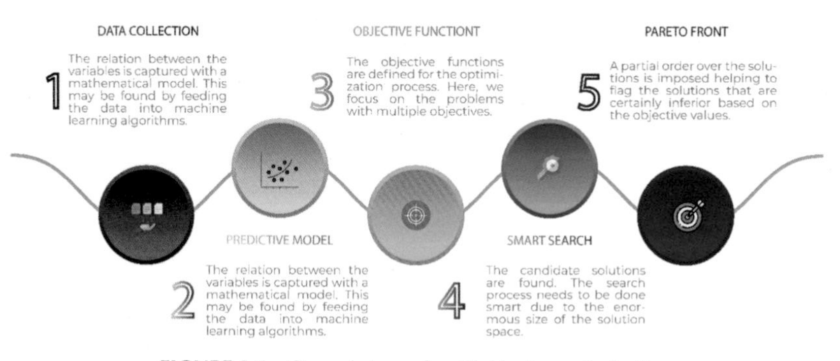

FIGURE 9.1 General steps of multiobjective optimization.

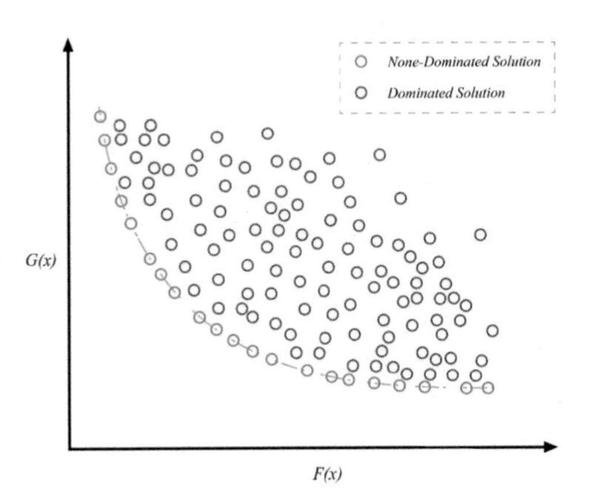

FIGURE 9.2 The general form of a Pareto Front.

simplicity and complexity. Finally, they compared the performance of one of these models with the M5P model tree. This comparison indicated that the ANN model had a half mean absolute percentage error for both train and test data.

Ultrahigh-performance fiber-reinforced concrete (UHPFRC) is interesting for the enhancement and seismic retrofitting of concrete structures. In cases like UHPFRC, many highly nonlinear variables make using conventional methods to mixture design almost impossible. To optimal mixture design of UHPFRC, Abellán-García and Guzmán-Guzmán [88] developed two RF-based models that predicted the energy absorption capacity and the maximum post-cracking strain. In the next stage, an R-coded multiobjective algorithm was implemented to reach optimum proportions of input variables that achieve the necessaries for seismic retrofitting applications at a lower cost. Finally, the proposed optimization method was compared with experimental results and revealed that this methodology could optimize new UHPFRC dosages.

To effectively use silica fume's advantages in concrete, we need an optimal mixture that trade-off among multiple objectives (strength, cost, and embodied CO_2) and consideration of a large number of variables under highly nonlinear constraints. In this context, Zhang et al. [89] developed a BPNN to model the relationship between inputs and SFC's uniaxial compressive strength (UCS). The BPNN's hyperparameters were tuned by the BAS algorithm and based on 10-fold cross-validation. In the next step, combining this model and two equations representing the cost and CO_2 emission using the weighted sum method, they constructed a single-objective function that MOBAS used to Search for optimal solutions. To choose the best solution among the Pareto front, they operated the TOPSIS method. Finally, they had a comparison with nondominated sorting genetic algorithm II (NSGA-II), multiobjective particle swarm optimization (MOPSO), and multiobjective differential evolution methods.

Traditional laboratory or statistics-based methods cannot consider multiple related variables and objectives (e.g., mechanical, economic, and environmental objectives) during optimizing mixtures. To address this issue, Zhang et al. [90] developed three models (RF, BPNN, and SVM) to predict UCS and splitting tensile strength of RAC. To improve the accuracy of models, hyperparameters were tuned by the firefly algorithm, and results showed that RF t and BPNN achieve the best prediction accuracy for predicting UCS and splitting tensile strength of RAC, respectively. The multi-objective firefly algorithm was adopted to search for the optimal RAC mixtures with minimum costs and CO_2 emissions, and maximum strength. In this respect, they obtained 50 nondominated mixture proportions of RAC for both the UCS–cost–CO_2 and STS–cost–CO_2 Pareto fronts by using RF and BPNN models as objective functions, respectively. Finally, they applied a multicriteria decision analysis method to select a final optimal solution closest to the ideal positive point.

Previously, to design the mixture proportion of concrete and pavement, in most of the research, just strength parameters were taken into account as the most critical characteristics, and other vital parameters such as cost, CO_2 emission, resistance, and durability against freeze and thaw cycles, abrasion resistance, and energy consumption were overlooked. In this respect, Shirzadi Javid et al. [86] compared four regression models (first polynomial, second polynomial, fractional, and partial polynomial–fractional type 2) to find the relation between inputs and outputs (flexural, abrasion freeze and thaw shrinkage, and slump), and results showed that second polynomial regression model was the most precise model. In the next step, they define a cost function that considered six criteria and operated four algorithms (GA, PSO, differential evolution, and interior–point optimizer) to achieve the optimal solution. Finally, the interior–point optimizer was the only one that reached the global optimum, and others achieved near-optimal solutions. In another research, Zhang et al. [91] developed several models to predict UCS of HPC mixtures and adopted the PSO algorithm to tune hyperparameters of models, and also developed several models to estimate UCS and slump of plastic concrete and compared the performance of models. Models that showed the best performance accepted as objective function beside cost function in the multi-objective optimization process. The MOPSO algorithm is used to find the Pareto front and among these nondominated points via the TOPSIS method final solutions were chosen. Among BPNN, RF, RT, KNN, LR, and SVR in both case studies BPNN had the best performance to predict the UCS, and RF was chosen to predict the slump in the second case study.

As previously mentioned, in order to enhance the service performances of asphalt pavement, many additives like polymers, diatomite, fibers, etc., were selected to mix with asphalt materials which makes asphalt mixture design challenging. Frequently, in mixture design, multiple objectives should optimize simultaneously. To address this issue, Liang et al. [92] developed the

Gaussian process regression-based machine learning method to model that represented the objective and constraint functions. In the next step, they defined three case studies and employed MOPSO to find optimum solutions for each case study. In the first case, basalt fiber-reinforced asphalt mixture Pareto front based on Marshall stability and cost was obtained. In the second case, diatomite and basalt fiber-reinforced asphalt mixture Pareto front based on high-temperature performance, low-temperature performance, and the cost was obtained. Finally, for the last case, rubber aggregate and basalt fiber concrete mixture Pareto front based on slump, flexural strength, compressive strength, and cost was obtained. The result showed that hybrid GPR-adaptive weight MOPSO could play a guidance role in the asphalt mixture design.

Multiobjective optimization methods can be employed in pavement maintenance. Gerami Matin et al. [93] employed GA, PSO [94], and combination of genetic algorithm and particle swarm optimization (GAPSO) as single-objective techniques and nondomination sorting genetic algorithm II (NSGAII) and MOPSO as multiobjective techniques. They were implemented to achieve optimized road maintenance planning that can minimize the life cycle cost of a road network and simultaneously maximize pavement condition. They utilized the pavement condition index and cost as objective functions and compared the algorithm's performance. Results showed that multiobjective algorithms performed better than single-objective algorithms, and among multiobjective algorithms, the NSGAII algorithm generally performs better than the MOPSO. In another research, Cao et al. [95] developed a multiobjective optimization model to maintain the low-noise pavement network system, which is a growing interest in the pavement management system. In this regard, they employed the nondominated sorting genetic algorithm II (NSGA-II) to reach an optimized solution that considers maximizing the average close proximity level reduction, minimizing the maintenance costs, and minimizing the greenhouse gas emissions.

6. Conclusions

Recently, many studies have discussed the application of machine learning in predicting the properties of construction materials, their damage detection and segmentation, mixture design, and multiobjective optimization as an alternative to traditional methods. A number of recent studies on applying machine learning in construction materials were reviewed in this paper. These studies have shown that while statistical and experimental methods cannot accurately model complex relationships between composite components, machine learning—based methods can do the task with much greater accuracy and can better deal well with inherent nonlinear relationships among mixture variables.

Here, an attempt has been made to look at the issue from a higher perspective and categorize the methods based on the applications instead of model's type. Based on research, it can be seen that the model's performance, regardless of the type of method, depends on the comprehensiveness of the data, hyperparameters tuning, and correct selection of input variables. Although several studies have been done so far, more studies need to be done for new emerging binders such as nanocomposites and geopolymer concrete. In addition, a lack of study on the usage of machine learning in multiobjective optimization exists. However, these few conducted studies showed that this approach has good potential to compete with other common methods such as response level.

In general, these models help to achieve a better understanding of relationships between variables. This insight can help in the evaluation and design, as well as damage identification to achieve sustainable and environmentally friendly materials. For example, the predictive models obtained from these methods can be applied to effectively use the industry's by-product or recycled aggregate in composites, resulting in less environmental pollution and resource sustainability or even help achieve an optimal mixture design with the lowest carbon dioxide production and cost for the required strength.

References

[1] E. Benhelal, G. Zahedi, E. Shamsaei, A. Bahadori, Global strategies and potentials to curb CO_2 emissions in cement industry, J. Clean. Prod. 51 (2013) 142–161.

[2] M. Asadnia, L.H. Chua, X. Qin, A. Talei, Improved particle swarm optimization–based artificial neural network for rainfall-runoff modeling, J. Hydrol. Eng. 19 (7) (2014) 1320–1329.

[3] M. Asadnia, M.S. Yazdi, A. Khorasani, An improved particle swarm optimization based on neural network for surface roughness optimization in face milling of 6061-T6 aluminum, Int. J. Appl. Eng. Res. 5 (19) (2010) 3191–3201.

[4] M. Asadnia, A.M. Khorasani, M.E. Warkiani, An accurate PSO-GA based neural network to model growth of carbon nanotubes, J. Nanomater. 2017 (2017).

[5] V. Chandwani, V. Agrawal, R. Nagar, Modeling slump of ready mix concrete using genetic algorithms assisted training of Artificial Neural Networks, Expert Syst. Appl. 42 (2) (2015) 885–893.

[6] H. Moayedi, B. Kalantar, L.K. Foong, D.T. Bui, A. Motevalli, Application of three metaheuristic techniques in simulation of concrete slump, Appl. Sci. 9 (20) (2019). Art. no. 4340.

[7] T.-D. Nguyen, T.-H. Tran, H. Nguyen, H. Nhat-Duc, A success history-based adaptive differential evolution optimized support vector regression for estimating plastic viscosity of fresh concrete, Eng. Comput. 37 (2) 1485–1498.

[8] M. Sonebi, A. Cevik, S. Grünewald, J. Walraven, Modelling the fresh properties of self-compacting concrete using support vector machine approach, Construct. Build. Mater. 106 (2016) 55–64.

[9] M.R. Kaloop, P. Samui, M. Shafeek, J.W. Hu, Estimating slump flow and compressive strength of self-compacting concrete using emotional neural networks, Appl. Sci. 10 (23) (2020) 1–17. Art. no. 8543.

[10] A. Kaveh, T. Bakhshpoori, S.M. Hamze-Ziabari, M5' and mars based prediction models for properties of self-compacting concrete containing fly ash, Period. Polytech. Civ. Eng. 62 (2) (2018) 281–294. Article.

[11] C. Kina, K. Turk, E. Atalay, I. Donmez, H. Tanyildizi, Comparison of extreme learning machine and deep learning model in the estimation of the fresh properties of hybrid fiber-reinforced SCC, Neural Comput. Appl. 33 (18) (2021) 11641–11659.

[12] O. Altay, M. Ulas, K.E. Alyamac, Prediction of the fresh performance of steel fiber reinforced self-compacting concrete using quadratic SVM and weighted KNN models, IEEE Access 8 (2020) 92647–92658. Art. no. 9093848.

[13] L. Chen, C.-H. Kou, S.-W. Ma, Prediction of slump flow of high-performance concrete via parallel hyper-cubic gene-expression programming, Eng. Appl. Artif. Intell. 34 (2014) 66–74.

[14] E. Gomaa, T. Han, M. ElGawady, J. Huang, A. Kumar, Machine learning to predict properties of fresh and hardened alkali-activated concrete, Cement Concr. Compos. 115 (2021). Art. no. 103863.

[15] P. Smarzewski, D. Barnat-Hunek, Mechanical and durability related properties of high performance concrete made with coal cinder and waste foundry sand, Construct. Build. Mater. 121 (2016) 9–17.

[16] A. Tavana Amlashi, P. Alidoust, M. Pazhouhi, K. Pourrostami Niavol, S. Khabiri, A.R. Ghanizadeh, AI-based formulation for mechanical and workability properties of eco-friendly concrete made by waste foundry sand, J. Mater. Civ. Eng. 33 (4) (2021). Art. no. 04021038.

[17] N. Sivakugan, R. Veenstra, N. Naguleswaran, Underground mine backfilling in Australia using paste fills and hydraulic fills, Int. J. Geosynth. Ground Eng. 1 (2) (2015).

[18] C. Qi, Q. Chen, X. Dong, Q. Zhang, Z.M. Yaseen, Pressure drops of fresh cemented paste backfills through coupled test loop experiments and machine learning techniques, Powder Technol. 361 (2020) 748–758.

[19] M.-C. Kang, D.-Y. Yoo, R. Gupta, Machine learning-based prediction for compressive and flexural strengths of steel fiber-reinforced concrete, Construct. Build. Mater. 266 (2021).

[20] Z. Tong, J. Gao, Z. Wang, Y. Wei, H. Dou, A new method for CF morphology distribution evaluation and CFRC property prediction using cascade deep learning, Construct. Build. Mater. 222 (2019) 829–838.

[21] D. Yuan, W. Jiang, Z. Tong, J. Gao, J. Xiao, W. Ye, Prediction of electrical conductivity of fiber-reinforced cement-based composites by deep neural networks, Materials 12 (23) (2019). Art. no. 3868.

[22] F. Farooq, W. Ahmed, A. Akbar, F. Aslam, R. Alyousef, Predictive modeling for sustainable high-performance concrete from industrial wastes: a comparison and optimization of models using ensemble learners, J. Clean. Prod. 292 (2021) 126032, 2021/04/10/2021.

[23] Q. Han, C. Gui, J. Xu, G. Lacidogna, A generalized method to predict the compressive strength of high-performance concrete by improved random forest algorithm, Construct. Build. Mater. 226 (2019) 734–742.

[24] J. Xu, X. Zhao, Y. Yu, T. Xie, G. Yang, J. Xue, Parametric sensitivity analysis and modelling of mechanical properties of normal- and high-strength recycled aggregate concrete using grey theory, multiple nonlinear regression and artificial neural networks, Construct. Build. Mater. 211 (2019) 479–491.

[25] F. Deng, Y. He, S. Zhou, Y. Yu, H. Cheng, X. Wu, Compressive strength prediction of recycled concrete based on deep learning, Construct. Build. Mater. 175 (2018) 562–569.

[26] A. Kandiri, F. Sartipi, M. Kioumarsi, Predicting compressive strength of concrete containing recycled aggregate using modified ann with different optimization algorithms, Art. no. 485, Appl. Sci. 11 (2) (2021) 1–19.

[27] U.K. Sevim, H.H. Bilgic, O.F. Cansiz, M. Ozturk, C.D. Atis, Compressive strength prediction models for cementitious composites with fly ash using machine learning techniques, Construct. Build. Mater. 271 (2021) 121584.

[28] M.R. Kaloop, D. Kumar, P. Samui, J.W. Hu, D. Kim, Compressive strength prediction of high-performance concrete using gradient tree boosting machine, Construct. Build. Mater. 264 (2020).

[29] E.M. Golafshani, A. Behnood, M. Arashpour, Predicting the compressive strength of normal and high-performance concretes using ANN and ANFIS hybridized with Grey Wolf Optimizer, Construct. Build. Mater. 232 (2020).

[30] A. Ashrafian, M.J. Taheri Amiri, M. Rezaie-Balf, T. Ozbakkaloglu, O. Lotfi-Omran, Prediction of compressive strength and ultrasonic pulse velocity of fiber reinforced concrete incorporating nano silica using heuristic regression methods, Construct. Build. Mater. 190 (2018) 479–494.

[31] J. Zhang, G. Ma, Y. Huang, J. sun, F. Aslani, B. Nener, Modelling uniaxial compressive strength of lightweight self-compacting concrete using random forest regression, Construct. Build. Mater. 210 (2019) 713–719.

[32] F. Naseri, F. Jafari, E. Mohseni, W. Tang, A. Feizbakhsh, M. Khatibinia, Experimental observations and SVM-based prediction of properties of polypropylene fibres reinforced self-compacting composites incorporating nano-CuO, Construct. Build. Mater. 143 (2017) 589–598.

[33] P.O. Awoyera, M.S. Kirgiz, A. Viloria, D. Ovallos-Gazabon, Estimating strength properties of geopolymer self-compacting concrete using machine learning techniques, J. Mater. Res. Technol. 9 (4) (2020) 9016–9028.

[34] Y. Zhang, R. Luo, R.L. Lytton, Characterizing permanent deformation and fracture of asphalt mixtures by using compressive dynamic modulus tests, J. Mater. Civ. Eng. 24 (7) (2012) 898–906.

[35] C. Zhang, S. Shen, X. Jia, Modification of the Hirsch dynamic modulus prediction model for asphalt mixtures, J. Mater. Civ. Eng. 29 (12) (2017).

[36] R. Dongré, et al., Field evaluation of Witczak and Hirsch Models for predicting dynamic modulus of hot-mix asphalt, in: 2005 Meeting of the Association of Asphalt Paving Technologists, Long Beach, CA vol. 74, 2005, pp. 381–442.

[37] J. Bari, et al., Development of a new revised version of the Witczak E Predictive Model for hot mix asphalt mixtures, in: Association of Asphalt Paving Technologists -Proceedings of the Technical Sessions 2006 Annual Meeting, Savannah, GA vol. 75, 2006, pp. 381–424.

[38] A. Behnood, E. Mohammadi Golafshani, Predicting the dynamic modulus of asphalt mixture using machine learning techniques: an application of multi biogeography-based programming, Construct. Build. Mater. 266 (2021).

[39] H. Gong, Y. Sun, Y. Dong, W. Hu, B. Han, P. Polaczyk, B. Huang, An efficient and robust method for predicting asphalt concrete dynamic modulus, Int. J. Pavement Eng. (2021) 1–12.

[40] G.S. Moussa, M. Owais, Pre-trained deep learning for hot-mix asphalt dynamic modulus prediction with laboratory effort reduction, Construct. Build. Mater. 265 (2020).

[41] H. Gong, et al., Improved estimation of dynamic modulus for hot mix asphalt using deep learning, Construct. Build. Mater. 263 (2020).

[42] A. Behnood, D. Daneshvar, A machine learning study of the dynamic modulus of asphalt concretes: an application of M5P model tree algorithm, Construct. Build. Mater. 262 (2020).

[43] F. Hussain, Y. Ali, M. Irfan, M. Ashraf, S. Ahmed, A data-driven model for phase angle behaviour of asphalt concrete mixtures based on convolutional neural network, Construct. Build. Mater. 269 (2021).

[44] B. Ghorbani, A. Arulrajah, G. Narsilio, S. Horpibulsuk, M. Win Bo, Thermal and mechanical properties of demolition wastes in geothermal pavements by experimental and machine learning techniques, Construct. Build. Mater. 280 (2021). Art. no. 122499.

[45] A. Ashrafian, A.H. Gandomi, M. Rezaie-Balf, M. Emadi, An evolutionary approach to formulate the compressive strength of roller compacted concrete pavement, Measurement 152 (2020).

[46] S. Debbarma, G.D. Ransinchung RN, Using artificial neural networks to predict the 28-day compressive strength of roller-compacted concrete pavements containing RAP aggregates, Road Mater. Pavement Des. (2020) 1−19.

[47] H. Majidifard, B. Jahangiri, W.G. Buttlar, A.H. Alavi, New machine learning-based prediction models for fracture energy of asphalt mixtures, Measurement 135 (2019) 438−451.

[48] A. Aksoy, E. Iskender, H. Tolga Kahraman, Application of the intuitive k-NN Estimator for prediction of the Marshall Test (ASTM D1559) results for asphalt mixtures, Construct. Build. Mater. 34 (2012) 561−569.

[49] H. Gong, Y. Sun, X. Shu, B. Huang, Use of random forests regression for predicting IRI of asphalt pavements, Construct. Build. Mater. 189 (2018) 890−897.

[50] E.F. Felix, R. Carrazedo, E. Possan, Carbonation model for fly ash concrete based on artificial neural network: development and parametric analysis, Construct. Build. Mater. 266 (2021).

[51] H. Lee, H.-S. Lee, P. Suraneni, Evaluation of carbonation progress using AIJ model, FEM analysis, and machine learning algorithms, Construct. Build. Mater. 259 (2020).

[52] P. Akpinar, I.D. Uwanuakwa, Investigation of the parameters influencing progress of concrete carbonation depth by using artificial neural networks, Mater. Construcción 70 (337) (2020). Art. no. e209.

[53] W.Z. Taffese, E. Sistonen, J. Puttonen, CaPrM: carbonation prediction model for reinforced concrete using machine learning methods, Construct. Build. Mater. 100 (2015) 70−82.

[54] R. Cai, et al., Prediction of surface chloride concentration of marine concrete using ensemble machine learning, Cement Concr. Res. 136 (2020).

[55] Q.F. Liu, M.F. Iqbal, J. Yang, X.Y. Lu, P. Zhang, M. Rauf, Prediction of chloride diffusivity in concrete using artificial neural network: modelling and performance evaluation, Construct. Build. Mater. 268 (2021). Art. no. 121082.

[56] W.Z. Taffese, E. Sistonen, Significance of chloride penetration controlling parameters in concrete: ensemble methods, Construct. Build. Mater. 139 (2017) 9−23.

[57] N.D. Hoang, C.T. Chen, K.W. Liao, Prediction of chloride diffusion in cement mortar using multi-gene genetic programming and multivariate adaptive regression splines, Meas. J. Int. Meas. Confed. 112 (2017) 141−149.

[58] P. Chen, W. Zheng, Y. Wang, W. Chang, Creep model of high-strength concrete containing supplementary cementitious materials, Construct. Build. Mater. 202 (2019) 494−506.

[59] Y. Yu, T.N. Nguyen, J. Li, L.F.M. Sanchez, A. Nguyen, Predicting elastic modulus degradation of alkali silica reaction affected concrete using soft computing techniques: a comparative study, Construct. Build. Mater. 274 (2021).

[60] H. Allahyari, A. Heidarpour, A. Shayan, V.P. Nguyen, A robust time-dependent model of alkali-silica reaction at different temperatures, Cement Concr. Compos. 106 (2020).

[61] C.K. Lau, H. Lee, V. Vimonsatit, W.Y. Huen, P. Chindaprasirt, Abrasion resistance behaviour of fly ash based geopolymer using nanoindentation and artificial neural network, Construct. Build. Mater. 212 (2019) 635−644.

[62] K. Mermerdaş, E. Güneyisi, M. Gesoğlu, T. Özturan, Experimental evaluation and modeling of drying shrinkage behavior of metakaolin and calcined kaolin blended concretes, Construct. Build. Mater. 43 (2013) 337–347.

[63] B. Kiani, S. Sajedi, A.H. Gandomi, Q. Huang, R.Y. Liang, Optimal adjustment of ACI formula for shrinkage of concrete containing pozzolans, Construct. Build. Mater. 131 (2017) 485–495.

[64] J. Liu, K. Yan, X. Zhao, Y. Hu, Prediction of autogenous shrinkage of concretes by support vector machine, Int. J. Pavement Res. Technol. 9 (3) (2016) 169–177.

[65] E. Yang, Y. Tang, L. Li, W. Yan, B. Huang, Y. Qiu, Research on the recurrent neural network-based fatigue damage model of asphalt binder and the finite element analysis development, Construct. Build. Mater. 267 (2021).

[66] H. Gong, Y. Sun, B. Huang, Gradient boosted models for enhancing fatigue cracking prediction in mechanistic-empirical pavement design guide, J. Transp. Eng. B Pavements 145 (2) (2019). Art. no. 04019014.

[67] H. Majidifard, B. Jahangiri, P. Rath, L. Urra Contreras, W.G. Buttlar, A.H. Alavi, Developing a prediction model for rutting depth of asphalt mixtures using gene expression programming, Construct. Build. Mater. 267 (2021).

[68] H. Gong, Y. Sun, Z. Mei, B. Huang, Improving accuracy of rutting prediction for mechanistic-empirical pavement design guide with deep neural networks, Construct. Build. Mater. 190 (2018) 710–718.

[69] A.K. Das, D. Suthar, C.K.Y. Leung, Machine learning based crack mode classification from unlabeled acoustic emission waveform features, Cement Concr. Res. 121 (2019) 42–57.

[70] Y. Dong, C. Su, P. Qiao, L. Sun, Microstructural crack segmentation of three-dimensional concrete images based on deep convolutional neural networks, Construct. Build. Mater. 253 (2020). Art. no. 119185.

[71] M. Szeląg, Fractal characterization of thermal cracking patterns and fracture zone in low-alkali cement matrix modified with microsilica, Cement Concr. Compos. 114 (2020).

[72] C. Liu, J. Li, J. Gao, Z. Gao, Z. Chen, Combination of pixel-wise and region-based deep learning for pavement inspection and segmentation, Int. J. Pavement Eng. (2021) 1–13.

[73] Y. Hou, et al., MobileCrack: object classification in asphalt pavements using an adaptive lightweight deep learning, J. Transp. Eng. B Pavements 147 (1) (2021).

[74] N.D. Hoang, Q.L. Nguyen, D. Tien Bui, Image processing-based classification of asphalt pavement cracks using support vector machine optimized by artificial bee colony, J. Comput. Civ. Eng. 32 (5) (2018). Art. no. 04018037.

[75] N.D. Hoang, Q.L. Nguyen, A novel method for asphalt pavement crack classification based on image processing and machine learning, Eng. Comput. 35 (2) (2019) 487–498.

[76] T.S. Tran, V.P. Tran, H.J. Lee, J.M. Flores, V.P. Le, A two-step sequential automated crack detection and severity classification process for asphalt pavements, Int. J. Pavement Eng. (2020) 1–15.

[77] Z. Tong, J. Huo, Z. Wang, High-throughput design of fiber reinforced cement-based composites using deep learning, Cement Concr. Compos. 113 (2020).

[78] R. Alsini, A. Almakrab, A. Ibrahim, X. Ma, Improving the outlier detection method in concrete mix design by combining the isolation forest and local outlier factor, Construct. Build. Mater. 270 (2021).

[79] H. Naseri, H. Jahanbakhsh, P. Hosseini, F. Moghadas Nejad, Designing sustainable concrete mixture by developing a new machine learning technique, J. Clean. Prod. 258 (2020).

[80] J. Gou, Z.-W. Fan, C. Wang, W.-P. Guo, X.-M. Lai, M.-Z. Chen, A minimum-of-maximum relative error support vector machine for simultaneous reverse prediction of concrete components, Comput. Struct. 172 (2016) 59–70.

[81] Z. Fan, R. Chiong, Z. Hu, Y. Lin, A fuzzy weighted relative error support vector machine for reverse prediction of concrete components, Comput. Struct. 230 (2020). Art. no. 106171.

[82] P. Ziolkowski, M. Niedostatkiewicz, Machine learning techniques in concrete mix design, Materials 12 (8) (2019). Art. no. 1256.

[83] S. Jafari, S.S. Mahini, Lightweight concrete design using gene expression programing, Construct. Build. Mater. 139 (2017) 93–100.

[84] E.M. Golafshani, A. Behnood, Estimating the optimal mix design of silica fume concrete using biogeography-based programming, Cement Concr. Compos. 96 (2019) 95–105.

[85] X. Ke, Y. Duan, A Bayesian machine learning approach for inverse prediction of high-performance concrete ingredients with targeted performance, Construct. Build. Mater. 270 (2021).

[86] A.A. Shirzadi Javid, H. Naseri, M.A. Etebari Ghasbeh, Estimating the optimal mixture design of concrete pavements using a numerical method and meta-heuristic algorithms, Iran. J. Sci. Technol. Trans. Civil Eng. 45 (2) (2021) 913–927.

[87] A. Kandiri, E. Mohammadi Golafshani, A. Behnood, Estimation of the compressive strength of concretes containing ground granulated blast furnace slag using hybridized multi-objective ANN and salp swarm algorithm, Construct. Build. Mater. 248 (2020).

[88] J. Abellán-García, J.S. Guzmán-Guzmán, Random forest-based optimization of UHPFRC under ductility requirements for seismic retrofitting applications, Construct. Build. Mater. 285 (2021). Art. no. 122869.

[89] J. Zhang, Y. Huang, G. Ma, B. Nener, Mixture optimization for environmental, economical and mechanical objectives in silica fume concrete: a novel frame-work based on machine learning and a new meta-heuristic algorithm, Resour. Conserv. Recycl. 167 (2021).

[90] J. Zhang, Y. Huang, F. Aslani, G. Ma, B. Nener, A hybrid intelligent system for designing optimal proportions of recycled aggregate concrete, J. Clean. Prod. 273 (2020).

[91] J. Zhang, Y. Huang, Y. Wang, G. Ma, Multi-objective optimization of concrete mixture proportions using machine learning and metaheuristic algorithms, Construct. Build. Mater. 253 (2020).

[92] C. Liang, et al., Machine learning approach to develop a novel multi-objective optimization method for pavement material proportion, Appl. Sci. 11 (2) (2021) 1–27. Art. no. 835.

[93] A. Gerami Matin, R. Vatani Nezafat, A. Golroo, A comparative study on using meta-heuristic algorithms for road maintenance planning: insights from field study in a developing country, J. Traffic Transp. Eng. 4 (5) (2017) 477–486.

[94] M. Farahnakian, M.R. Razfar, M. Moghri, M. Asadnia, The selection of milling parameters by the PSO-based neural network modeling method 57 (1) (2011) 49–60.

[95] R. Cao, Z. Leng, J. Yu, S.C. Hsu, Multi-objective optimization for maintaining low-noise pavement network system in Hong Kong, Transp. Res. Transp. Environ. 88 (2020). Art. no. 102573.

Chapter 10

The AI-assisted removal and sensor-based detection of contaminants in the aquatic environment

Sweta Modak[1], Hadi Mokarizadeh[2], Elika Karbassiyazdi[3], Ahmad Hosseinzadeh[3], Milad Rabbabni Esfahani[1]

[1]*Department of Chemical and Biological Engineering, The University of Alabama, Tuscaloosa, AL, United States;* [2]*Department of Chemical Engineering, Amirkabir University of Technology, Tehran, Iran;* [3]*Centre for Technology in Water and Wastewater, University of Technology Sydney, Sydney, NSW, Australia*

1. Introduction

Contamination of limited fresh and accessible water resources with a wide range of pollutants, including heavy metals [1], microbial species [2,3], disinfection by-products (DBPs) [4,5], pharmaceutical substances [6], and organic contaminants [7–9], is a critical challenge facing humanity growth and development. Prolonged exposure to these contaminants poses serious health risks to humans and could adversely affect organs. For instance, a recent study revealed a higher mortality risk for COVID-19 in a population severely exposed to per- and polyfluoroalkyl substances (PFAS) [10]. Thus, scientists have tried to develop various technologies such as adsorption, ion exchange, membrane filtration, advanced oxidation, and biological treatment to deliver clean water to people in need and repurpose wastewater throughout human history [11,12]. Computer science, machine learning (ML), and sensors have been developed to innovate water and wastewater treatment in recent years [13].

PFAS are a family of synthetic organic chemicals divided mainly into two classes, polymeric and nonpolymeric. These compounds, particularly the perfluoroalkyl acids, are exceptionally stable to metabolic and environmental degradation. They reduce the surface tension of water, and hence, they possess strong surfactant properties that make them ideal lubricants as well as water and oil repellents [14]. PFAS are distributed ubiquitously and detectable in

Artificial Intelligence and Data Science in Environmental Sensing
https://doi.org/10.1016/B978-0-323-90508-4.00005-8

environmental media, edible plants, wildlife, and humans. Human exposure to PFAS can occur through dietary intake, food packaging, drinking water, house dust, and consumer products [15,16]. Technical advances made in analytical chemistry during the past two decades have afforded the identification of an enormous body of PFAS in consumer and industrial products, as well as drinking and groundwater, wastewater, landfill leachate, and soil [17–19]. Some PFAS are highly toxic and affect human health on continuous long-term exposure (Fig. 10.1) [20].

Based on the United States Environment Protection Agency (USEPA) published health advisories, the third Unregulated Contaminant Monitoring Rule (UCMR3), the existence of six types of PFAS (perfluorooctanesulfonic acid [PFOS], perfluorooctanoic acid [PFOA], perfluorohexane sulfonic acid [PFHxS], perfluorononanoic acid [PFNA], perfluorodecanoic acid [PFDA], and 2-(N-methyl-perfluorooctane sulfonamido) acetic acid [MeFOSAA]) was reported in drinking water. The PFOS and PFOA were the most frequently detected compounds across all system sizes and had detection frequencies of 0.79% and 1.03%, respectively. The highest concentration of PFOS was found to be 7000 ng/L. The minimum reporting level (MRL), mean, and maximum

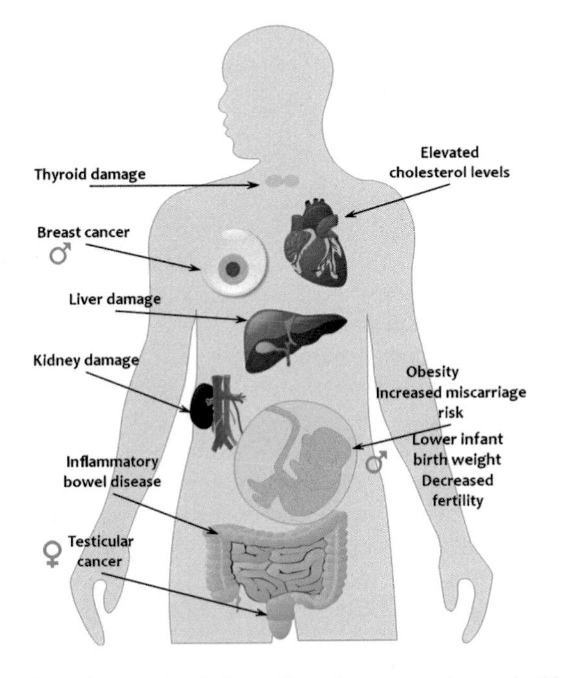

FIGURE 10.1 Effect of per- and polyfluoroalkyl substances on human health. *Adopted and edited from S.E. Fenton, et al., Per- and polyfluoroalkyl substance toxicity and human health review: current state of knowledge and strategies for informing future research, Environ. Toxicol.Chem., 40 (3) (2021) 606–630, https://doi.org/10.1002/etc.4890.*

TABLE 10.1 UCMR3 results for perfluorooctanoic acid (PFOA) and perfluorooctanesulfonic acid (PFOS) concentrations by drinking water utility system size and source water type [21].

	Small systems – surface water			Large systems – surface water		
	Concentration (ng/L)			Concentration (ng/L)		
	MRL	Mean	Maximum	MRL	Mean	Maximum
PFOA	20	–	–	20	31	100
PFOS	40	54	59	40	77	400
	Small systems – groundwater			**Large systems – groundwater**		
	Concentration (ng/L)			**Concentration (ng/L)**		
	MRL	**Mean**	**Maximum**	**MRL**	**Mean**	**Maximum**
PFOA	20	100	206	20	45	349
PFOS	40	158	300	40	200	7000

concentrations of PFOS and PFOA in groundwater and surface water across small (serving <10,000 people) and large (serving >10,000 people) water systems are specified in Table 10.1 [21]. The analysis of PFAS is mainly based on high-performance liquid chromatography (HPLC) coupled with tandem mass spectroscopy (MS); however, gas chromatography (GC-MS) and LC-MS are also used for detecting selected PFAS.

2. AI-assisted techniques for PFAS detection and removal

Recently, the artificial intelligence (AI) process has appeared as a promising tool to assess the detection and removal of PFAS compounds. In the pioneering work in this area, Akbar Reza et al. [22] examined the application of unsupervised ML to automatically assess C–F bond in terms of bond dissociation energies delivering an effective removal strategy for PFAS compounds (Fig. 10.2). Exact estimates for dissociation energies of C–F bond were reported using different ML algorithms such as least absolute shrinkage, random forest (RF), feedforward neural networks, and selection operator regression. The ML approach was highly efficient, needing a short time to train the data (<10 min) and calculate the C–F bond dissociation energy of a new compound (<1 s). The sole requirement to produce trustworthy outcomes was chemical connectivity information in the structure of PFAS, making the approach cheap and feasible [22].

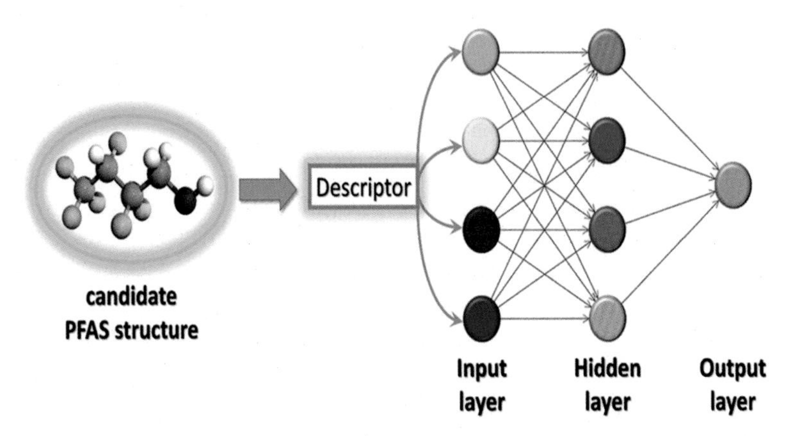

FIGURE 10.2 Schematic of the Feedforward Neural Network used to predict per- and polyfluoroalkyl substances (PFAS) bond dissociation energies. *Reprinted with permission from Reza © 2019 ACS Publication.*

Chang and colleagues (2019) built ML-based quantitative structure–activity relationship models (called QSAR) to forecast PFAS potential hazards, including bioactivity [23]. Authors developed the first PFAS-specific database covering the bioactivity data for a considerable number of PFAS compounds (1012) for 26 bioassays. Five diverse ML models, including conventional models (such as logistic regression, multitask neural network [MNN], and RF) and advanced graph-based models (such as graph convolutional network and weave model), were used to train the built database. The best act was observed for MNN and graph-based models with 0.916 as the average of the best area-under-the-curve score for each bioassay [23]. Kibbey and colleagues [24] shed some light on the importance of source allocation to deliver visions for environmental behaviors of specific PFAS compounds. To allocate PFAS source, the authors assessed supervised ML classifiers comprising three conventional ones (K-neighbors, support vector machines [SVMs], and extra trees) and one multilayer perceptron feedforward deep neural network tested. A large database of concentrations of PFAS species (1197 samples from worldwide sites) was used for this study. Compared to SVMs and K-neighbors methods, the deep neural network and extra trees showed improved performance at the classification of samples from different sources. Because these approaches act according to totally unlike principles but deliver similar forecasts, supervised ML was concluded as an encouraging tool for forensic source allocation [24]. Build on this observation, Kibbey and colleagues [25] later assessed a series of 12 classifiers, including 11 conventional classifiers and 1 deep neural network ensemble to source allocate PFAS compounds in water samples, emphasizing difference between an aqueous film-forming foam (AFFF) and non-AFFF compounds. A large dataset containing 8040 samples was used to evaluate different preprocessing methods and the influence of feature selection on classification performance. The outcomes proved that supervised ML based on composition could be practical to develop patterns to distinguish PFAS sources [25]. Su and co-workers [26] defined a database framework capable of discovering systematics in structure–function relationships accompanying developing PFAS chemistries. The developed context plots data based on the SMILES approach wherein functionality data, which include physicochemical and bioactivity property, are used to encode molecular structure. The resultant three-dimensional "PFAS-Map" could be used as an unsupervised visualization tool to categorize novel PFAS chemistries in an automatic manner using existing PFAS classification principles. There are additional examples of how to use the created PFAS-Map to (1) analyze fundamental physical characteristics of PFAS that have yet to be tested, (2) highlight features in in the current material classification schemes, and (3) fuse data from various sources [26].

3. Sensors for detection of PFAS

Different types of sensors have been used for the detection of PFAS in the aquatic environment. These sensors can be categorized into three groups, including electrochemical, optical, and biosensors. Each of these sensors has advantages and disadvantages based on their detection mechanism, as summarized below.

3.1 Electrochemical sensors

There are different electrochemical sensors, including potentiometric sensors, conductometric sensors, impedimetric sensors, and voltammetric sensors [27–29]. Among these sensors, voltammetric and potentiometric sensors are highly used for the detection of PFAS. The key point for achieving enhanced sensor performance is the functionalization of electrodes for effective interaction with the target analyte through ion exchange or complexation. One promising functionalization approach is the use of molecularly imprinted polymers (MIPs), which provide a polymeric matrix on the surface of the electrode with voids, or recognition sites, that are complementary to the shape, size, and functional groups of the target analyte [30,31]. In recent years, MIPs have been applied to detect PFOS in water [30,32], though the lack of electrochemical activity of PFOS is a primary challenge. As a solution to this problem, Karimian et al. [30] utilized ferrocenecarboxylic acid (FcCOOH), which acted as an electroactive reporter molecule competing with PFOS (nonelectroactive) for the MIP sites. The mechanism of sensing in the presence of PFOS is that the voltammetric signal decreases in correlation with the PFOS concentration, and the signal is developed. It was found that the voltammetric signal of FcCOOH, as the reporter molecule, was inversely proportional to the concentration of PFOS in the solution. Tran et al. [32] developed a photoelectrochemical PFOS sensor that consists of molecularly imprinted polyacrylamide on vertically aligned TiO_2 nanotubes that detected PFOS by measuring the increases in photocurrent resulted from interactions between PFOS and the MIP coating. Unlike the results obtained from Karimian et al. [30], in which the observed signal was inversely proportional to the concentration of PFOS (differential pulse voltammetry [DPV]), the sensor developed by Tran et al. [32] showed a direct correlation between PFOS concentration and the photocurrent observed from different sensing mechanisms (photocurrent response). Though efficient, this as-prepared sensor has a limit of detection (LOD) of 86 ng/mL, which is higher than concentrations typically found in the natural environment.

The two significant limitations of electrochemical affinity sensors are the transduction step and low sensitivity. The transduction step requires the target analyte to reach the recognition element, which results in long detection times (hours). In addition, to improve an often-compromised transducer signal, bulky and expensive instrumentation is typically required. Cheng et al. [33] addressed the need for an ultrasensitive detection technique suited for first-response devices by embedding metal–organic framework (MOF) capture probes on a microfluidic platform between interdigitated microelectrodes to increase the sensitivity of the observed impedance change. The synergistic effect provided by the mesoporous probes and the microelectrodes ensured penetration of the electric field across the entire platform, allowing for interaction with the PFOS at a molecular level and capturing minute changes in interfacial charge transport at any position within the channel. Furthermore, the authors found that this approach led to a significant increase in the signal-to-noise ratio. Other researchers have recently used MOFs for various sensing and ion extraction purposes [34–37]. Ranaweera et al. [38] reported a bubble nucleation–based electrochemical method that detects concentrations of PFAS based on their high surface activity [38]. This new solution removes the complications associated with the low electroactivity of PFAS. The technique consists of applying a sub-50-nm Pt nanoelectrode to an acidic solution, which causes hydrogen evolution reactions (HERs), resulting in a measurable current upon negatively scanning the nanoelectrode potential until it reaches a peak value (ipeak). The sudden drop in the HER current past the peak corresponds to the formation of an H_2 gas bubble at the electrode, thereby blocking the electrode surface. The presence of PFAS in the solution reduces the surface tension of the gas–liquid interface, thus reducing the nucleation barrier due to H_2 bubble nuclei stabilization. Therefore, an electrochemical transducer reports changes in the surface tension of the gas–liquid interface, which is proportional to the PFAS concentration due to their effect on the stabilization of gas nuclei. Potentiometric detection of fluorosurfactants has also been recently demonstrated. For instance, Fang et al. [39] utilized the MIP technique and pencil lead as an electrode material to detect concentrations of PFOA, PFOS, and 1,1,2H,2H-perfluorooctanesulfonic acid (6:2FTS) in the range of 10 μM−10 mM. It was discovered during experimentation that the selectivity of the PFOA-MIP for PFOA was higher than others due to its small recognition site. Another successful demonstration of potentiometric detection was performed by Chen et al. [40], who utilized ion-selective electrodes (ISEs) with fluorous anion liquid exchange membranes (LIX) to detect perfluorooctanoate (PFO^-) and PFOS with a low LOD of 0.07 μg/L. However, when these ISEs were applied to a native New Jersey lake, it was discovered that the presence of other perfluorinated anions that differed only in their number of carbon atoms hindered the selectivity of this method. This finding suggests while these electrochemical sensing technologies offer low LODs for their respective target analytes, their sensing is often limited to only one analyte or may have interference with other similar molecules and can therefore primarily be used as prescreening tools (i.e., first-response devices) rather than for multianalyte detection in water bodies.

3.2 Optical and fluorescence sensors

Fluorescence quantification of PFOS has served as another successful means of PFAS detection. Table 10.2 shows a summary of the various optical and fluorescence sensor for PFAS detection. Proposed by Feng et al. [41], an MIP fluorescence sensor in the form of MIP-capped silicon dioxide (SiO_2) nanoparticles anchored with a fluorescent dye was able to detect PFOS concentrations in water as low as 5.57 μg/L. Upon the binding of PFOS to the recognition sites of this sensor, fluorescence quenching occurred due to the electron transfer between the dye (fluorescein 6-isothiocyanate [FITC]) and PFOS, thereby reducing the fluorescence emission of the dye, which was easily measured using a fluorescence spectrophotometer. A drawback to this method is that PFOS binding on the surface only occurs under extremely acidic conditions (pH 3.5), increasing operational costs and potential environmental hazards. Additionally, the basic amine groups on the surface of the sensor tend toward protonation in acidic environments reducing the selectivity in solutions that contain various perfluoroalkyl homologs [41]. Like fluorescence sensors, optical sensing techniques often utilize organic dyes, though they are less reliant on analytical devices, such as fluorescence spectrophotometers, as detection can usually be performed with the naked eye. Optical sensors can rapidly and efficiently detect anions; therefore, they have recently been considered as one of the most practical detecting methods for PFAS. Generally, PFOA and PFOS are anionic surfactants that can interact with a cationic dye (e.g., methylene blue or ethyl violet) to form an ion pair [42]. Based on this principle, a study by Cennamo et al. [43] presented a D-shaped plastic optical fiber (POF) platform based on a specific MIP, which demonstrated an LOD of 0.5 ppb for PFAS. Various optical sensors (e.g., polymer-coated gold nanoparticle [AuNP] and self-assembled monolayer−AuNP sensors) were also reported for PFOA detection based on the colorimetric

TABLE 10.2 Summary of the various optical and fluorescence sensors for per- and polyfluoroalkyl substances detection.

Matrix	Detector	Working range	LOD	Note	References
D-shaped POF a	Optical density	0–200 ppb	0.21 ppb	D-shaped POF was characterized using a straightforward and low-cost experimental setup based on an LED and two photodetectors	[43]
Polymer-AuNP b	Naked eye	–	100 ppm	PFOA detached polystyrene from AuNP surface	[44]
SAM-AuNP	Naked eye	10–1000 ppb	10 ppb	The colorimetric assay for the detection of PFCs, but the long chain of PFCs (>7) is discerned	[45]
QD c-bioassay	Fluorescence	2.7–7.5 ppt	2.5 ppt	Bioassay based on PFOS binding to PPARα	[49]
Bio-AuNP	Optical density	50 ppt–500 ppb	5 ppt	Bioassay based on the silver enhancement of AuNP and interaction among ligands, PPARα, and PPRE	[50]
MIP-C3N4	Electrochemiluminescence	0.02–400 ppb	0.01 ppb	PFOA is efficiently oxidized by the electrogenerated (SO^{4-}); thus, this sensor is highly sensitive to PFOA	[47]
MPA d-QD	Fluorescence	200–16,000 ppb	120 ppb	PFOA strongly quenched the fluorescence emission of the MPA-CdS QDs	[48]
App-based	Smartphone camera	10–1000 ppb	0.5 ppb	PFOA sensing requires SPE e pretreatment of samples	[42]
SPR-POF-MIP	Optical density	–	<1 ppb	PFBS sensing	[51]

reaction, where LODs of 100 ppm and 10 ppb were achieved, respectively [44,45]. Nanoparticle-based sensors used for PFAS identification have emerged due to their unique behaviors (aggregation, disaggregation, adsorption, and desorption), small size (1–100 nm), and outstanding sensitivity in the form of portable, cheap, and reliable devices [46]. Another type of detection method for PFAS involves a combined carbon nitride (C_3N_4) nanosheet with an MIP, which demonstrated a higher PFOA sensing sensitivity and a low LOD (10 ppt) using the electrochemiluminescence method [47]. Quantum dots (QDs) have also been developed to detect PFOA with an LOD of 120 ppb when fluorescent detection was employed [48]; however, the illumination detection using a photomultiplier tube limited the on-site application of this method. Optical sensing methods still lack certain characteristics (such as not portable) and have limitations (like selectivity), thus necessitating the need for further research and development to increase their selectivity and sensitivity for PFAS detection such that they compare to laboratory-based results.

4. Biosensors

A biosensor is an analytical device that converts the biochemical signals (sensed by a biological element, which serves as a receptor) into a measurable electrical signal by a physicochemical transducer converter. A general schematic diagram for the working principle of biosensors has been shown in Fig. 10.3. The major characteristics of biosensors are high sensitivity, short response time, relatively small size, specificity, simplicity of operation, and relatively low cost [52–54].

An electrochemical biosensor that detects PFOS based on the biocatalysis process of an enzymatic biofuel cell (BFC) was reported by Zhang et al. [56].

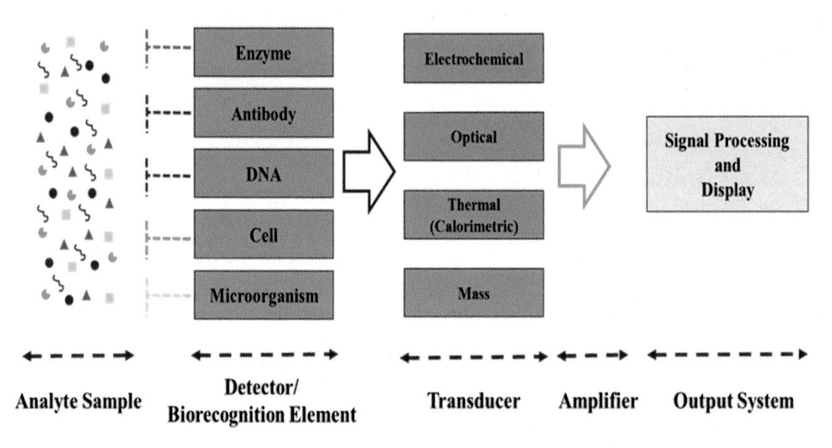

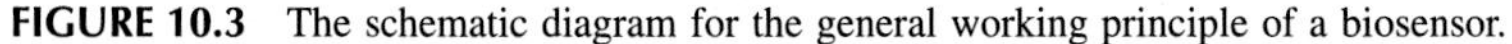
FIGURE 10.3 The schematic diagram for the general working principle of a biosensor.

They developed the one-compartment BFC comprising multiwalled carbon nanohorn–modified glassy carbon electrodes (GCEs), which were used for both the bio-anode and bio-cathode substrate, and glutamic dehydrogenase and bilirubin oxidase, which were used as the bio-anode and bio-cathode biocatalysts, respectively. Within this design, the presence of PFOS affects the bioactivity of the biocatalysts at both the bio-anode and bio-cathode, resulting in a decrease in the open-circuit voltage of the BFC. The biosensor showed an acceptable correlation of $R^2 = 0.976$ between PFOS and the reduction in voltage. In addition, this study incorporated the potential effects on PFOS detection of four perfluorinated chemicals (PFOA, non-afluorobutanesulfonic acid potassium, perfluorooctanesulfonamide, and heptadecafluorononanoic acid) with similar structures to PFOS and two types of chemicals (SMNBS and SDS) that may co-exist in micropolluted environments. It was shown that the electrochemical biosensor exhibited acceptable selectivity (relative standard deviation from 3.6% to 7.7%) for PFOS, even in real water samples obtained from a local river and reservoir in Dalian, China. Cennamo et al. [55] took a different approach for biosensors by developing a configuration that included a platform functionalized with a bioreceptor. The proposed biosensor was characterized by a surface plasmon resonance (SPR) platform based on POFs, together with a bioreceptor for the detection of PFOA and PFOS. In this bioassay method, PFOA compounds are covalently attached to an immunological protein carrier (bovine serum albumin [BSA]) with high affinity and selectivity. By increasing the PFOA concentration and thus the produced antibodies, a decrease in the refractive index value of the receptor layer was observed. The assay's LOD was less than 0.21 ppb (in seawater). The study expanded their investigation by studying the interactions between the produced antibodies and PFOS; it was found that PFOS exhibited a similar response to PFOA such that the antibodies could also be used to monitor PFOS molecules [55]. Other studies have demonstrated similar results with the bioassay method in which PFOS binds to a specific enzyme (e.g., peroxisomal proliferator-activated receptor alpha [PPAR-α]), where a detection range of 2.5–7.5 ppt at a wavelength of 605 nm was achieved [49,50].

5. Disinfection by-products

Water is treated with various disinfectants to protect the public from bacteria and viruses. These disinfectants react with naturally occurring organic and inorganic matter present in water and form DBPs. Disinfectants like chlorine, chloramines, ozone, and chlorine dioxide produce DBPs like chlorate, chlorite, bromate, trihalomethanes (THMs), and haloacetic acids (HAAs) in drinking water [56]. Different DBPs are generated depending on the type of disinfectant and the type of inorganic and organic matter present in the water, the water temperature and pH, and the concentration of precursors and disinfectants.

To date, more than 600 DBPs are identified in drinking water [57]. Researchers are detecting many new DBPs (like idoacid DBPs, halo phenols, and polar halogenated DBPs), which possess higher toxicity [58–60]. Some of the DBPs are carcinogenic and affect health adversely [61–63]. The consumption of DBPs like THM4 may cause bladder cancer [64], gestational age in childbirth [65,66], and miscarriages [67]. The removal of precursors before disinfection can prevent the formation of DBPs. This can be achieved using enhanced coagulation, powdered activated carbon, and granulated activated carbon [68]. Moreover, the DBPs can be removed using filtration techniques (microfiltration and ultrafiltration) and biological activated carbon [69,70].

5.1 AI-assisted techniques for disinfection by-products removal

Regarding the diversity of the processes for water treatment and the DBPs' precursor variety, there are complex relationships between the precursors and DBPs formation. Therefore, AI-assisted procedures can considerably discover such complexities and ambiguities to help avoid DBPs formation, removal, and other aspects of these pollutants [13]. AI-assisted techniques have been applied for DBPs studies in terms of their formation analysis and modeling [71–73], the risk assessment of DBPs [71], internal connection between the DBPs' precursors [72], and the identification of 15 main DBPs [72]. According to the outcomes of the principal component analysis (PCA) procedure, there is a consistent relationship between the proportions of the organic substances in water and the generation potential of DBPs during the disinfection processes [72]. PCA [71–75], artificial neural network (ANN) [73,76–78], kernel partial least squares regression [79], SVM [77,79], gene expression programming (GEP) [77], linear regression [78,80], supervised principal component regression [81], boosted regression tree [81], and autoencoder [75] have ever been applied for DBP studies in water, among which the ANN and PCA were the most widely used procedures.

THMs, as very harmful compounds with potential carcinogenic effects, are considered as one of the essential DBPs generated in chlorinated waters containing organic matter. Singh and Gupta [77] applied three supervised ML procedures, i.e., ANN, SVM, and GEP, to model and predict the generation of THMs in waster owing to chlorination. Contact time, bromide concentration, temperature, water pH, and Cl_2/dissolved organic carbon (DOC) were regarded as independent variables. SVM model demonstrated better performance than ANN and GEP with R^2 and root mean square errors proportions of (0.998 and 0.7); (0.998 and 0.09); and (0.990 and 3.07) correspondingly. Based on the sensitivity analysis, initial pH, contact time, and temperature were recognized as the essential variables in THMs formation (Fig. 10.4) [77]. In 2017, an MLR model was used to model and predict the chloroform yields in chlorinated water containing organic matters. In 19 descriptors used, the developed MLR model could predict chloroform

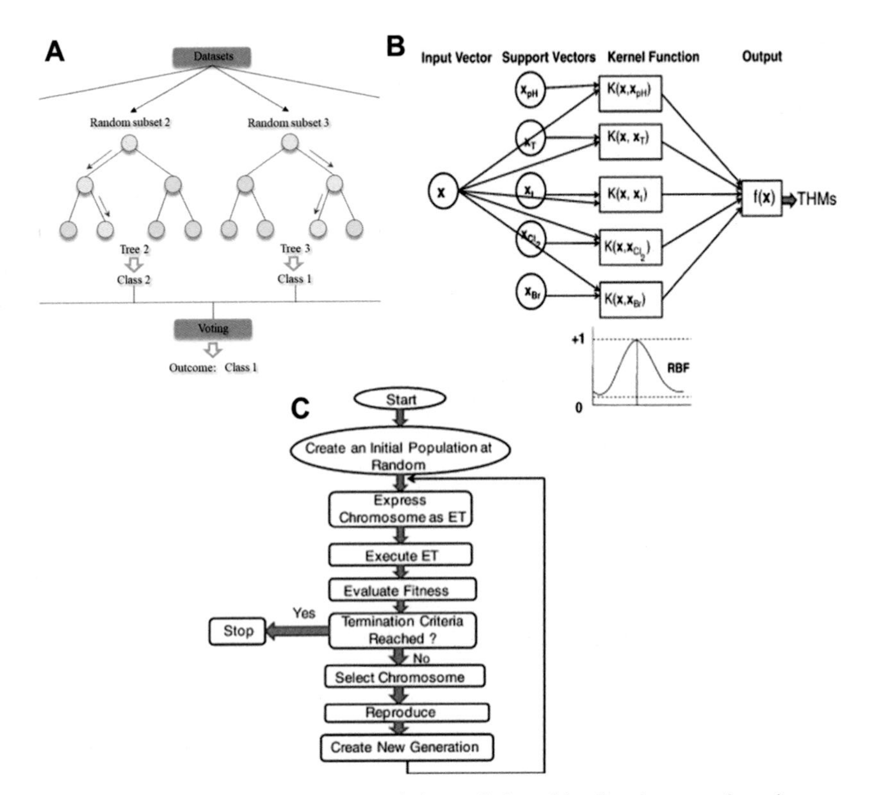

FIGURE 10.4 The example structures of the applied machine learning procedures in water treatment process. (A) Random forest (RF), (B) support vector machine (SVM), and (C) gene expression programming (GEP).

formation as the response variable with standard error and R^2 of 8.83% and 0.91, respectively [82]. Then, Cordero et al. [83] used chemical predictors of DBPs precursors as attributes in SVM, RF, and ANN to predict the potential formation of HAAs, i.e., trichloroacetic acid (TCAA) and dichloroacetic acid (DCAA). The developed RF was introduced as the most accurate model to forecast the TCAA with R^2 of 0.58 and $Q^2 > 0.50$; however, the model for DCAA was less accurate with R^2 of 0.41 and $Q^2 > 0.35$. The most significant predictors recognized for the potential formation of TCAA were electron topological descriptors associated with atomic electronegativity distribution and electrostatic interaction, along with hydrophilicity and the aromatic bond proportions [83]. Hong et al. [84] applied two different procedures, i.e., a hybrid gray relational analysis (GRA)—radial basis function artificial neural network (RBF ANN) along with a single RBF ANN to predict THMs level in a natural water distribution system. In their study, ammonia, nitrite, residual free chlorine, bromide, DOC, UVA_{254}, pH, and temperature were considered as inputs of the models. RBF ANN procedure was recognized as a robust model with more accuracy than the linear and log-linear regression models. In addition, RBF ANNs applying fewer inputs based on GRA could demonstrate outstanding strength in output forecast.

5.2 Sensors for detection of DBPs

Electrochemical biosensors have gained noticeable research interest in environmental monitoring due to their rapid response, excellent sensitivity, affordability, and potential portability [85–87]. The electrochemical biosensing of DBPs in drinking water generally depends on three main aspects: (1) recognition molecules, (2) electrodes and their modifications, and (3) signal readout techniques. Superior sensitivity and specificity of electrochemical sensors are achieved with the help of well-covered recognition molecules. These recognition molecules are considered as bridges for electron transfer between electrodes and target molecules. Some of the recognition molecules used for the determination of various DBPs are redox proteins such as hemoglobin (Hb) [88–102] and myoglobin (Mb) [103–115], metal complexes including porphyrin [116], enzymes like horseradish peroxidase [117,118], and phthalocyanine [119] and MIPs [120,121]. The electrodes for electrochemical biosensing of DBPs can be classified into two main groups: metal and carbon electrodes. Only a few studies reported electrochemical detection of DBPs using metal electrodes, which usually required an electrode assembly without using either signal amplification or recognition molecules. As compared with metal electrodes, carbon materials (e.g., graphite) are more cost-effective and suitable for mass production due to their ubiquitous source. Therefore, they are often preferred as electrode materials and received far more attention in the electrochemical detection of DBPs. The electrochemical detection of DBP in water was mainly carried out using current-related monitoring techniques, such as amperometry [102], cyclic voltammetry (CV) [114], DPV [93], and square wave voltammetry [101]. Ceto et al. [121] fabricated a selective MIP as a recognition element, entrapped it into electrosynthesized polypyrrole (PPy) matrix and used it for detection based on electrochemical impedance spectroscopy results. The MIP sensor demonstrated the rapid, selective, and reliable detection of N-nitrosodimethylamine (NDMA) in water samples. Finally, this developed sensor showed a liner detection range from 10 to 230 g/L and an LOD of 0.85 g/L toward NDMA. In another study, a highly sensitive sensor for TCAA detection based on iron(II) phthalocyanine (PcFe) and a Zn-based MOF (ZIF-8) composite was fabricated by Zeng et al. [122]. They found that the TCAA oxidized the reduction product of PcFe and accelerated the electron transfer on the electrode interface, thus augmenting the reduction peak current of PcFe. The authors also deduced that using ZIF-8 as the support may help minimize the agglomeration and loss of PcFe from the electrode. Fang et al. [123] focused on detecting highly genotoxic and cytotoxic trichloroacetamide (TCAM) in drinking water.

They reported a novel electrochemical sensor based on the triangular silver nanoprism (NPR) and MoS_2 nanosheet (AgNPR@MoS_2) heterostructure. The 2D plate-like AgNPR triangle catalyzed the dechlorination reaction for TCAM detection, and further, the heterostructure of AgNPR@MoS_2 promoted the charge transfer rate, the electrocatalytic capacity, and the stability of the electrode. This sensor showed two linear sensing ranges of 0.5−10 μM and 10−80 μM, with a detection limit of 0.17 μM. The combination of electrochemical sensors and chemometrics for the analysis of HAAs in water samples was explored by Ceto et al. [124]. The voltammetric measurements on a bare gold electrode were combined with chemometric processing for fingerprints extraction of different HAAs. Lastly, a fast Fourier transform was used to preprocess the cyclic and square wave voltammograms to provide the coefficients used as inputs for an ANN model.

5.3 Heavy metals

The term heavy metal generally refers to any metallic elements with an atomic number greater than 23 and density higher than 5 g/cm^3, such as mercury (Hg), cadmium (Cd), arsenic (As), chromium (Cr), thallium (TL), and lead (Pb) [125,126]. These elements are toxic or poisonous at low concentrations due to their ability to accumulate over time in plants, animals, and water [127]. Heavy metals toxicity is challenging since they are soluble in water and easily absorbed by living organisms [128]. Recommended limits for heavy metals and other elements in drinking water based on USEPA and European Union (EU) guidelines were tabulated in Table 10.3. Heavy metals mobility in water is mainly influenced by the pH of water, presence and concentration of carbonates and phosphates, the presence of hydrated forms of manganese (Mn) and iron (Fe), and the content of organic matter [129,130]. Consequently, the growing concern for sustainability and environmental preservation leads to the development of different detection methods such as spectroscopic, voltammetric, potentiometric, impedimetric, and biosensing techniques to monitor the heavy metals presence and concentration in the environment [131−135]. Recently, sensors have been used for the detection of heavy metals due to the advantages such as low cost and simplicity of operation compared to the conventional analytical methods, including cold vapor atomic absorption spectrometry and inductively coupled plasma mass spectrometry [136]. Different types of sensors such as biosensors, thermal sensors, and electrochemical sensors have been used for detection and monitoring of different types of heavy metals.

6. AI-assisted techniques for removal of heavy metal

ML for heavy metals has gained massive attention over the last 2 years (Fig. 10.5) [139]. It is essential to gain insights into different parameters

TABLE 10.3 Recommended limits of heavy metals and other elements in drinking water based on Environmental Protection Agency (EPA) and European Union (EU) guidelines [137,138].

Parameter	Unit	EPA limit	EU limit
Antimony	ppb	6	5
Arsenic	ppb	10	10
Cadmium	ppb	5	5
Chromium	ppb	100	25
Copper	ppb	1.3	2
Lead	ppb	15	5
Mercury	ppb	2	1
Nickel	ppb	100	20
Selenium	ppb	50	10
Uranium	ppb	30	30

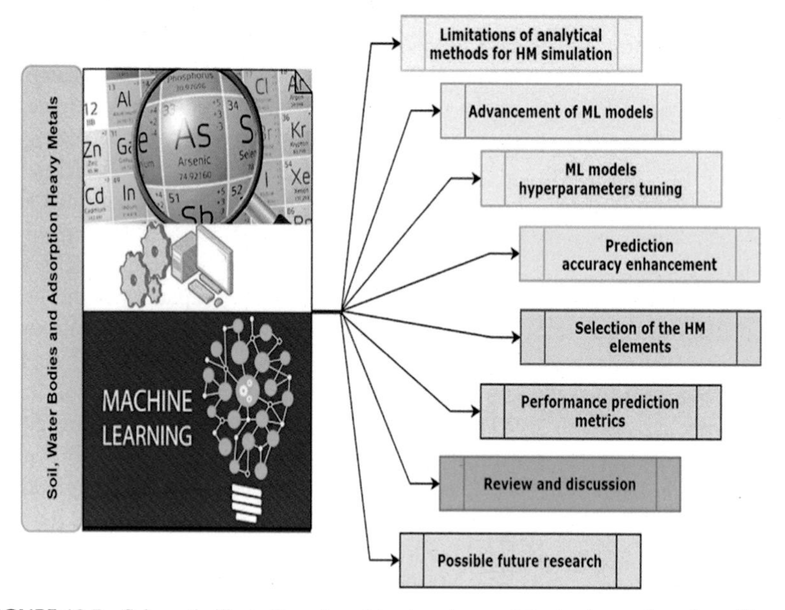

FIGURE 10.5 Schematic illustration of machine learning models simulates adsorption of heavy metals. *Reprinted with permission from Yaseen © 2021 Elsevier Ltd.*

(adsorption process, initial concentration, etc.) to improve removal efficiency when dealing with wastewater containing heavy metals. ML is one of the best tools to model and learn adsorption behavior in heavy metals [140]. Herein, we summarized the recent literature in this area.

Zhu et al. [141] investigated two different models of ML, including ANN and RF, to forecast the adsorption capacity of biochar for heavy metals (lead, cadmium, nickel, arsenic, copper, and zinc) based on environmental conditions, metal sources, and adsorbent characteristics (surface area and pore size). The RF model exhibited better accuracy ($R^2 = 0.973$), predictive performance, and generalization ability for adsorption efficiency than the ANN model. The biochar characteristics that contribute to the adsorption effectiveness were cation exchange capacity and pH_{H2O} of biochars, which accounted for almost 66% of the biochar characteristics. The surface area of the biochars delivered only 2% of adsorption efficiency. Collectively, the authors signified the benefit of the established model used, which could significantly decrease experiment capacity [141]. El Hanandeh et al. [142] formulated a multi-input multi-output (MIMO) model for the adsorption process of three heavy metals, including copper, nickel, and lead, on biochar. The authors accounted for the mutual interactions of crucial process parameters on the adsorption capacity in multisolute systems. Investigated ML models such as ANN models, radial basis, and gradient boosting algorithms to model the MIMO process provided highly accurate predictions ($R^2 > 0.99$). The developed generalized regression network delivered the best match to the experimental data while allowing operators to forecast the adsorption system response to changes in the operations to access a tool for process optimization [142]. Wanyonyi et al. [143] used ML and molecular simulation techniques to investigate the adsorption of hazardous heavy metal ions on 242 zeolites, emphasizing topology, pores, and chemical characteristics of zeolites adsorption efficiency. The simulation findings showed that only the zeolites with large pore sizes (16.45 Å) could adsorb the cations significantly. Boruta algorithm, built around RF classier, was used to expose the impact of parameters like total energy, pore size, and average volume on the loading capacity of the heavy metals. Boruta algorithm revealed that total energy plays a crucial role in the adsorption of heavy metals. It was also observed that the pore size of the best performing zeolite frameworks is accountable for the adsorption of larger size heavy metals (Pb^{2+}) [143]. Ke et al. [144] studied the use of 20 AI models to forecast the feasibility and efficiency of biochar for removal of heavy metals from wastewater and water, according to five ML algorithms (SVM, RF, ANN, M5Tree, and Gaussian process [GP] algorithms). The individual models were bagged with each other to generate 20 intelligent models, including SVM, RF, M5Tree, GP, ANN, BA-SVM, BARF, BA-M5Tree, BA-GP, BA-ANN, SVM-RF, GP-ANN, SVM-M5Tree, SVM-GP, SVM-ANN, RF-ANN, RF-M5Tree, RF-GP, M5Tree-GP, and M5Tree-ANN, and used to evaluate the impact of biochar characteristics (surface area and pore size), metal sources, and environmental conditions such as temperature and pH on biochar adsorption performance. A dataset of adsorption efficiency of heavy metal was collected and processed with 353 experimental tests and different performance indexes such as color intensity applied to evaluate the models. It was discovered that AI models could predict the adsorption efficiency of heavy metals onto biochar with a high degree ($R^2 > 0.813$) of accuracy. The efficiency of joint models is higher than that of individual models, with the SVM-ANN model being the most effective [144]. Zhao and co-workers [145] proposed Kriging models and Kernel extreme learning machine (KELM) to forecast heavy metals (ions and single ions) adsorption efficiency on biochar. Data collection and output value fixation tried to advance the precision of model fitting with the best R^2 value of 0.919 and 0.980 for KELM and Kriging, respectively. Conducted analysis displayed a strong relationship between adsorption efficiency and pH solute, and KELM and Kriging delivered some helpful predictive frameworks in selecting suitable biochar for particular treatment scenarios, which diminished the amount of required adsorption of metal-biochar experiments [145].

6.1 Sensors for detection of heavy metals

Electrochemical biosensors, as the first proposed and commercialized biosensors, are preferable due to their enhanced sensitivity, fast response, simpler approach, low power cost, and ease of adaptability for the in-situ multielement analysis of heavy metals. The main disadvantages of these kinds of sensors are poor reproducibility, poor selectivity, and stability [127,146,147]. Recently, various types of nanomaterials, including nanoparticles due to their stability and carrier capacity, nanorods due to their capability of detection sensitivity, nanowires, carbon nanotubes due to their enhanced surface area, electrical and thermal conductivity, and QDs due to their color tunability, have been widely explored and employed for the fabrication of electrochemical biosensors [148–152]. Nanoparticles in electrochemical-based biosensors are typically utilized as signal amplifiers to enhance the sensitivity and obtain lower detection limits [153,154]. He et al. [154] developed a novel electrochemical biosensor for Hg^{2+} detection and used three signal amplification strategies, including a DNA dual cycle, organic-inorganic hybrid nanoflowers ($Cu_3(PO_4)_2$ HNFs), and a gold nanoparticle (AuNP) probe to improve the performance of the sensor. To improve the sensitivity of the sensor, the authors introduced nicking endonuclease and exonuclease III (Exo III) for target recycling to realize signal circle amplification. The authors noted that the fabricated electrochemical biosensor showed enhanced sensitivity and stability. The biosensor showed a great linear response range from 1 fM to 10 nM with an LOD of 0.19 fM (S/N = 3), which was approximately eight orders of magnitude more sensitive than the allowed Hg^{2+} levels in drinking water. They also noted that the introduced signal amplification strategies could be easily adjusted and extended to identify other hazardous heavy metals and nucleic acids [154]. In another study, Gupta et al. [155] employed carbon nanotubes microelectrodes with parts per trillion (ppt) LOD of heavy metals in drinking water. They noted carbon nanotubes served as an ultrasensitive platform for the detection of heavy metals, especially lead. They fabricated microsensor with excellent LOD for individual Cu^{2+} (6.0 nM (376 ppt)), Pb^{2+} (0.45 nM (92 ppt)), and Cd^{2+} (0.24 nM (27 ppt)) in tap water and 0.32 nM (20 ppt), 0.26 nM (55 ppt), and 0.25 nM (28 ppt) in simulated drinking water, respectively. They also mentioned that the microelectrode detected Pb^{2+} ions well below the USEPA limits in a broad range of water quality conditions reported for temperature (5–45 and 55°C) and conductivity (55–600 μS/cm) [155].

Optical biosensors have also been used for the detection of heavy metals. Noor Ul Amin et al. [156] developed a novel method by using 3 μg/mLBSA as a probe and liquid crystal as a transducer (signal reporter) for Cu^{2+} and Pb^{2+} ions detection in water at neutral pH. In the developed biosensor, the BSA served as a multidentate ligand for multiple metal ion detection. Based on their results, a dark optical image was observed in the absence of heavy metal ions and a bright optical image in their presence. For the developed biosensor, the LOD was approximately 1 nM. They also noted that the developed sensor was more sensitive to copper ions detection than other heavy metal ions [156]. In another study, two new ratiometric fluorescent nanosensors with dual- and three-emission carbon dot nanohybrids were developed for the detection of Pb^{2+} and Hg^{2+} in water, respectively. The authors reported the low LOD for the Pb^{2+} (~0.14 nM) and Hg^{2+} (~0.22 nM) ions, which were far lower than the recommended limits (Pb^{2+}, 48 nM; Hg^{2+}, 5 nM) for Pb^{2+} and Hg^{2+} in drinking water [157].

Mass-based biosensors or gravimetric biosensors are mainly categorized into surface acoustic wave biosensors, microcantilever-based biosensors, and piezoelectric sensors, which are more commonly used for heavy metal detection among mass-based sensors. The changes in pressure, temperature, acceleration, strain, or force could be measured and converted to an electrical signal by piezoelectric sensors [158]. Quartz crystals, among the various materials with a piezoelectric effect, have been commonly utilized in piezoelectric biosensors due to the advantages like availability, high-temperature resistance, and chemical stability in aqueous solution [127,147]. Ballen et al. [159] developed a novel cantilever nanosensor and nanobiosensor based on receptor molecules including urease enzyme, graphene oxide (GO), and hybrid GO/urease for detection of cadmium in the ppt range. The authors noted that the device developed with complex receptor molecules exhibited high sensitivity (14.7467 nm/ppb), low detection limits (from 0.01831 ppb [GO/urease nanobiosensor] to 0.03776 ppb [urease nanobiosensor]), and higher quantification (from 0.06103 ppb [GO/urease nanobiosensor] to 0.12586 ppb [urease nanobiosensor]) in the ppt range than the device developed with the simple receptor molecules [159]. In another study, an ultrasensitive cantilever nanobiosensor based on GO and urease was used for cadmium detection in river water. The authors reported high sensitivity (0.0147 nm/ppt) and detection limit (18 ppt) for the developed device, which was functionalized with GO and urease [160].

6.2 Antibiotics, endocrine-disrupting chemicals/pharmaceuticals

The emerging water contaminants such as pharmaceuticals and endocrine-disrupting chemicals (EDCs) have been detected in the surface water, groundwater, domestic and municipal, and industrial effluents of most countries [6,161]. The universal consumption and improper disposal of these contaminants lead to the presence of these compounds in drinking water sources and water and wastewater treatment plants [162,163]. Globally, about 100,000 tons of different pharmaceutical drugs have been used by humans every year [164]. EDCs are a concern due to their potential threats to aquatic ecosystems and their harmful effects on male and female reproductive health, metabolic disorders, thyroid homeostasis, and increase risk of hormone-sensitive cancers. Most conventional water and wastewater treatment plants, including coagulation/flocculation, sedimentation, and sand filtration, are ineffective in removing trace amounts of EDCs compounds, and therefore improved detection technique will help their removal and waste management [6,165].

6.2.1 AI-assisted techniques for removal of antibiotics, endocrine-disrupting chemicals/pharmaceuticals

In terms of application of the AI-assisted procedures in antibiotics, EDCs, and pharmaceutics removal from aqueous solutions, there are only some limited studies that used supervised ANN procedure to model and predict the performance of various water and wastewater treatment processes [166]. The decomposition of three antibiotics, i.e., cloxacillin, ampicillin, and amoxicillin by Fenton process, was modeled using ANN. A network with a 5-14-1 structure (Fig. 10.6) showing 14 neurons in a hidden layer was developed and applied to predict Fenton process performance in the mentioned antibiotics with five various inputs, including reaction time, H_2O_2/COD, H_2O_2/Fe^{2+}, pH, and antibiotics concentration. The developed model showed considerable strength in the prediction of COD removal by the Fenton process with a determination coefficient and the mean square error (MSE) of 0.997 and 0.000376 consecutively. According to the sensitivity analysis, the H_2O_2/Fe^{2+} ratio demonstrated higher relative importance in the degradation process with 25.8% [167].

Andaluri et al. [168] utilized the ANN procedure to model the performance of the sonochemistry process in the elimination of seven different

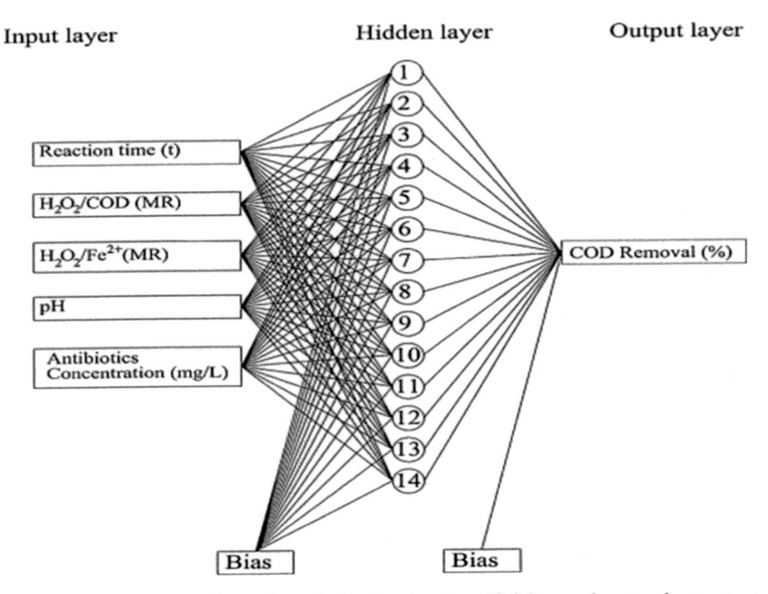

FIGURE 10.6 Schematic illustration of the developed artificial neural network structure to model the behavior of the Fenton process in antibiotics removal [167].

estrogen hormones from water, i.e., equilin, estrone, 17α-dihydroequilin, 17α-ethinylestradiol, estriol, 17β-estradiol, and 17α-estradiol. Reactor design factors, ultrasound amplitude, power intensity, and power density were regarded as the independent variables. The area-specific power density, together with the volume-specific power intensity, was identified as the most critical factor. Optimization of the process could improve fivefold the energy consumption efficiency in the mentioned process. Furthermore, 17α-ethinylestradiol (EE2) elimination from wastewaters by photocatalytic degradation process was modeled by ANN procedure with five inputs: conductivity, organic content, EE2 concentration, TiO_2, and reaction time. The developed model with a structure of 5-13-1 could mimic well with the real removal values of EE2 with MSE and R^2 of 7.77×10^{-4} and 0.994, respectively [169]. In addition, the ANN has been applied to model the performance of sequencing batch reactor-adsorbent process in the elimination of pharmaceutics from synthetic wastewater with two inputs of contact time and pharmaceutics concentrations and demonstrated the considerable performance of this AI-assisted procedure with $R^2 < 0.99$ [170]. Yu et al. [171] used a backpropagation ANN physical modeling procedure (BP-ANN-P) to model the performance of a hybrid photoelectrocatalytic process in the decomposition of Norfloxacin, which is an antibiotic in saline water. The developed model structure was 4-11-1, in which contact time, current density, initial pH, and Norfloxacin concentration were four inputs, and the only output was total organic carbon removal. The model predictions could mimic well with the actual results with 89.5% accuracy and −0.04159 to 0.03856 errors. It is observed that despite the considerable capabilities of the AI-assisted procedures in different aspects of the processes, only a narrow line of this field, which is the study of the relationships between dependent and independent variables, has been regarded, and the other aspects of this field should be more considered.

6.3 Sensors for detection of heavy metals antibiotics, endocrine-disrupting chemicals/pharmaceuticals

Generally, the HPLC, liquid chromatography coupled with mass spectrometry (LC-MS), liquid chromatography coupled with electrochemical detection (LC-ED), gas chromatography (GC), gas chromatography coupled with mass spectrometry (GC-MS), and capillary electrophoresis commonly are utilized for the determination of different EDCs and antibiotics. Although these approaches offer remarkable selectivity and detection limits, they are not appropriate for the rapid processing of multiple samples and real-time detection. These detection processes also are time-consuming and required highly skilled operators. Among the alternative methods for EDCs and antibiotics detection, different electrochemical sensors, optical sensors, and biosensors present advantages such as high sensitivity, selectivity, and simplicity [172,173].

Qian et al. [174] developed a selective and sensitive electrochemical sensor based on the five different GO-based nanomaterials including GO, thermally reduced graphene oxide (TrGO), boron-doped partially reduced graphene oxide (B-rGO), nitrogen-doped partially reduced graphene oxide (N-rGO), and fluorine-doped partially reduced graphene oxide (F-GO) for the detection of naproxen. The strongest activity toward the electrochemical oxidation of naproxen was achieved for GO/GCEs compared to GCE and other modified GO electrodes. The authors noted that the GO-based electrochemical sensor showed a wide linear range (10 μM^{-1} mM), high sensitivity (0.60 $\mu A/\mu M cm^2$), high selectivity (no notable response was detected to the interference/competitive species and the GO/GCE showed 91% of the oxidation signal compared to pure naproxen), and a strong anti-interference capacity over potential interfering species that may exist in a biological system for the detection of naproxen [174]. In another study, a GCE modified with branch-like poly(eriochrome black T) film (P(EBT)/GCE) was developed for the detection of chloramphenicol (CAP) in pharmaceutical samples. The branch-like P(EBT) film was deposited by electropolymerization using CV. The CV curve at the bare GCE showed that CAP produced only a cathodic peak at −0.72 V, while at the P(EBT)/GCE, peak of CAP was observed at a more positive region (−0.61 V). The observed cathodic peak attributed to the irreversible reduction of the nitro group of CAP. Additionally, the CV curve of CAP at the P(EBT)/GCE showed a couple of redox peaks located at −0.01 V and −0.0 9 V, which was attributed to the reversible redox of hydroxylamine group. The P(EBT)/GCE exhibited two linear ranges (0.01−0.10 and 0.10−4.0 μmol/L) with high sensitivity (5.2 μA/μM). The authors also noted 3 nmol/L for LOD and 11 nmol/L for quantitation [175]. Wang et al. [176] developed a paper-based antibiotic sensor via wax printing and surface modification for the detection of tetracycline (TET) and CAP. The authors noted that the low LOD (0.5 ng/mL and 0.05 ng/m) with wide linear detecting ranges (as 1 ~ 103 ng/mL and 0.1 ~ 100 ng/mL) was observed for the developed sensor for the detection of the TET and CAP. In another study, a novel 3D MOF as fluorescence probe was developed for selective detection of dimetridazole, metronidazole, and antibiotics as well as 2,6-dichloro-4-nitroaniline pesticide in water. The solvothermal synthesized sensor via MOF exhibited a strong fluorescence effect, undisturbed by the wide pH range (pH = 3−12) [176]. The MIPs were developed via high-affinity synthetic receptors coupled with an SPR sensor to detect glycopeptide antibiotics in milk samples. The nano MIP-SPR sensor facilitated vancomycin quantification with the LODs of 4.1 and 17.7 ng/mL using direct and competitive assays, respectively. The author also noted that the developed optical sensor enabled the detection of vancomycin in the range of 10−1000 ng/mL [177].

References

[1] F. Yang, C. Tang, M. Antonietti, Natural and artificial humic substances to manage minerals, ions, water, and soil microorganisms, Chem. Soc. Rev. (2021), https://doi.org/10.1039/D0CS01363C.

[2] E. Yakamercan, A. Aygün, Fate and removal of pentachlorophenol and diethylhexyl phthalate from textile industry wastewater by sequencing batch biofilm reactor: effects of hydraulic and solid retention times, J. Environ. Chem. Eng. 9 (4) (2021) 105436, https://doi.org/10.1016/j.jece.2021.105436, 2021/08/01.

[3] F. Ejeian, et al., Biosensors for wastewater monitoring: a review, Biosens. Bioelectron 118 (2018) 66−79, https://doi.org/10.1016/j.bios.2018.07.019.

[4] M.A. Shannon, P.W. Bohn, M. Elimelech, J.G. Georgiadis, B.J. Mariñas, A.M. Mayes, Science and technology for water purification in the coming decades, Nature 452 (7185) (2008) 301−310, https://doi.org/10.1038/nature06599, 2008/03/01.

[5] I.B. Gomes, J.-Y. Maillard, L.C. Simões, M. Simões, Emerging contaminants affect the microbiome of water systems—strategies for their mitigation, NPJ Clean Water 3 (1) (2020) 39, https://doi.org/10.1038/s41545-020-00086-y, 2020/09/18.

[6] S. Alipoori, et al., Polymer-based devices and remediation strategies for emerging contaminants in water, ACS Appl. Polym. Mater. 3 (2) (2021) 549−577, https://doi.org/10.1021/acsapm.0c01171.

[7] E.M. Sunderland, X.C. Hu, C. Dassuncao, A.K. Tokranov, C.C. Wagner, J.G. Allen, A review of the pathways of human exposure to poly- and perfluoroalkyl substances (PFASs) and present understanding of health effects, J. Expo. Sci. Environ. Epidemiol. 29 (2) (2019) 131−147, https://doi.org/10.1038/s41370-018-0094-1, 2019/03/01.

[8] S. Foorginezhad, et al., Recent advances in sensing and assessment of corrosion in sewage pipelines, Process Saf. Environ. Prot. 147 (2020) 192−213, https://doi.org/10.1016/j.psep.2020.09.009.

[9] P. Srivastava, R. Abbassi, A.K. Yadav, V. Garaniya, M. Asadnia, A review on the contribution of an electron in electroactive wetlands: electricity generation and enhanced wastewater treatment, Chemosphere (2020) 126926, https://doi.org/10.1016/j.chemosphere.2020.126926.

[10] D. Catelan, et al., Exposure to perfluoroalkyl substances and mortality for COVID-19: a spatial ecological analysis in the Veneto region (Italy), Int. J. Environ. Res. Publ. Health 18 (5) (2021) 2734, https://doi.org/10.3390/ijerph18052734.

[11] C. Santhosh, V. Velmurugan, G. Jacob, S.K. Jeong, A.N. Grace, A. Bhatnagar, Role of nanomaterials in water treatment applications: a review, Chem. Eng. J. 306 (2016) 1116–1137, https://doi.org/10.1016/j.cej.2016.08.053, 2016/12/15.

[12] M.R. Esfahani, et al., Nanocomposite membranes for water separation and purification: fabrication, modification, and applications, Separ. Purif. Technol. 213 (2019) 465–499.

[13] L. Li, S. Rong, R. Wang, S. Yu, Recent advances in artificial intelligence and machine learning for nonlinear relationship analysis and process control in drinking water treatment: a review, Chem. Eng. J. 405 (2021) 126673, https://doi.org/10.1016/j.cej.2020.126673, 2021/02/01.

[14] Organisation for Economic Cooperation and Development, Working towards a Global Emission Inventory of PFASs: Focus on PFCAs - Status Quo and the Way Forward, 2015.

[15] J.L. Domingo, M. Nadal, Human exposure to per- and polyfluoroalkyl substances (PFAS) through drinking water: a review of the recent scientific literature, Environ. Res. 177 (October 2019) 108648, https://doi.org/10.1016/j.envres.2019.108648.

[16] M. Haukas, U. Berger, H. Hop, B. Gulliksen, G.W. Gabrielsen, Bioaccumulation of per- and polyfluorinated alkyl substances (PFAS) in selected species from the Barents Sea food web, Environ. Pollut. 148 (1) (July 2007) 360–371, https://doi.org/10.1016/j.envpol.2006.09.021.

[17] S.F. Nakayama, et al., Worldwide trends in tracing poly- and perfluoroalkyl substances (PFAS) in the environment, Trac. Trends Anal. Chem. 121 (2019), https://doi.org/10.1016/j.trac.2019.02.011.

[18] A. Moller, et al., Distribution and sources of polyfluoroalkyl substances (PFAS) in the River Rhine watershed, Environ. Pollut. 158 (10) (October 2010) 3243–3250, https://doi.org/10.1016/j.envpol.2010.07.019.

[19] J.R. Lang, B.M. Allred, J.A. Field, J.W. Levis, M.A. Barlaz, National estimate of per- and polyfluoroalkyl substance (PFAS) release to U.S. Municipal Landfill Leachate, Environ. Sci. Technol. 51 (4) (February 21, 2017) 2197–2205, https://doi.org/10.1021/acs.est.6b05005.

[20] S.E. Fenton, et al., Per- and polyfluoroalkyl substance toxicity and human health review: current state of knowledge and strategies for informing future research, Environ. Toxicol. Chem. 40 (3) (March 2021) 606–630, https://doi.org/10.1002/etc.4890.

[21] B.C. Crone, et al., Occurrence of per- and polyfluoroalkyl substances (PFAS) in source water and their treatment in drinking water, Crit. Rev. Environ. Sci. Technol. 49 (24) (June 2019) 2359–2396, https://doi.org/10.1080/10643389.2019.1614848.

[22] A. Raza, et al., A machine learning approach for predicting defluorination of per- and polyfluoroalkyl substances (PFAS) for their efficient treatment and removal, Environ. Sci. Technol. Lett. 6 (10) (2019) 624–629, https://doi.org/10.1021/acs.estlett.9b00476, 2019/10/08.

[23] W. Cheng, C.A. Ng, Using machine learning to classify bioactivity for 3486 per- and polyfluoroalkyl substances (PFASs) from the OECD list, Environ. Sci. Technol. 53 (23) (2019) 13970–13980, https://doi.org/10.1021/acs.est.9b04833, 2019/12/03.

[24] T.C.G. Kibbey, R. Jabrzemski, D.M. O'Carroll, Supervised machine learning for source allocation of per- and polyfluoroalkyl substances (PFAS) in environmental samples, Chemosphere 252 (2020) 126593, https://doi.org/10.1016/j.chemosphere.2020.126593, 2020/08/01.

[25] T.C.G. Kibbey, R. Jabrzemski, D.M. O'Carroll, Source allocation of per- and polyfluoroalkyl substances (PFAS) with supervised machine learning: classification performance and the role of feature selection in an expanded dataset, Chemosphere 275 (2021) 130124, https://doi.org/10.1016/j.chemosphere.2021.130124, 2021/07/01.

[26] A. Su, K. Rajan, A database framework for rapid screening of structure-function relationships in PFAS chemistry, Sci. Data 8 (1) (2021) 14, https://doi.org/10.1038/s41597-021-00798-x, 2021/01/18.

[27] M. Asadnia, et al., Mercury (II) selective sensors based on AlGaN/GaN transistors, Anal. Chim. Acta 943 (2016) 1–7, https://doi.org/10.1016/j.aca.2016.08.045.

[28] M. Asadnia, et al., Ca^{2+} detection utilising AlGaN/GaN transistors with ion-selective polymer membranes, Anal. Chim. Acta 987 (2017) 105–110, https://doi.org/10.1016/j.aca.2017.07.066.

[29] T.M. Sanders, et al., Description of ionophore-doped membranes with a blocked interface, Sens. Actuators B Chem. 250 (2017) 499–508, https://doi.org/10.1016/j.snb.2017.04.143.

[30] N. Karimian, A.M. Stortini, L.M. Moretto, C. Costantino, S. Bogialli, P. Ugo, Electrochemosensor for trace analysis of perfluorooctanesulfonate in water based on a molecularly imprinted poly(o-phenylenediamine) polymer, ACS Sens. 3 (7) (July 27, 2018) 1291–1298, https://doi.org/10.1021/acssensors.8b00154.

[31] M. Mahmud, et al., Recent progress in sensing nitrate, nitrite, phosphate, and ammonium in aquatic environment, Chemosphere (2020) 127492, https://doi.org/10.1016/j.chemosphere.2020.127492.

[32] T. Tran, et al., Molecularly imprinted polymer modified TiO_2 nanotube arrays for photoelectrochemical determination of perfluorooctane sulfonate (PFOS), Sensor. Actuator. B Chem. 190 (2014) 745–751, https://doi.org/10.1016/j.snb.2013.09.048.

[33] Y.H. Cheng, et al., Metal-organic framework-based microfluidic impedance sensor platform for ultrasensitive detection of perfluorooctanesulfonate, ACS Appl. Mater. Interfaces 12 (9) (March 4, 2020) 10503–10514, https://doi.org/10.1021/acsami.9b22445.

[34] M. Abdollahzadeh, et al., Low humid transport of anions in layered double hydroxides membranes using polydopamine coating, J. Membr. Sci. 624 (2021) 118974, https://doi.org/10.1016/j.memsci.2020.118974.

[35] H. Ahmadi, et al., Incorporation of natural lithium-ion trappers into graphene oxide nanosheets, Adv. Mater. Technol. (2020) 2000665, https://doi.org/10.1002/admt.202000665.

[36] A. Razmjou, M. Asadnia, E. Hosseini, A.H. Korayem, V. Chen, Design principles of ion selective nanostructured membranes for the extraction of lithium ions, Nat. Commun. 10 (1) (2019) 1–15, https://doi.org/10.1038/s41467-019-13648-7.

[37] A. Razmjou, et al., Lithium ion-selective membrane with 2D subnanometer channels, Water Res. 159 (2019) 313–323, https://doi.org/10.1016/j.watres.2019.05.018.

[38] R. Ranaweera, C. Ghafari, L. Luo, Bubble-nucleation-based method for the selective and sensitive electrochemical detection of surfactants, Anal. Chem. 91 (12) (June 18, 2019) 7744–7748, https://doi.org/10.1021/acs.analchem.9b01060.

[39] C. Fang, Z. Chen, M. Megharaj, R. Naidu, Potentiometric detection of AFFFs based on MIP, Environ. Technol. Innov. 5 (2016) 52–59, https://doi.org/10.1016/j.eti.2015.12.003.

[40] L.D. Chen, et al., Fluorous membrane ion-selective electrodes for perfluorinated surfactants: trace-level detection and in situ monitoring of adsorption, Anal. Chem. 85 (15) (August 6, 2013) 7471–7477, https://doi.org/10.1021/ac401424j.

[41] H. Feng, N. Wang, T. TranT, L. Yuan, J. Li, Q. Cai, Surface molecular imprinting on dye–(NH_2)–SiO_2 NPs for specific recognition and direct fluorescent quantification of perfluorooctane sulfonate, Sensor. Actuator. B Chem. 195 (2014) 266–273, https://doi.org/10.1016/j.snb.2014.01.036.

[42] C. Fang, X. Zhang, Z. Dong, L. Wang, M. Megharaj, R. Naidu, Smartphone app-based/portable sensor for the detection of fluoro-surfactant PFOA, Chemosphere 191 (January 2018) 381–388, https://doi.org/10.1016/j.chemosphere.2017.10.057.

[43] N. Cennamo, et al., A simple and low-cost optical fiber intensity-based configuration for perfluorinated compounds in water solution, Sensors 18 (9) (September 8, 2018), https://doi.org/10.3390/s18093009.

[44] M. Takayose, K. Akamatsu, H. Nawafune, T. Murashima, J. Matsui, Colorimetric detection of perfluorooctanoic acid (PFOA) utilizing polystyrene-modified gold nanoparticles, Anal. Lett. 45 (18) (2012) 2856–2864, https://doi.org/10.1080/00032719.2012.696225.

[45] H. Niu, S. Wang, Z. Zhou, Y. Ma, X. Ma, Y. Cai, Sensitive colorimetric visualization of perfluorinated compounds using poly(ethylene glycol) and perfluorinated thiols modified gold nanoparticles, Anal. Chem. 86 (9) (May 6, 2014) 4170–4177, https://doi.org/10.1021/ac403406d.

[46] M. Al Amin, et al., Recent advances in the analysis of per- and polyfluoroalkyl substances (PFAS)—a review, Environ. Technol. Innov. 19 (2020), https://doi.org/10.1016/j.eti.2020.100879.

[47] S. Chen, A. Li, L. Zhang, J. Gong, Molecularly imprinted ultrathin graphitic carbon nitride nanosheets-Based electrochemiluminescence sensing probe for sensitive detection of perfluorooctanoic acid, Anal. Chim. Acta 896 (October 8, 2015) 68–77, https://doi.org/10.1016/j.aca.2015.09.022.

[48] Q. Liu, A. Huang, N. Wang, G. Zheng, L. Zhu, Rapid fluorometric determination of perfluorooctanoic acid by its quenching effect on the fluorescence of quantum dots, J. Lumin. 161 (2015) 374–381, https://doi.org/10.1016/j.jlumin.2015.01.045.

[49] W. Xia, et al., Sensitive bioassay for detection of PPARalpha potentially hazardous ligands with gold nanoparticle probe, J. Hazard Mater. 192 (3) (September 15, 2011) 1148–1154, https://doi.org/10.1016/j.jhazmat.2011.06.023.

[50] J. Zhang, et al., A rapid and high-throughput quantum dots bioassay for monitoring of perfluorooctane sulfonate in environmental water samples, Environ. Pollut. 159 (5) (May 2011) 1348–1353, https://doi.org/10.1016/j.envpol.2011.01.011.

[51] N. Cennamo, et al., Water monitoring in smart cities exploiting plastic optical fibers and molecularly imprinted polymers. The case of PFBS detection, in: Presented at the IEEE Int. Symp. Meas. Networking, 2019.

[52] V. Naresh, N. Lee, A review on biosensors and recent development of nanostructured materials-enabled biosensors, Sensors 21 (4) (2021) 1109, https://doi.org/10.3390/s21041109.

[53] A.M. Shrivastav, U. Cvelbar, I. Abdulhalim, A comprehensive review on plasmonic-based biosensors used in viral diagnostics, Commun. Biol. 4 (1) (2021) 1–12, https://doi.org/10.1038/s42003-020-01615-8.

[54] A. Hashem, M.M. Hossain, M. Al Mamun, K. Simarani, M.R. Johan, Nanomaterials based electrochemical nucleic acid biosensors for environmental monitoring: a review, Appl. Surf. Sci. Adv. 4 (2021) 100064, https://doi.org/10.1016/j.apsadv.2021.100064.

[55] N. Cennamo, et al., A High Sensitivity Biosensor to detect the presence of perfluorinated compounds in environment, Talanta 178 (February 1, 2018) 955–961, https://doi.org/10.1016/j.talanta.2017.10.034.

[56] S.D. Richardson, T.V. Caughran, A.D. Thruston, T.W. Collette, K.M. Schenck, B.W. Lykins, Identification of new drinking water disinfection by-products from ozone, chlorine dioxide, chloramine and chlorine, Water Air Soil Pollut. (1999), https://doi.org/10.1533/9780857090324.2.46.

[57] S.E. Hrudey, J.W.A. Charrois, Disinfection by Products and Human Health, IWA Publ., London, UK, 2012.

[58] M.J. Plewa, E.D. Wagner, S.D. Richardson, A.D. Thruston, Y.T. Woo, A.B. McKague, Chemical and biological characterization of newly discovered iodoacid drinking water disinfection byproducts, Environ. Sci. Technol. 38 (18) (2004) 4713–4722, https://doi.org/10.1021/es049971v.

[59] J. Han, X. Zhang, J. Liu, X. Zhu, T. Gong, Characterization of halogenated DBPs and identification of new DBPs trihalomethanols in chlorine dioxide treated drinking water with multiple extractions, J. Environ. Sci. 58 (August 2017) 83–92, https://doi.org/10.1016/j.jes.2017.04.026.

[60] R. Selvam, S. Muniraj, T. Duraisamy, V. Muthunarayanan, Identification of disinfection by-products (DBPs) halo phenols in drinking water, Appl. Water Sci. 8 (5) (2018), https://doi.org/10.1007/s13201-018-0771-1.

[61] X.F. Li, W.A. Mitch, Drinking water disinfection byproducts (DBPs) and human health effects: multidisciplinary challenges and opportunities, Environ. Sci. Technol. 52 (4) (February 20, 2018) 1681–1689, https://doi.org/10.1021/acs.est.7b05440.

[62] N.U. Benson, O.A. Akintokun, A.E. Adedapo, Disinfection byproducts in drinking water and evaluation of potential health risks of long-term exposure in Nigeria, J Environ. Public Health 2017 (2017) 7535797, https://doi.org/10.1155/2017/7535797.

[63] S.D. Richardson, M.J. Plewa, E.D. Wagner, R. Schoeny, D.M. Demarini, Occurrence, genotoxicity, and carcinogenicity of regulated and emerging disinfection by-products in drinking water: a review and roadmap for research, Mutat. Res. 636 (1–3) (Nov-Dec 2007) 178–242, https://doi.org/10.1016/j.mrrev.2007.09.001.

[64] N. Costet, et al., Water disinfection by-products and bladder cancer: is there a European specificity? A pooled and meta-analysis of European case-control studies, Occup. Environ. Med. 68 (5) (May 2011) 379–385, https://doi.org/10.1136/oem.2010.062703.

[65] J.M. Wright, A. Evans, J.A. Kaufman, Z. Rivera-Nunez, M.G. Narotsky, Disinfection by-product exposures and the risk of specific cardiac birth defects, Environ. Health Perspect. 125 (2) (February 2017) 269–277, https://doi.org/10.1289/EHP103.

[66] J. Grellier, et al., Exposure to disinfection by-products, fetal growth, and prematurity: a systematic review and meta-analysis, Epidemiology 21 (3) (May 2010) 300–313, https://doi.org/10.1097/EDE.0b013e3181d61ffd.

[67] K. Waller, S.H. Swan, G. DeLorenze, B. Hopkins, Trihalomethanes in drinking water and spontaneous abortion, Epidemiology 9 (1998) 134–140.

[68] I. Kristiana, C. Joll, A. Heitz, Powdered activated carbon coupled with enhanced coagulation for natural organic matter removal and disinfection by-product control: application in a Western Australian water treatment plant, Chemosphere 83 (5) (April 2011) 661–667, https://doi.org/10.1016/j.chemosphere.2011.02.017.

[69] M.A. Zazouli, L.R. Kalankesh, Removal of precursors and disinfection by-products (DBPs) by membrane filtration from water; a review, J Environ. Health Sci. Eng. 15 (2017) 25, https://doi.org/10.1186/s40201-017-0285-z.

[70] Q. Lin, F. Dong, Y. Miao, C. Li, W. Fei, Removal of disinfection by-products and their precursors during drinking water treatment processes, Water Environ. Res. 92 (5) (May 2020) 698–705, https://doi.org/10.1002/wer.1263.
[71] W. Wang, J. Yu, W. An, M. Yang, Occurrence and profiling of multiple nitrosamines in source water and drinking water of China, Sci. Total Environ. 551 (2016) 489–495.
[72] J. Fu, W.-N. Lee, C. Coleman, K. Nowack, J. Carter, C.-H. Huang, Removal of disinfection byproduct (DBP) precursors in water by two-stage biofiltration treatment, Water Res. 123 (2017) 224–235.
[73] P. Kulkarni, S. Chellam, Disinfection by-product formation following chlorination of drinking water: artificial neural network models and changes in speciation with treatment, Sci. Total Environ. 408 (19) (2010) 4202–4210.
[74] C.M. Villanueva, et al., Concentrations and correlations of disinfection by-products in municipal drinking water from an exposure assessment perspective, Environ. Res. 114 (2012) 1–11.
[75] N.M. Peleato, R.L. Legge, R.C. Andrews, Neural networks for dimensionality reduction of fluorescence spectra and prediction of drinking water disinfection by-products, Water Res. 136 (2018) 84–94.
[76] B. Legube, B. Parinet, K. Gelinet, F. Berne, J.-P. Croue, Modeling of bromate formation by ozonation of surface waters in drinking water treatment, Water Res. 38 (8) (2004) 2185–2195.
[77] K.P. Singh, S. Gupta, Artificial intelligence based modeling for predicting the disinfection by-products in water, Chemometr. Intell. Lab. Syst. 114 (2012) 122–131, https://doi.org/10.1016/j.chemolab.2012.03.014, 2012/05/15.
[78] J. Milot, M.J. Rodriguez, J.B. Sérodes, Contribution of neural networks for modeling trihalomethanes occurrence in drinking water, J. Water Resour. Plann. Manag. 128 (5) (2002) 370–376.
[79] S. Platikanov, J. Martín, R. Tauler, Linear and non-linear chemometric modeling of THM formation in Barcelona's water treatment plant, Sci. Total Environ. 432 (2012) 365–374.
[80] J. Milot, M.J. Rodriguez, J.-B. Sérodes, Modeling the susceptibility of drinking water utilities to form high concentrations of trihalomethanes, J. Environ. Manag. 60 (2) (2000) 155–171.
[81] B.F. Trueman, S.A. MacIsaac, A.K. Stoddart, G.A. Gagnon, Prediction of disinfection by-product formation in drinking water via fluorescence spectroscopy, Environ. Sci. Water Res. Technol. 2 (2) (2016) 383–389.
[82] T. Bond, N. Graham, Predicting chloroform production from organic precursors, Water Res. 124 (2017) 167–176, https://doi.org/10.1016/j.watres.2017.07.063, 2017/11/01.
[83] J.A. Cordero, K. He, K. Janya, S. Echigo, S. Itoh, Predicting formation of haloacetic acids by chlorination of organic compounds using machine-learning-assisted quantitative structure-activity relationships, J. Hazard Mater. 408 (2021) 124466, https://doi.org/10.1016/j.jhazmat.2020.124466, 2021/04/15.
[84] H. Hong, et al., Radial basis function artificial neural network (RBF ANN) as well as the hybrid method of RBF ANN and grey relational analysis able to well predict tri-halomethanes levels in tap water, J. Hydrol. 591 (2020) 125574, https://doi.org/10.1016/j.jhydrol.2020.125574, 2020/12/01.
[85] W. Zhang, B. Jia, H. Furumai, Fabrication of graphene film composite electrochemical biosensor as a pre-screening algal toxin detection tool in the event of water contamination, Sci. Rep. 8 (1) (July 16, 2018) 10686, https://doi.org/10.1038/s41598-018-28959-w.

[86] W. Zhang, et al., A 3D graphene-based biosensor as an early microcystin-LR screening tool in sources of drinking water supply, Electrochim. Acta 236 (2017) 319–327, https://doi.org/10.1016/j.electacta.2017.03.161.

[87] G. Hernandez-Vargas, J.E. Sosa-Hernandez, S. Saldarriaga-Hernandez, A.M. Villalba-Rodriguez, R. Parra-Saldivar, H. Iqbal, Electrochemical biosensors: a solution to pollution detection with reference to environmental contaminants, Biosensors 8 (2) (March 24, 2018), https://doi.org/10.3390/bios8020029.

[88] W. Sun, et al., Electrodeposited graphene and silver nanoparticles modified electrode for direct electrochemistry and electrocatalysis of hemoglobin, Electroanalysis 24 (10) (2012) 1973–1979, https://doi.org/10.1002/elan.201200103.

[89] M. Najafi, S. Darabi, A. Tadjarodi, M. Imani, Determination of trichloroacetic acid (TCAA) using CdO nanoparticles modified carbon paste electrode, Electroanalysis 25 (2) (2013) 487–492, https://doi.org/10.1002/elan.201200462.

[90] W. Chen, et al., Investigation on direct electrochemical and electrocatalytic behavior of hemoglobin on palladium-graphene modified electrode, Mater. Sci. Eng. C Mater. Biol. Appl. 80 (November 1, 2017) 135–140, https://doi.org/10.1016/j.msec.2017.05.129.

[91] W. Zhao, Application of ionic liquid-graphene-NiO hollowsphere composite modified electrode for electrochemical investigation on hemoglobin and electrocatalysis to trichloroacetic acid, Int. J. Electrochem. Sci. (2017) 4025–4034, https://doi.org/10.20964/2017.05.06.

[92] L. Kong, Electrochemistry of hemoglobin-ionic liquid-graphene-SnO_2 nanosheet composite modified electrode and electrocatalysis, Int. J. Electrochem. Sci. (2017) 2297–2305, https://doi.org/10.20964/2017.03.66.

[93] W. Chen, et al., Boron-doped Graphene quantum dots modified electrode for electrochemistry and electrocatalysis of hemoglobin, J. Electroanal. Chem. 823 (2018) 137–145, https://doi.org/10.1016/j.jelechem.2018.06.001.

[94] T. Zhan, Z. Tan, X. Wang, W. Hou, Hemoglobin immobilized in g-C3N4 nanoparticle decorated 3D graphene-LDH network: direct electrochemistry and electrocatalysis to trichloroacetic acid, Sensor. Actuator. B Chem. 255 (2018) 149–158, https://doi.org/10.1016/j.snb.2017.08.048.

[95] W. Sun, Y. Guo, X. Ju, Y. Zhang, X. Wang, Z. Sun, Direct electrochemistry of hemoglobin on graphene and titanium dioxide nanorods composite modified electrode and its electrocatalysis, Biosens. Bioelectron. 42 (April 15, 2013) 207–213, https://doi.org/10.1016/j.bios.2012.10.034.

[96] W. Sun, et al., Direct electrochemistry with enhanced electrocatalytic activity of hemoglobin in hybrid modified electrodes composed of graphene and multi-walled carbon nanotubes, Anal. Chim. Acta 781 (June 5, 2013) 41–47, https://doi.org/10.1016/j.aca.2013.04.010.

[97] W. Sun, et al., Electrochemical biosensor based on graphene, Mg_2Al layered double hydroxide and hemoglobin composite, Electrochim. Acta 91 (2013) 130–136, https://doi.org/10.1016/j.electacta.2012.12.088.

[98] W. Sun, L. Dong, Y. Deng, J. Yu, W. Wang, Q. Zhu, Application of N-doped graphene modified carbon ionic liquid electrode for direct electrochemistry of hemoglobin, Mater. Sci. Eng. C Mater. Biol. Appl. 39 (June 1, 2014) 86–91, https://doi.org/10.1016/j.msec.2014.02.029.

[99] W. Sun, et al., Direct electrochemistry and electrocatalysis of hemoglobin on three-dimensional graphene modified carbon ionic liquid electrode, Sensor. Actuator. B Chem. 219 (2015) 331–337, https://doi.org/10.1016/j.snb.2015.05.015.

[100] F. Shi, et al., Application of graphene-copper sulfide nanocomposite modified electrode for electrochemistry and electrocatalysis of hemoglobin, Biosens. Bioelectron. 64 (February 15, 2015) 131–137, https://doi.org/10.1016/j.bios.2014.08.064.

[101] T. Zhan, X. Wang, X. Li, Y. Song, W. Hou, Hemoglobin immobilized in exfoliated Co2Al LDH-graphene nanocomposite film: direct electrochemistry and electrocatalysis toward trichloroacetic acid, Sensor. Actuator. B Chem. 228 (2016) 101–108, https://doi.org/10.1016/j.snb.2015.12.095.

[102] D. Qian, et al., Voltammetric sensor for trichloroacetic acid using a glassy carbon electrode modified with Au@Ag nanorods and hemoglobin, Microchim. Acta 184 (7) (2017) 1977–1985, https://doi.org/10.1007/s00604-017-2175-6.

[103] C. Ruan, et al., Electrochemical myoglobin biosensor based on graphene–ionic liquid–chitosan bionanocomposites: direct electrochemistry and electrocatalysis, Electrochim. Acta 64 (2012) 183–189, https://doi.org/10.1016/j.electacta.2012.01.005.

[104] G. Li, et al., Electrodeposited nanogold decorated graphene modified carbon ionic liquid electrode for the electrochemical myoglobin biosensor, J. Solid State Electrochem. 17 (8) (2013) 2333–2340, https://doi.org/10.1007/s10008-013-2098-z.

[105] X. Chen, Fabrication of myoglobin-sodium alginate-graphene composite modified carbon ionic liquid electrode via the electrodeposition method and its electrocatalysis toward trichloroacetic acid, Int. J. Electrochem. Sci. (2017) 11633–11645, https://doi.org/10.20964/2017.12.65.

[106] W. Zheng, Effect of carboxyl graphene on direct electrochemistry of myoglobin and electrocatalytic investigation, Int. J. Electrochem. Sci. (2017) 4341–4350, https://doi.org/10.20964/2017.05.02.

[107] Z. Wen, Electrodeposited ZnO@three-dimensional graphene composite modified electrode for electrochemistry and electrocatalysis of myoglobin, Int. J. Electrochem. Sci. (2017) 2306–2314, https://doi.org/10.20964/2017.03.09.

[108] W. Sun, et al., Fabrication of graphene-platinum nanocomposite for the direct electrochemistry and electrocatalysis of myoglobin, Mater. Sci. Eng. C Mater. Biol. Appl. 33 (4) (May 1, 2013) 1907–1913, https://doi.org/10.1016/j.msec.2012.12.077.

[109] W. Sun, et al., Electrodeposited nickel oxide and graphene modified carbon ionic liquid electrode for electrochemical myglobin biosensor, Thin Solid Films 562 (2014) 653–658, https://doi.org/10.1016/j.tsf.2014.05.002.

[110] F. Shi, et al., Application of three-dimensional reduced graphene oxide-gold composite modified electrode for direct electrochemistry and electrocatalysis of myoglobin, Mater. Sci. Eng. C Mater. Biol. Appl. 58 (January 1, 2016) 450–457, https://doi.org/10.1016/j.msec.2015.08.049.

[111] W. Wang, et al., Electrochemistry of multilayers of graphene and myoglobin modified electrode and its biosensing, J. Chin. Chem. Soc. 63 (3) (2016) 298–302, https://doi.org/10.1002/jccs.201500378.

[112] W. Wang, et al., Electrochemistry and electrocatalysis of myoglobin on electrodeposited ZrO_2 and graphene-modified carbon ionic liquid electrode, J. Iran. Chem. Soc. 13 (2) (2015) 323–330, https://doi.org/10.1007/s13738-015-0740-7.

[113] X. Wang, L. Liu, W. Zheng, W. Chen, G. Li, W. Sun, Electrochemical behaviors of myoglobin on graphene and Bi film modified electrode and electrocatalysis to trichloroacetic acid, Int. J. Electrochem. Sci. 11 (2016) 1821–1830.

[114] X. Chen, et al., A novel biosensor based on electro-co-deposition of sodium alginate-Fe3O4-graphene composite on the carbon ionic liquid electrode for the direct

electrochemistry and electrocatalysis of myoglobin, Polym. Bull. 74 (1) (2016) 75–90, https://doi.org/10.1007/s00289-016-1698-z.

[115] S. Kang, Electrochemical behaviors of myoglobin on ionic liquid- graphene-cobalt oxide nanoflower composite modified electrode and its electrocatalytic activity, Int. J. Electrochem. Sci. (2017) 2184–2193, https://doi.org/10.20964/2017.03.64.

[116] W. Tu, J. Lei, H. Ju, Functionalization of carbon nanotubes with water-insoluble porphyrin in ionic liquid: direct electrochemistry and highly sensitive amperometric biosensing for trichloroacetic acid, Chemistry 15 (3) (2009) 779–784, https://doi.org/10.1002/chem.200801758.

[117] W. Sun, Y. Guo, T. Li, X. Ju, J. Lou, C. Ruan, Electrochemistry of horseradish peroxidase entrapped in graphene and dsDNA composite modified carbon ionic liquid electrode, Electrochim. Acta 75 (2012) 381–386, https://doi.org/10.1016/j.electacta.2012.05.018.

[118] W. Zheng, et al., Direct electron transfer of horseradish peroxidase at Co_3O_4–graphene nanocomposite modified electrode and electrocatalysis, J. Iran. Chem. Soc. 14 (4) (2017) 925–932, https://doi.org/10.1007/s13738-016-1042-4.

[119] M. Kurd, A. Salimi, R. Hallaj, Highly sensitive amperometric sensor for micromolar detection of trichloroacetic acid based on multiwalled carbon nanotubes and Fe(II)-phtalocyanine modified glassy carbon electrode, Mater. Sci. Eng. C Mater. Biol. Appl. 33 (3) (April 1, 2013) 1720–1726, https://doi.org/10.1016/j.msec.2012.12.085.

[120] R.W. Kibechu, M.A. Mamo, T.A.M. Msagati, S. Sampath, B.B. Mamba, Synthesis and application of reduced graphene oxide and molecularly imprinted polymers composite in chemo sensor for trichloroacetic acid detection in aqueous solution, Phys. Chem. Earth 76–78 (2014) 49–53, https://doi.org/10.1016/j.pce.2014.09.008.

[121] X. Cetó, C.P. Saint, C.W.K. Chow, N.H. Voelcker, B. Prieto-Simón, Electrochemical detection of N-nitrosodimethylamine using a molecular imprinted polymer, Sensor. Actuator. B Chem. 237 (2016) 613–620, https://doi.org/10.1016/j.snb.2016.06.136.

[122] Z. Zeng, X. Fang, W. Miao, Y. Liu, T. Maiyalagan, S. Mao, Electrochemically sensing of trichloroacetic acid with iron(II) phthalocyanine and Zn-based metal organic framework nanocomposites, ACS Sens. 4 (7) (July 26, 2019) 1934–1941, https://doi.org/10.1021/acssensors.9b00894.

[123] X. Fang, et al., Ultrasensitive detection of disinfection byproduct trichloroacetamide in drinking water with Ag nanoprism@MoS2 heterostructure-based electrochemical sensor, Sensor. Actuator. B Chem. 332 (2021), https://doi.org/10.1016/j.snb.2021.129526.

[124] X. Cetó, C. Saint, C.W.K. Chow, N.H. Voelcker, B. Prieto-Simón, Electrochemical fingerprints of brominated trihaloacetic acids (HAA3) mixtures in water, Sensor. Actuator. B Chem. 247 (2017) 70–77, https://doi.org/10.1016/j.snb.2017.02.179.

[125] T. Deblonde, C. Cossu-Leguille, P. Hartemann, Emerging pollutants in wastewater: a review of the literature, Int. J. Hyg. Environ. Health 214 (6) (2011) 442–448, https://doi.org/10.1016/j.ijheh.2011.08.002.

[126] M.B. Tahir, S. Arif, M. Sagir, A. Batool, Semiconductor-based photocatalytic nanomaterials for environmental applications, Encycl. Renew. Sustain. Mater. 2 (2020) 320–325, https://doi.org/10.1016/B978-0-12-803581-8.11560-7.

[127] A. Odobašić, I. Šestan, S. Begić, Biosensors for determination of heavy metals in waters, in: Biosensors for Environmental Monitoring, IntechOpen, 2019.

[128] M. Jaishankar, T. Tseten, N. Anbalagan, B.B. Mathew, K.N. Beeregowda, Toxicity, mechanism and health effects of some heavy metals, Interdiscip. Toxicol. 7 (2) (2014) 60–72, https://doi.org/10.2478/intox-2014-0009.

[129] A. Kubier, R.T. Wilkin, T. Pichler, Cadmium in soils and groundwater: a review, J. Appl. Geochem. 108 (2019) 104388, https://doi.org/10.1016/j.apgeochem.2019.104388.

[130] A.G. Caporale, A. Violante, Chemical processes affecting the mobility of heavy metals and metalloids in soil environments, Curr. Pollut. Rep. 2 (1) (2016) 15–27, https://doi.org/10.1007/s40726-015-0024-y.

[131] C. Krantz-Rülcker, M. Stenberg, F. Winquist, I. Lundström, Electronic tongues for environmental monitoring based on sensor arrays and pattern recognition: a review, Anal. Chim. Acta 426 (2) (2001) 217–226, https://doi.org/10.1016/S0003-2670(00)00873-4.

[132] Z. Wei, Y. Yang, J. Wang, W. Zhang, Q. Ren, The measurement principles, working parameters and configurations of voltammetric electronic tongues and its applications for foodstuff analysis, J. Food Eng. 217 (2018) 75–92, https://doi.org/10.1016/j.jfoodeng.2017.08.005.

[133] Y. Vlasov, A. Legin, A. Rudnitskaya, Electronic tongues and their analytical application, Anal. Bioanal. Chem. 373 (3) (2002) 136–146, https://doi.org/10.1007/s00216-002-1310-2.

[134] A. Riul Jr., C.A. Dantas, C.M. Miyazaki, O.N. Oliveira Jr., Recent advances in electronic tongues, Analyst 135 (10) (2010) 2481–2495, https://doi.org/10.1039/C0AN00292E.

[135] M. Li, H. Gou, I. Al-Ogaidi, N. Wu, Nanostructured sensors for detection of heavy metals: a review, ACS Sustain. Chem. Eng. 1 (7) (2013) 713–723, https://doi.org/10.1021/sc400019a.

[136] N. Verma, M. Singh, Biosensors for heavy metals, Biometals 18 (2) (2005) 121–129, https://doi.org/10.1007/s10534-004-5787-3.

[137] EPA, 2018 Edition of the Drinking Water Standards and Health Advisories Tables, 2018.

[138] A.G.-M. Ferrari, P. Carrington, S.J. Rowley-Neale, C.E. Banks, Recent advances in portable heavy metal electrochemical sensing platforms, Environ. Sci.: Water Res. Technol. 6 (10) (2020) 2676–2690, https://doi.org/10.1039/D0EW00407C.

[139] Z.M. Yaseen, An insight into machine learning models era in simulating soil, water bodies and adsorption heavy metals: review, challenges and solutions, Chemosphere 277 (2021) 130126, https://doi.org/10.1016/j.chemosphere.2021.130126, 2021/08/01.

[140] E. Z-Flores, M. Abatal, A. Bassam, L. Trujillo, P. Juárez-Smith, Y. El Hamzaoui, Modeling the adsorption of phenols and nitrophenols by activated carbon using genetic programming, J. Clean. Prod. 161 (2017) 860–870, https://doi.org/10.1016/j.jclepro.2017.05.192, 2017/09/10.

[141] X. Zhu, X. Wang, Y.S. Ok, The application of machine learning methods for prediction of metal sorption onto biochars, J. Hazard Mater. 378 (2019) 120727, https://doi.org/10.1016/j.jhazmat.2019.06.004, 2019/10/15.

[142] A. El Hanandeh, Z. Mahdi, M.S. Imtiaz, Modelling of the adsorption of Pb, Cu and Ni ions from single and multi-component aqueous solutions by date seed derived biochar: comparison of six machine learning approaches, Environ. Res. 192 (2021) 110338, https://doi.org/10.1016/j.envres.2020.110338, 2021/01/01.

[143] F.S. Wanyonyi, T.T. Fidelis, G.K. Mutua, F. Orata, A.M. Pembere, Role of pore chemistry and topology in the heavy metal sorption by zeolites: from molecular simulation to machine learning, Comput. Mater. Sci. 195 (2021) 110519, https://doi.org/10.1016/j.commatsci.2021.110519, 2021/07/01.

[144] B. Ke, et al., Predicting the sorption efficiency of heavy metal based on the biochar characteristics, metal sources, and environmental conditions using various novel hybrid machine learning models, Chemosphere 276 (2021) 130204, https://doi.org/10.1016/j.chemosphere.2021.130204, 2021/08/01.

[145] Y. Zhao, Y. Li, D. Fan, J. Song, F. Yang, Application of kernel extreme learning machine and Kriging model in prediction of heavy metals removal by biochar, Bioresour. Technol. 329 (2021) 124876, https://doi.org/10.1016/j.biortech.2021.124876, 2021/06/01.

[146] P. Gautam, S. Suniti, A. Kumari, D. Madathil, B. Nair, A review on recent advances in biosensors for detection of water contamination, Int. J. Environ. Sci. 2 (3) (2012) 1565–1574, https://doi.org/10.6088/ijes.002020300041.

[147] L. Eddaif, A. Shaban, J. Telegdi, Sensitive detection of heavy metals ions based on the calixarene derivatives-modified piezoelectric resonators: a review, Int. J. Environ. Anal. Chem. 99 (9) (2019) 824–853, https://doi.org/10.1080/03067319.2019.1616708.

[148] K.R. Singh, V. Nayak, T. Sarkar, R.P. Singh, Cerium oxide nanoparticles: properties, biosynthesis and biomedical application, RSC Adv. 10 (45) (2020) 27194–27214, https://doi.org/10.1039/D0RA04736H.

[149] A. Zamora-Galvez, E. Morales-Narváez, C.C. Mayorga-Martinez, A. Merkoçi, Nanomaterials connected to antibodies and molecularly imprinted polymers as bio/receptors for bio/sensor applications, Appl. Mater. Today 9 (2017) 387–401, https://doi.org/10.1016/j.apmt.2017.09.006.

[150] A. Merkoçi, Electrochemical biosensing with nanoparticles, FEBS J. 274 (2) (2007) 310–316, https://doi.org/10.1111/j.1742-4658.2006.05603.x.

[151] S.A. Lim, M.U. Ahmed, Electrochemical immunosensors and their recent nanomaterial-based signal amplification strategies: a review, RSC Adv. 6 (30) (2016) 24995–25014, https://doi.org/10.1039/C6RA00333H.

[152] H. Bhardwaj, G. Sumana, C.A. Marquette, Gold nanobipyramids integrated ultrasensitive optical and electrochemical biosensor for Aflatoxin B1 detection, Talanta 222 (2021) 121578, https://doi.org/10.1016/j.talanta.2020.121578.

[153] A.S. Maghsoudi, S. Hassani, K. Mirnia, M. Abdollahi, Recent advances in nanotechnology-based biosensors development for detection of arsenic, lead, mercury, and cadmium, Int. J. Nanomed. 16 (2021) 803, https://doi.org/10.2147/IJN.S294417.

[154] W. He, et al., A novel electrochemical biosensor for ultrasensitive Hg^{2+} detection via a triple signal amplification strategy, Chem. Commun. 57 (5) (2021) 619–622, https://doi.org/10.1039/D0CC07268K.

[155] P. Gupta, et al., Parts per trillion detection of heavy metals in as-is tap water using carbon nanotube microelectrodes, Anal. Chim. Acta 1155 (2021) 338353, https://doi.org/10.1016/j.aca.2021.338353.

[156] H.M. Siddiqi, Y. Kun Lin, Z. Hussain, N. Majeed, N. Amin, Bovine serum albumin protein-based liquid crystal biosensors for optical detection of toxic heavy metals in water, Sensors 20 (1) (2020) 298, https://doi.org/10.3390/s20010298.

[157] Z. Liu, et al., Ratiometric fluorescent sensing of Pb^{2+} and Hg^{2+} with two types of carbon dot nanohybrids synthesized from the same biomass, Sens. Actuators B Chem. 296 (2019) 126698, https://doi.org/10.1016/j.snb.2019.126698.

[158] N. Bereli, D. Çimen, H. Yavuz, A. Denizli, Sensors for the detection of heavy metal contaminants in water and environment, Nanosens. Environ. Food Agric. (2021) 1–21, https://doi.org/10.1007/978-3-030-63245-8_1.

[159] S.C. Ballen, J. Steffens, C. Steffens, Stability characteristics of cantilever nanobiosensors with simple and complex molecules for determination of cadmium, Sens. Actuators, A Phys. 324 (2021) 112686, https://doi.org/10.1016/j.sna.2021.112686.

[160] S.C. Ballen, G.M. Ostrowski, J. Steffens, C. Steffens, Graphene oxide/urease nanobiosensor applied for cadmium detection in river water, IEEE Sens. J. 21 (8) (2021) 9626–9633, https://doi.org/10.1109/JSEN.2021.3056042.

[161] J.O. Tijani, O.O. Fatoba, L.F. Petrik, A review of pharmaceuticals and endocrine-disrupting compounds: sources, effects, removal, and detections, Water Air Soil Pollut. 224 (11) (2013) 1–29, https://doi.org/10.1007/s11270-013-1770-3.

[162] V. Chander, et al., Pharmaceutical compounds in drinking water, J. Xenobiot. 6 (1) (2016) 1–7, https://doi.org/10.4081/xeno.2016.5774.

[163] J. Derksen, G. Rijs, R. Jongbloed, Diffuse pollution of surface water by pharmaceutical products, Water Sci. Technol. 49 (3) (2004) 213–221, https://doi.org/10.2166/wst.2004.0198.

[164] K. Kümmerer, Pharmaceuticals in the Environment: Sources, Fate, Effects and Risks, Springer Science & Business Media, 2008.

[165] L. Zhao, et al., Nanomaterials for treating emerging contaminants in water by adsorption and photocatalysis: systematic review and bibliometric analysis, Sci. Total Environ. 627 (2018) 1253–1263, https://doi.org/10.1016/j.scitotenv.2018.02.006.

[166] M. Asadnia, L.H. Chua, X. Qin, A. Talei, Improved particle swarm optimization–based artificial neural network for rainfall-runoff modeling, J. Hydrol. Eng. 19 (7) (2014) 1320–1329.

[167] E.S. Elmolla, M. Chaudhuri, M.M. Eltoukhy, The use of artificial neural network (ANN) for modeling of COD removal from antibiotic aqueous solution by the Fenton process, J. Hazard Mater. 179 (1) (2010) 127–134, https://doi.org/10.1016/j.jhazmat.2010.02.068, 2010/07/15.

[168] G. Andaluri, E.V. Rokhina, R.P.S. Suri, Evaluation of relative importance of ultrasound reactor parameters for the removal of estrogen hormones in water, Ultrason. Sonochem. 19 (4) (2012) 953–958, https://doi.org/10.1016/j.ultsonch.2011.12.005, 2012/07/01.

[169] Z. Frontistis, et al., Photocatalytic (UV-A/TiO_2) degradation of 17α-ethynylestradiol in environmental matrices: experimental studies and artificial neural network modeling, J. Photochem. Photobiol. Chem. 240 (2012) 33–41, https://doi.org/10.1016/j.jphotochem.2012.05.007, 2012/07/15.

[170] A. Mojiri, J. Zhou, M. Vakili, H. Van Le, Removal performance and optimisation of pharmaceutical micropollutants from synthetic domestic wastewater by hybrid treatment, J. Contam. Hydrol. 235 (2020) 103736, https://doi.org/10.1016/j.jconhyd.2020.103736, 2020/11/01.

[171] H. Yu, Z. Zhang, L. Zhang, H. Dong, H. Yu, Improved Norfloxacin degradation by urea precipitation Ti/SnO_2–Sb anode under photo-electro catalysis and kinetics investigation by BP-neural-network-physical modeling, J. Clean. Prod. 280 (2021) 124412, https://doi.org/10.1016/j.jclepro.2020.124412, 2021/01/20.

[172] F.-D. Munteanu, A.M. Titoiu, J.-L. Marty, A. Vasilescu, Detection of antibiotics and evaluation of antibacterial activity with screen-printed electrodes, Sensors 18 (3) (2018) 901, https://doi.org/10.3390/s18030901.

[173] R. Chauhan, J. Singh, T. Sachdev, T. Basu, B. Malhotra, Recent advances in mycotoxins detection, Biosens. Bioelectron. 81 (2016) 532–545, https://doi.org/10.1016/j.bios.2016.03.004.

[174] L. Qian, A.R. Thiruppathi, R. Elmahdy, J. van der Zalm, A. Chen, Graphene-oxide-based electrochemical sensors for the sensitive detection of pharmaceutical drug naproxen, Sensors 20 (5) (2020) 1252, https://doi.org/10.3390/s20051252.

[175] K. Kaewnu, K. Promsuwan, P. Kanatharana, P. Thavarungkul, W. Limbut, A simple and sensitive electrochemical sensor for chloramphenicol detection in pharmaceutical samples, J. Electrochem. Soc. 167 (8) (2020) 087506, https://doi.org/10.1149/1945-7111/ab8ce5.

[176] X. Wang, et al., Paper-based antibiotic sensor (PAS) relying on colorimetric indirect competitive enzyme-linked immunosorbent assay for quantitative tetracycline and chloramphenicol detection, Sens. Actuators B Chem. 329 (2021) 129173, https://doi.org/10.1016/j.snb.2020.129173.

[177] Z. Altintas, Surface plasmon resonance based sensor for the detection of glycopeptide antibiotics in milk using rationally designed nanoMIPs, Sci. Rep. 8 (1) (2018) 1−12, https://doi.org/10.1038/s41598-018-29585-2.

Chapter 11

Recent progress in biosensors for wastewater monitoring and surveillance

Pratiksha Srivastava[1], Yamini Mittal[3,4], Supriya Gupta[3,4], Rouzbeh Abbassi[2], Vikram Garaniya[1]

[1]*Australian Maritime College, College of Sciences and Engineering, University of Tasmania, Launceston, TAS, Australia;* [2]*School of Engineering, Faculty of Science and Engineering, Macquarie University, Sydney, NSW, Australia;* [3]*Environment and Sustainability Department, CSIR-Institute of Minerals and Materials Technology, Bhubaneswar, Odisha, India;* [4]*Academy of Scientific and Innovative Research (AcSIR), CSIR-Human Resource Development Centre, CSIR-HRDC Campus, Ghaziabad, India*

1. Introduction

Water quality assessment and pollutant monitoring have increased enormous attention [1–3]. Water pollution is found to be intrinsically linked to public health issues and habitat degradation [4,5]. Wastewater is a composition of various harmful components produced and discharged from different sources such as domestic, industrial, and agriculture activities [5]. The deterioration of water quality with increased level of pollutants such as heavy metal, pesticide, and noncommunicable and infectious diseases has gained ample attention in recent years [4]. The timely monitoring of these threats can save public health as well as habitat degradation. The concentration of wastewater constituents and indication of any infectious diseases is important in determining treatment methods, designing, functioning, and reusing water or waste [5–7]. Over the time, varying concentration of pollutants needs to be monitored to apply treatment processes or monitor public health. The wastewater quality parameters are mainly tested in laboratory-based sophisticated methods, which require a longer time, rapid sampling, and money for the analysis. These factors indicated the necessity of rapid and on-site monitoring of multi-pollutants in wastewater along with an indication of any public health threat of infectious diseases. Therefore, biosensors have emerged as a potential alternative for on-site monitoring of wastewater pollutants [8–10].

Artificial Intelligence and Data Science in Environmental Sensing
https://doi.org/10.1016/B978-0-323-90508-4.00010-1

Biosensors have received substantial attention in recent years as a powerful analytical tool and their excellent characteristics in wastewater and public health monitoring [4,11]. This testing is rapid, simple, low cost, onsite, and only requires a small device to monitor multipollutants simultaneously [4]. Biosensors are devices established on biochemical reactions assisted by biological receptors such as microbes, enzymes, and antibodies and typically identify targets established on electrical, thermal, and other signals [4,12]. A typical working of a biosensor is represented in Fig. 11.1. Biosensors are important in determining wastewater pollutants such as organics, inorganics, toxic pollutants, microorganisms, and noncommunicable and infectious diseases. Considering wide range of its application, bioelectrochemical, photoelectrochemical (PEC), and enzymatic biosensors for sensing various pollutants in wastewater are discussed in this chapter.

Bioelectrochemical systems (BES) such as microbial fuel cell (MFC)-dependent biosensors use microbes as a catalyst for oxidation of organics and inorganics to produce electric current [13,14]. In general, MFC act as a biosensor, where microbes in the anode act as a biological recognition factor and proton exchange membrane (PEM) and conductive materials as a transducer [15,16]. The output energy represents a function of microbial respiration and fuel concentration in BES. The microbes in MFC can sense the concentration change of any pollutants in the wastewater and can indicate through output energy [15,17,18]. Hence, MFC has been utilized for the biosensing of various pollutants of wastewater. Moreover, to detect the target analyte without the influence of other sample constituents, enzymatic biosensors hold the most extensive commercial market [19–22]. Enzyme biosensors are also based on the theory of biocatalysis like bioelectrochemical-based biosensors. In the abundance of suitable substrate, enzymes catalyze electrochemical reactions and estimate electric variations on the transducer.

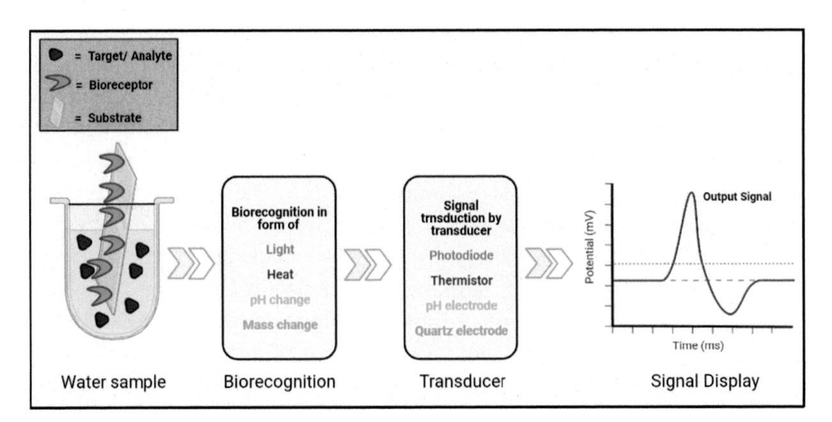

FIGURE 11.1 Working components and mechanism of a biosensor for water monitoring.

Nevertheless, PEC-based biosensors have also become a sustainable technology since it works on photoelectrochemistry. Since the discovery of the photoelectric effect in 1839 by Becquerel [23,24], the PEC has emerged with various applications such as photocatalysis, photovoltaics [25], and sensing. The PEC is based on the principle of charge separation and consequent charge transfer due to photon-to-electricity conversion [26,27]. Due to the charge transfer in PEC, possibility of bioanalysis emerged as a new platform for wastewater pollutant monitoring [26]. These PEC sensors are based on the charge transfer process upon light irradiation using light as an excitation source between the analyte, photoactive material, and electrode [27]. Bioelectrochemical, enzymatic, and photoelectric biosensors are individually discussed in detail in the chapter.

2. Principles and working of BES as a biosensor

Biosensors are sensing devices efficient for detecting and quantifying the target molecules by quantifying the signals generated from biological recognition events [28]. In general, the biosensor consists of a bioreceptor present on the surface of the transducer. The bioreceptor binds the target molecule/analyte to form an intermediate complex that results in characteristic changes, and the transducer translates the incurred changes into desired signals which are then evaluated and displayed to the user through an interface (Fig. 11.1). Thus, depending upon the recognition element, biosensors can be classified as immunosensors, aptasensors, genosensors, and enzymatic biosensors, when aptamers, nucleic acids, antibodies, and enzymes are used, respectively. They are also categorized on the basis of transduction principles such as surface plasmon resonance biosensors and optical fiber based on optical signals, amperometric biosensors, potentiometric biosensors, impedimetric biosensors, and conductimetric biosensor based on electrochemical signals and quartz crystal microbalance biosensors based on piezoelectric signals [29].

Among all these biosensors, BES, for instance, MFCs and microbial electrolysis cells (MECs), has demonstrated promising results in water quality monitoring [30]. MFCs and MECs emerged as strong competitors in organic and inorganic pollutants sensing over cumbersome conventional technologies, since conventional technologies require extensive processing time and information gaps on biological responses [30,31]. MFCs and MECs are beneficial in terms of sensing due to their (1) longevity with barely any maintenance requirement, (2) mechanical and electrical design simplicity, and (3) low cost of construction [32,33]. This section focuses on the principle and functioning of MFCs and MECs as a biosensor.

2.1 Microbial fuel cell as a sensor

MFCs can directly convert chemical energy present in wastewater (organic matter) into bioelectricity utilizing metabolic processes of microorganisms (i.e., renewable catalysts). It consists of an anodic and cathodic compartment separated by cation exchange membrane (CEM) or PEM [14], as shown in Fig. 11.2. In the anodic compartment, electroactive bacteria (EABs) form biofilm over the electrode surface, which metabolizes organic matter present in the wastewater to form CO_2, electron (e^-), and proton (H^+) [31]. These generated electrons are transferred to the anode electrode by extracellular electron transfer through direct (electron shuttling) or mediated pathways (redox mediators, primary metabolites, and their intermediates) [34,35], as shown in Fig. 11.2. Further, both electrons and H^+ reach the cathode electrode through a wire and CEM/PEM, respectively, for reduction reaction [36,37]. The electron produced at the anode electrode by EABs can be straightaway linked with the metabolic activity of the developed biofilm, as depicted in Fig. 11.2. Any disturbance to this metabolic activity (such as an increase in organic load) can be sensed/detected provided all process parameters, for example, pH, temperature, humidity, and conductivity, are set as constant [31,38]. MFC sensors are based on electroactive biofilm on the anode surface, which acts as a recognition element for in situ monitoring and eliminates external equipment's need to act as transducers [38].

In MFCs, the electron flow rate to the anode gets interrupted by the specific disruption to the anode electroactive biofilm and transduced as a considerable change in the electricity production of MFCs [31]. Furthermore, BES's sensitivity is defined as a change in the electrical signal output (ΔI in mA) per unit change of analyte concentration (ΔI in mM) in the anodic compartment, which is inversely dependable on electrode surface area [30], as shown in Eq. (11.1).

$$\text{Sensitivity} = \frac{\Delta I}{\Delta C} \times \frac{1}{A} \quad (11.1)$$

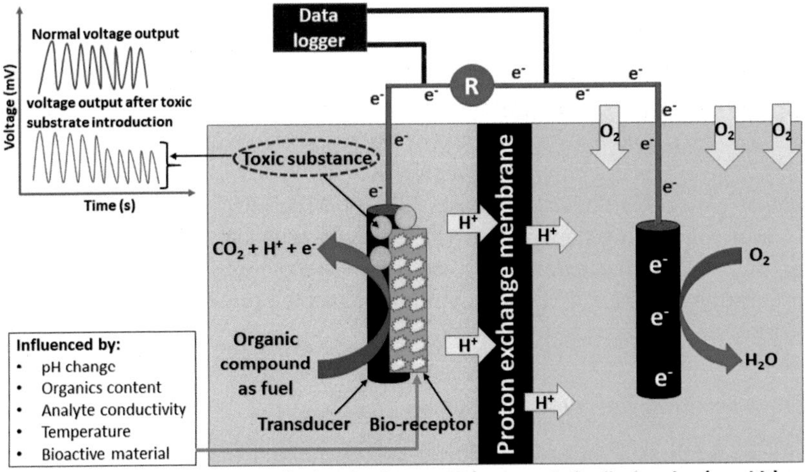

FIGURE 11.2 Schematic diagram and operation of microbial fuel cells as a biosensor.

The sensitivity equation plays a crucial role in determining toxic, bioactive compounds and other parameters in wastewater such as chemical oxygen demand (COD), biological oxygen demand (BOD), and nitrogen through the shift in the current generation [31]. Thus, electroactive biofilm present at the anode electrode surface can function as a low-cost quantitative biosensor against other biosensors (Fig. 11.3).

MFCs are advantageous in real-time applications and can be employed as in situ environmental or bioprocess monitoring devices [30]. However, before employment of MFCs as biosensors, few points can be taken into consideration, including (1) it shall provide linear response with the concentration with appropriate sensitivity and fast response, (2) capable of detecting a wide variety of analytes, and (3) eliminate low-power output MFCs as it might disturb the sensitivity of biosensor by reducing the unit variation in current per unit of the analyte [39]. Generally, anode potential ranging between −350 and −400 mV v/s Ag/AgCl electrode is determined as the most stable operating voltage in MFCs at which constant current density can be attained [40]. So far, MFCs as biosensors are employed in the detection of (1) pollutants such as acetate, assimilable organic carbon, COD, volatile fatty acids (VFAs), BOD, and dissolved oxygen, (2) metal ion such as iron, chromium, and magnesium sensor, (3) toxicity compounds such as acid toxicity and copper toxicity in wastewater, (4) hazardous chemicals such as fumarate sensor and acetaldehyde sensor, and (5) water quality assessment purpose [31]. Di Lorenzo et al. [41] demonstrated an air cathode single-chambered MFCs as a biosensor against double-chambered MFC in measuring range, where linearity

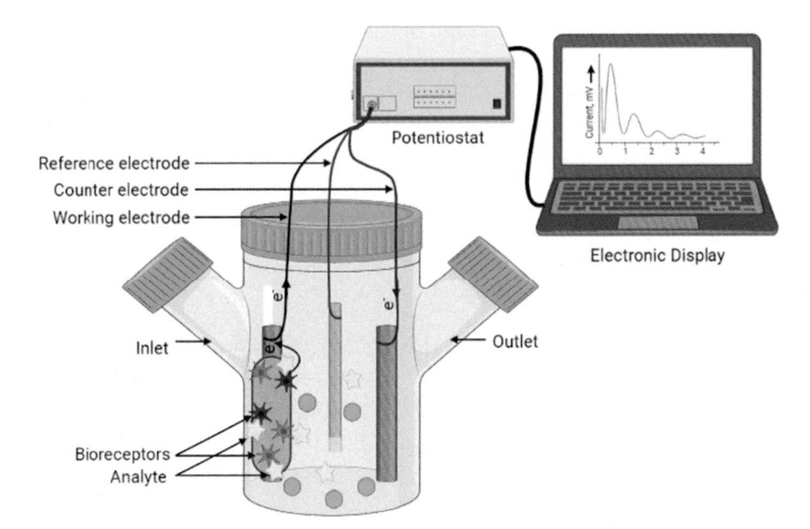

FIGURE 11.3 Schematic of three-electrode microbial electrolysis cell biosensor for water monitoring.

was observed at 350 and 500 mg/L COD calibrated against current and charge, respectively. The increase in measuring range was due to better oxygen supply in air cathode MFCs. Furthermore, Sun et al. [42] operated innovatively designed MFCs to monitor acetate in the anaerobic digestion process. Wherein acetate ions were first fed to the cathode compartment of CW-MFC and then traveled through the membrane to the anode compartment, where it acted as a sole substrate for pre-enriched electroactive biofilm for the current generation. The study observed this method to be highly selective for acetate with the least interference from VFAs and showed a linear detection range of 0.5–20 mM within 5 h of reaction time. VFAs are a very crucial parameter for monitoring bioprocesses occurring in MFCs. One such study correlated VFA concentration with current (columbic efficiency) and electrochemical response (cyclic voltammetry) using an MFC biosensor [43]. The study revealed slow response through columbic efficiency parameter of >20 h for 20 mg/L of VFA concentration. While oxidation peak at constant scan rate demonstrated linear correlation between VFA concentration of <40 mg/L and produced peak current in 1–2 min of response time.

Recently, Adekunle et al. [44] determined the sensitivity of a mining rock drainage for total heavy metal concentration in an MFC biosensor. A significant difference in electroactive microbial population was observed in MFC biosensors exposed with heavy metal compared to the nonexposed. The results indicated the capability of the MFC biosensor as a real-time detection tool for long-term online heavy metal toxicity monitoring (quantitative or qualitative) in remote locations. Lately, Khan et al. [45] demonstrated zinc detection in MFC biosensors by using engineered *Escherichia coli* BL21, which helped in Zn^{2+} sensing through riboflavin and porin production. The obtained results revealed a positive linear relationship of Zn^{2+} (0–400 μM) with high levels of riboflavin production through engineered strains. Linear correlation of voltage (160, 183, 260, 292, and 342 mV) with Zn^{2+} concentration (0, 100, 200, 300, and 400 μM, respectively) was observed at pH of 9, 37°C temperature, and 1000 Ω of internal resistance [45]. Thus, utilization of MFCs as biosensors is efficacious with the potential benefit of simple operation, rapid and on-site biosensing capability.

2.2 Microbial electrolysis cell as a sensor

Like MFC, MEC is a biocatalytic device that depends on the oxidation of organic matter for the generation of metabolic electrons, accepted by solid state anode, which subsequently transferred to the cathode by an external circuit for resource recovery by cathodic reduction reactions. However, unlike MFC, MEC is impressed with an external voltage that is subsequently utilized by the microorganisms to overcome the anodic and cathodic overpotentials; and utilize the opportunity to generate current to their maximum potential

[32,46]. A pollutant-specific stable potential is poised in MEC for the oxidation or reduction in the anode or cathode, respectively, with respect to the reference electrode (RE). The resultant potential or current of the cell is determined against the applied external resistance. The resultant electrical outputs achieved in MEC are more stable as compared to MFCs. As shown in Fig. 11.3, the MEC biosensor consists of a working electrode (WE), a counter electrode (CE), and an RE assembled in a three-electrode cell setup and connected to a potentiostat. The working electrode acts as the transducer that consists of a bioreceptor and is dedicated to collect the electrons generated during the oxidation of the target pollutant at a constant potential applied with respect to the RE. The transduction principle in MEC is electrochemical, and generated electric signals are analyzed using a potentiostat to determine the concentration of the target pollutant.

The MEC biosensors have evolved as a modification strategy to MFC biosensors. MFC biosensors were challenging to operate continuously due to noncontrollable anode potential, which was overcome by providing applied potential [47]. Moreover, the linear correlation between BOD and current, i.e., substrate utilization rate in MFC biosensors, is obtained at only low BOD concentrations. It is already known that after the substrate concentration is increased up to a specific limit, the microbial substrate utilization rate reaches a saturation level following Monod growth kinetics. Chang et al. [48] obtained the linearity up to 100 mg/L, whereafter the higher concentrations were measured by model fitting or increasing the hydraulic retention time (HRT). Moreover, Modin and Wilén [46] observed a good correlation curve ($R^2 = 0.97$) with MEC biosensors working at the HRT of 20 h with a higher BOD concentration (acetate as substrate) of 1280 mg/L. The correlation as however found to fluctuate with a decrease in HRT. At the HRT of 15, 10, and 5 h, a good linear correlation ($R^2 = 0.99$) was achieved at the concentration of 640 and 320 mg/L, respectively. The detection limit of MEC biosensors for BOD concentration was yet higher than MFC biosensors and ranged from 32 to 1280 mg/L. It was due to the higher BOD oxidation rate influenced by applied potential [46]. Previous studies have also reported about simplification of biosensor design with impressed potential. The selection of applied voltage in MEC needs to be taken care of while recording current since high applied voltage can result in water electrolysis that will further contribute to the current output, resulting in false detection [46]. A high sensitivity marine MEC biosensor was used for the early warning of biofouling potential in seawater with a rapid response within 10 min using an amperometric signal [47]. The reported detection limit was 10 μM of acetate at the poised anode potential of −300 mV ($R^2 = 0.99$). The sensitivity and recovery time was improved further with higher poised anode potential [47]. The advantages of MEC biosensor include low detection limits, high reproducibility, and a wide linear response range with several target compounds [28].

3. Biosensor for various pollutant monitoring

Many types of biosensors are developed for monitoring various pollutant levels in wastewater. They are broadly based on two analytical techniques, electrochemical or optical [49–51]. Among which wastewater treatment researchers significantly adapt electrochemical methods. The simple operation strategy, fast response, low cost, high selectivity and sensitivity, and high diversity in fabrication have made these electrochemical biosensors more widespread [52–54]. This section includes different biosensors for pollutant monitoring.

3.1 Organic pollutants

Organic pollutants are one of the most common and vital wastewater parameters to access water quality [15]. Organic pollutants can have natural or anthropogenic sources, and municipal, industrial, and agricultural wastewater can have a broad spectrum of it [5]. Organic pollutants in wastewater are generally measured in terms of BOD and COD, which identifies biological and chemical oxygen consumed in the oxidation of organic pollutants. In conventional systems, the determination of these parameters is involved with the complex procedure, time-consuming, low detection, and requires costly and hazardous reagent [1,15]. Similarly, biodegradable organic matter (BOM) in wastewater is another factor to determine the wastewater treatment processes or management and assessment of wastewater [55]. BOD is the key parameter for the quantification of BOM of wastewater. However, the BOD test requires 5–7 days for the analysis, trained manpower, and careful sampling in conventional systems. Thus, it is essential to use alternative solutions to avoid unnecessary delays due to conventional techniques. The rapid detection techniques in the form of biosensors hence become popular in wastewater treatment industries. Several biosensors based on the principle of photocatalytic and electrochemical theory have been investigated for wastewater monitoring [1,26]. These biosensor techniques are proved to be very effective and efficient for monitoring; however, BES-based biosensors are more effective and popular in organics detection in wastewater [56]. The organic detection in BES is based on a linear relationship between amount of organics and current output. The operational stability of over 5 years has been found in BES-based biosensors with minimum maintenance [41,48,57,58]. Gao et al. [56] have stated that three different approaches can be adapted to determine electrical signals in BES-based biosensors. The first approach is based on maximum current output. Yamashita et al. [59] developed an open-type BES biosensor for in situ estimation of BOD during intermittent aeration. The study observed that current output was similar in aerating and nonaerating situations and BOD concentration had a logarithmic correlation ($R^2 > 0.9$) with the current. Similarly, Kharkwal et al. [60] investigated real wastewater for continuous monitoring of BOD in MnO_2 cathode-based BES biosensor. The voltage output and BOD concentration have a reasonable correlation. The second approach is grounded on open-circuit voltage. Wang et al. [61] investigated BES biosensor based on open-circuit voltage and stated that the sensor was stable and sensitive with organic matter concentration. The third approach is coulombic efficiency or yield, which is precisely related to the concentration of pollutants. A linear relationship between concentration and current output is used for biosensors [58]. Kim et al. [58] also demonstrated the practical application of BES biosensor based on coulombic efficiency operated for 60 days.

Furthermore, Yao et al. [1] have used thermal biosensors for COD determination. Thermal biosensors measure the heat generated during the oxidation of organics in wastewater. In this biosensor, the change in enthalpy of organics in proportion to the height of the thermometric peak is measured. It is also reported that thermistor is insensitive to the electrochemical, optical, and other properties of wastewater; thus, a degree of interference is very low, and the detection efficiency is very high [1]. Other organics present in wastewater can originate from pesticides, phenolic compounds, and pharmaceuticals, which can also be determined by biosensors [4].

3.2 Nitrogen pollutants

Nitrogen pollutants are frequently present in wastewater in normal to high concentrations, and it is the leading cause of 70% of eutrophication. The ammonium ions concentration in wastewater is problematic to detect due to cation and pH interferences [52]. The optical sensor for ammonium biosensor based on polydimethylsiloxane (PDMS)-nitroprusside composite obtained reproducible results. This type of sensor was based on encapsulation of nitroprusside on the PDMS matrix, though it has a low detection limit and a longer response time. Raud et al. [62] immobilized nitrifying microbes (*Nitrosomonas* sp.) on an agarose gel matrix as a biosensor for ammonium-nitrogen. The life of the biosensor was 14 days, and the response time observed was 15–25 min. A bioenzyme sensor for ammonium determination was constructed by the amperometric method by immobilizing glutamate oxidase (GXD) and glutamate dehydrogenase (GIDH) on a Clark-type oxygen electrode, and enzymes were entrapped on a Teflon membrane by a poly-carbamoyl sulfonate [63]. GXD consumes dissolved oxygen, and it acts as a crucial substance for enzymatic reactions. Thus, the maximum rate of dissolved oxygen consumption detected by GXD is used to determine ammonium [63].

However, BES biosensor operates without a transducer signal; thus, it is considered a rapid, real-time monitoring device [64]. In BES biosensor, ammonium acts as an electron donor for electricity generation, via which the ammonium concentration is monitored. Ammonium oxidation in BES supplies electrons in anaerobic and aerobic environments, giving signals in terms of current output [36,65]. Also, nitrifying bacteria use ammonium in the anaerobic chamber of BES to produce organic compounds, which heterotrophs can use to produce electricity [64]. These are the important paths to detect ammonium concentration in wastewater.

However, some wastewater may contain other forms of nitrogen, such as nitrate and nitrite. Larsen et al. [66] developed a microscale biosensor to monitor nitrate/nitrite in activated sludge plant. The biosensor was developed based on nitrate/nitrite diffusion by a tip membrane into a thick mass of microbes, where it gets converted into nitrous oxide and detected by electrochemical detection. To detect nitrate/nitrite, majorly ion-selective electrodes or ionophores are used. It is a class of components that facilitate certain ion transport across membranes by forming reversible complexes with an analyte ion [52].

3.3 Toxic pollutants

Wastewater occasionally contains a high load of pollutants depending on the source of wastewater. The pollutant concentration, such as heavy metals or organics, higher than the usual concentration, is considered a toxic pollutant or shock pollutant [31,67]. Since wastewater treatment plants are usually designed to treat average concentration, increased concentration of pollutants creates operational difficulties [67]. The key player of biological treatment processes, i.e., microbes, gets damaged, which stops wastewater treatment operation [31]. Thus, biosensors were employed for early detection of shock load to allow intervention to minimize operational difficulties.

Aptamers are the key player as a recognition element for heavy metal detection in wastewater [4]. It is a particular nucleic acid sequence to determine heavy metal ions [68]. To determine other metallic pollutants, nucleic acid sequences can be recognized with screening systematic evolution of ligands by exponential enrichment (SELEX) technology, which can be utilized as an aptamer for specific pollutants [4,69]. The other most adapted biosensor technique to detect other pollutants is a microbial biosensor. BES biosensor based on microbes−electrode interaction has shown promising results for determining "shock" or "toxicity" in wastewater. BES gives an early indication if the pollutant concentration increases in the systems; with an increase in pollutant concentration, the decrease in power is used to indicate system failure [31]. Xu et al. [70] developed a miniature BES for detecting nickel and chromium (Cr) shock in wastewater. They observed an excellent response for Cr shock with 5−20 mg/L. Moreover, it took 80 h for the system to recover and act normally after 10 mg/L of shock. Similarly, Shen et al. [71] determined shear rate effect on BES for toxic shock. The results indicated that flow rate was a major factor for getting a good toxic response from the system. However, 85% of the inhibition effect was observed with 7 ppm of Cu^{2+}. Thus, BES sensors are a good choice for early detection of any potential effects on wastewater treatment plants.

4. Photoelectrochemical biosensors

Amid several biosensors employment for wastewater monitoring purposes, PEC biosensors have also received considerable attention regarding detection of abnormal changes or sensing of pollutants found in the water environment. A typical PEC biosensor involves a PEC reaction and a specific recognition process. PEC biosensor generally consists of three indispensable elements: (1) light source for excitation, (2) WE, CE, and electrolytic solution/wastewater as detection system, and (3) signal acquisition system. It works on the principle of converting a photo to an electrical signal, which generates a photocurrent signal after irradiation of light to the working electrode (photoactive material) [72,73], as depicted in Fig. 11.4A. The overall mechanism involved in the working of PEC biosensors involves (1) light adsorption by the photoactive material, (2) generation of photoexcited carriers, i.e., electrons (e^-) and holes (h^+), and (3) charge separation to the respective terminals (e^- on working electrode and h^+ to the electrolyte) [26,74]. A suitable light source excites electrons present in the photoactive material from valence band (VB) to conduction band (CB), which eventually produces a photo e^- - h^+ couple. Subsequently, produced photo e^- injects from the CB of photoactive material to the conductive working electrode. Concurrently, high energy photo h^+ is scavenged by the targeted biomolecules/pollutants present in the electrolytic solution/wastewater and oxidizes the targeted biomolecules, which contributes an electron to the photoactive material [72]. However, oxidation of biomolecules depends on the kinetic energy of photo h^+; the oxidation potential of target analyte present in electrolyte should be less than VB energy of bioactive material for effective photocatalytic oxidation. Meanwhile, photo e^- is transported to the CE, reducing protons to hydrogen (Fig. 11.4A). Conclusively, under irradiation of light, enhancement in photocurrent quantifies the amount of target analytes [26], as represented in Fig. 11.4B.

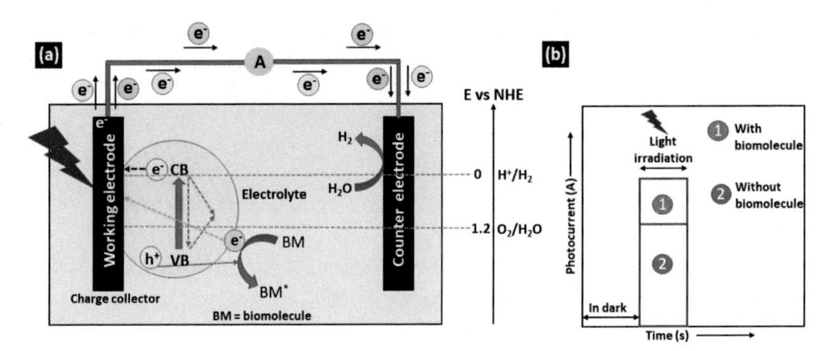

FIGURE 11.4 (A) Schematic representation of mechanism followed by a typical photoelectrochemical biosensor, (B) output photocurrent response in presence and absence of targeted biomolecule.

This suggests detection of ultralow pollutant concentration in wastewater, which is hardly possible by other electrochemical biosensors. Target can directly or indirectly impact/change the properties of electrolyte environment or photoactive material resulting in variation of photocurrent generation and quantitative estimation [73].

Several salient features of PEC biosensor contribute to its improved performance such as the following: (1) it combines three different aspects including photocatalysis, biocatalysis, and electrocatalysis, which synergistically contribute their advantages in terms of characteristically fast response, miniaturization feasibility, and high sensitivity, (2) exploration of specific biomolecules for interaction with targeted analyte lead to high selectivity and sensitivity in PEC biosensors, and (3) separation of excitation source (light) from recognition signal (electric current) contributes to low background signal and high sensitivity [72,73].

Progressively, PEC biosensor has been explored for online monitoring and portability aspect for sensing of pollutant in a water environment. In this regard, an extended gate field-effect transistor (EGFET)-dependent PEC sensor was developed to detect COD as a water quality assessment parameter. The extended gate comprises 3D TiO_2 nanotubes with Ti mesh electrodes modified with platinum NPs. The promising structure of EGFET was able to detect the lowest COD concentration of 0.12 mg/L and a broad COD detection range from 1.44 to 672 mg/L with 1 mL/s of continuous flow rate. COD detection results in real wastewater show that the PEC sensor is good consensus with the standard dichromate method of COD detection [75]. Furthermore, a portable COD sensor based on the photoelectrochromic principle was introduced [76], where a tiny indium tin oxide (ITO) electrode was altered with photocatalytic $TiO_2/g\text{-}C_3N_4$ material and an electrochromic Prussian blue material. Beneath the illumination of light, an organic pollutant in wastewater was oxidized by $TiO_2/g\text{-}C_3N_4$ to Prussian blue. Thus, the COD value was detected by the change in the ITO electrode color. With good stability and repeatability, this sensor showed a flexible COD detection range of 0.025–750 mg/L [76]. Lately, a novel concept came into focus, where living microalgae was explored for the detection of a toxic chemical such as metal ions in wastewater [77]. Herein, photocurrent was amplified through exited photosynthetic activity from Chlorella (living algal species), and Cu electrode and Cu NPs formed nanoactivity by almost two orders of magnitude. Accordingly, microalgal sensors were employed for heavy and light metal detection with a breakthrough detection limit of 50 nM [77].

Furthermore, a PEC-based enzymatic biosensor is also considerably used in wastewater pollutant detection. It inherits PEC bioanalysis' sensitivity and the enzymes' selectivity simultaneously.

4.1 Photoelectrochemical enzymatic biosensors

PEC enzymatic biosensors work on the principle of electrical signal tracking before and after the enzymatic transformation, where photoelectrode, RE, and CE work in conjunction within an electrolyte/water environment. At first, enzymes are confined onto the surface of photoelectrode, and then subsequent biocatalytic events are recorded through electrical signals [78], as depicted in Fig. 11.5. Enzymes can be incredibly advantageous as biorecognition elements owing to their (1) high efficiency to generate numerous detectable products and greatly amplify the signal, (2) fast detection of individual substrate owing to their exquisite specificity for a particular substrate, and thus (3) this inherent selectivity avoid disturbance through interfering species in the sample [79].

For rapid and sensitive recognition of organophosphates (OPs), a visible light-based PEC biosensor paired with enzyme inhibition was constructed on dual-functional $Cd_{0.5}Zn_{0.5}S$-reduced graphene oxide ($Cd_{0.5}Zn_{0.5}S$-rGO) nanocomposite. $Cd_{0.5}Zn_{0.5}S$-rGO was immobilized with AChE, and as visible

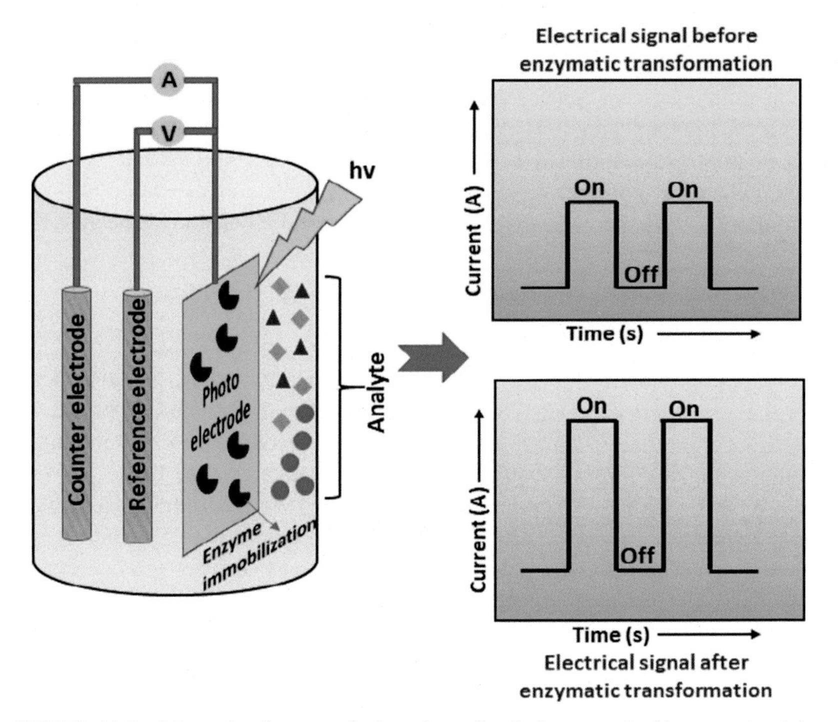

FIGURE 11.5 Schematic diagram of photoelectrochemical enzymatic biosensor involving enzyme immobilization on the photoelectrode surface (working as a recognition element) for biocatalytic transformation and detection through electrical signal enhancement.

light illumination takes place, it hydrolyzes acetylthiocholine chloride into thiocholine. The hydrolysis assists photocurrent enhancement of the photoelectrode enzyme, and further as OPs come in contact with the enzyme, inhibition takes place, which declines photocurrent response. Results revealed a wide range of OPs detection, i.e., 0.001−1 μg and the lowest detection limit as 0.3 ng/mL. It has been successfully demonstrated for OPs pesticides detection in water and can be useful for sensing other hazardous compounds [80]. Furthermore, a PEC enzymatic biosensor composed of TiO_2 nanotubes array (TNA), HRP enzyme, and CdY was employed to detect commonly used herbicide asulam in real wastewater. HRP was assembled to the inner wall of TNA covalently, which catalyzes H_2O_2 and reduces $S_2O_3^{2-}$ to S^{2-}, and further generated S^{2-} reacted with CdY to form quantum dots under visible light illumination. A decrease in the photocurrent will be observed owing to the inhibitory effect of asulam present in wastewater toward HRP. PEC TNA/HRP biosensor has shown successful implementation in real wastewater with a detection limit of 0.02−2.0 ng/mL and a low detection limit of 4.1 pg/mL [81].

Further development of PEC enzymatic biosensors could include their expansion in industrial applications. Advancement in enzyme immobilization, accessibility, surface coverage, stability, and activity of the enzyme molecules along with custom engineering of biomolecules, new progress in material science, and nanotechnology could contribute to new detection principles and signal amplification. Besides, the synergistic effect could permit detection even at very low concentrations. Application of nucleic acids−based enzymes will allow specific and sensitive detection of several biochemical species [82].

5. Biosensors as a perspective to monitor infectious disease outbreak

Currently, public health and hygiene are attracting ample attention from governing authorities due to the unfortunate global pandemic. Despite the existing diagnostic techniques, wastewater-based epidemiology is blooming as an influential tool for the on-site and on-time detection of borne diseases caused by the pathogenic contamination of water. Waterborne diseases hold a great share in the global mortality rate arising every year [4,5]. The sewage drained out from the household activities and food industries are overloaded with organics and nutrients and are conducive for the growth of pathogenic microorganisms such as bacteria, viruses, parasites, protozoa, etc. [4] (Fig. 11.6). The known causes of waterborne diseases are lack of suitable water monitoring systems, water infrastructure, poor sanitation, and public knowledge. Therefore, the availability of real-time water monitoring devices such as biosensors with appreciably short response times will be the safeguard for combating waterborne illness and decreasing the mortality rate. It would be possible to monitor the changing water quality and take on-time attentive measures to design the treatment processes with the view of human vulnerability and control the widespread of diseases [4,5,28].

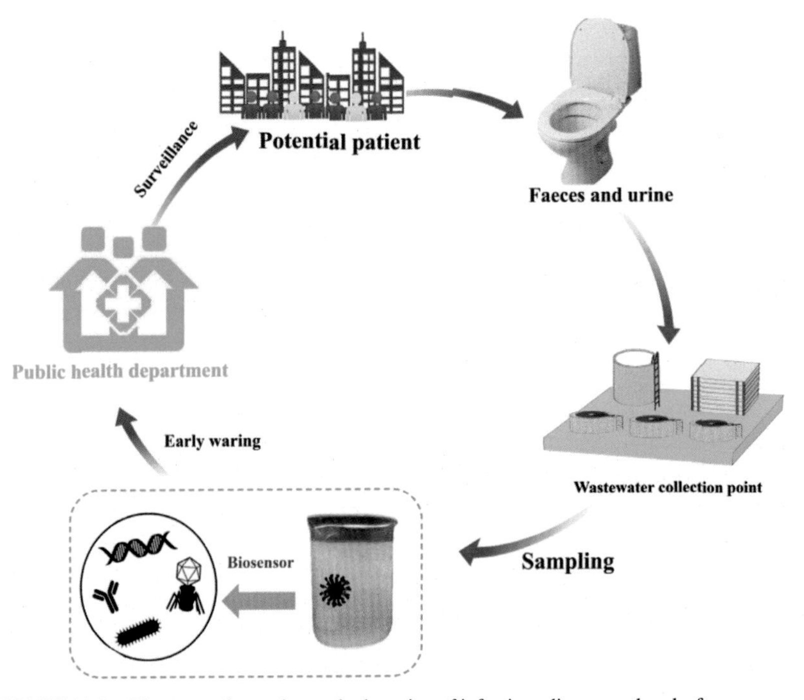

FIGURE 11.6 Biosensors for testing early detection of infectious disease outbreaks from sewage. *Taken from K. Mao, K. Zhang, W. Du, W. Ali, X. Feng, H. Zhang, The potential of wastewater-based epidemiology as surveillance and early warning of infectious disease outbreaks, Curr. Opin. Environ. Sci. Health 17 (2020) 1-7.*

The unique features of biosensors, such as high selectivity, sensitivity, economical, fast, real-time analytical technique, and miniaturization, have been attracting the scientific community's attention for the surveillance and quick notice of infectious disease epidemics and transmission. It eliminates the unnecessary human intervention for sewage sampling and analysis for the qualitative and quantitative inferences. Furthermore, its application avoids the requirement of a technical platform for sophisticated and rigorous analysis [5]. Depending upon the transducer principle, optical and electrochemical biosensors are most commonly designed to diagnose microorganisms as a whole or biomarkers involved in infectious disease outbreaks [4]. Researchers have used two different mechanisms to sense the presence of pathogens in the water testing sample, i.e., detection of specific nucleic acid sequences of a microorganism, and the acknowledgment of an element, such as antibodies, bacterial aptamers, host−guest relationships between bacteriophages, and alternation with chemical groups [4,84]. An aptamer modified silver nanoparticle was developed for *Staphylococcus aureus*' electrochemical dual aptamer−based sandwich recognition [84]. The second mechanism of detecting specific nucleic acid sequences of a microorganism involves nucleic acid–based polymerase chain reaction (PCR) analysis; however, the complex wastewater generally contains certain PCR inhibitors that hinder their detection [85]. A LAMP-based biosensor was then developed for the on-site detection of *Toxoplasma gondii* and *Entamoeba histolytica* in wastewater [85]. Later on, paper-based devices were developed using the microfluidic technique for nucleic acid testing. The chemiluminescence immunosensor was developed to detect *Escherichia coli* O157: H7 by applying a two-step sandwich measurement strategy by Zhang et al. [78]. Other optical biosensors developed for the detection of *E. coli* O157: H7 utilized fluorescently labeled specific aptamer [86], aptamers adsorbed on the surface of AuNPs [87], and carboxyl functionalized graphene quantum dots (cf-GQDs) [88]. Güner et al. [89] developed a voltametric immunosensor by substrate alteration of the graphite electrode through a bionanocomposite of polypyrrole (PPy)/AuNP/MWCNT/chitosan for the detection of *E. coli* O157: H7.

The metabolically active *Legionella pneumophila* was detected in environmental water samples using various optical biosensors. Foudeh et al. [90] detected *L. pneumophila*'s presence by the RNA detector probe embedded on the biochip gold surface, which was based on the determination of bacterial RNA. Manera et al. [91] utilized the gold substrate, which was functionalized in conjunction with an antibody solution against *L. pneumophila* and a self-assembled monolayer of protein A. The LOD of 10^3 CFU/mL was obtained in a response time of 30 min. An electrochemical biosensor was developed for *L. pneumophila* detection dependent on principle of grating coupling screen plasmon resonance that showed improved LOD of 10 CFU/mL and appreciable specificity [92].

Much of the work has been done for the easy and accurate detection of SARS-CoV-2 using biosensors since the year 2020. Xu et al. [93] reviewed the rapid biosensing of SARS-CoV-2 using facile biosensors and stated that the target components used for the COVID-19 detection could be spike (S), membrane (M), and envelop (E) proteins and viral RNA, as well as the host response, generated antibodies (IgG and IgM). For the quick detection of COVID-19 with 100% specificity and sensitivity for SARS-CoV-2, a manifold reverse transcription loop–mediated isothermal amplification conjugated by a nanoparticle lateral flow biosensor (mRT-LAMP-LFB) was developed with a limit of detection of 12 copies per reaction in 1 h [94]. The global pandemic situation due to COVID-19 had been observed as an emergency, and WHO emphasized testing and continuous testing until the vaccination came into the role as a way to protect mass populations from severity/death. Meanwhile, several biosensors were designed and developed to combat the urgent testing situation with gradual improvements of the improved limit of detection and reduced response time. Qiu et al. [95] established a dual-functional plasmonic biosensor that combined the localized surface plasmon resonance and plasmonic photothermal sensing transduction. It provides accurate sensitivity for viral sequences including E genes, ORF1ab, and RdRp from SARS-CoV-2 by LOD with low concentration of 0.22 pM and high precision as a substitute for clinical COVID-19 detection. For rapid detection of SARS-CoV-2 RNA in 10 min, a DNA nanoscaffold hybrid chain reaction–based fluorescence biosensor was developed [96]. Another approach of COVID-19 detection was the development of immunosensor for the estimation of surface antigens and viruses. In this direction, a graphene field-effect transistor (Gr-FET) was reported with a low response time of 2 min and LOD down to 0.2 pM [97]. The biomarkers were also targeted as an interesting technique for the estimation of SARS-CoV-2. The several biomarkers studied are mitochondrial reactive oxygen species [98], homocysteine and angiotensin II [99], and cytokine (e.g., IL-6) [100].

6. Conclusions, future trends, and prospective of biosensors

The field of biosensors is progressive as they have an obvious advantage over the existing analytical methods. Biosensors are rapid, accurate, economical, and easy-to-use analytical devices that require minimum human intervention and no technical handling obligation. Researchers have so far designed and tested several biosensors for water quality monitoring parameters such as dissolved oxygen, COD, BOD, heavy metals, toxins, pathogens, biomarkers, antibiotic resistance genes, etc. Few among them have already made their way to market. However, some have been tested on a laboratory scale, but the real-time application is yet to be explored. For instance, an MFC biosensor using bioanode and biocathode to detect pollutants in water and wastewater is highly promising with long stability and high linearity. The bioanode biosensor offers the great advantage of analyzing the treated streams of anaerobic digestors for organic content. However, the response time and sensitivity still need to be improved for its commercialization. Similarly, biocathode biosensors can be directly applied to clean drinking water to monitor toxins without high organic load shock interference. A thoughtful insight into the mass-transfer phenomenon, which changes with the biofilm formation, is still required [31].

The major challenge in wastewater management is the complexity of wastewater due to spatial and temporal variations. The presence of several components together in the complex wastewater matrix demands a multi-constituent detectable sensor with a more extensive detection range and limit of detection in the lower concentration range. The use of nanomaterials to integrate multiple sensors into a single device would be a helpful future development for miniaturizing [101–104]. The use of nanomaterials in biosensors further improves the sensitivity and reliability and decreases the response time. Integration of different sensing components such as potentiostats with several electronic equipment, low-power sources, etc., can also be considered.

The presence of pathogens and other biomarkers in the wastewater and water streams has attracted much attention currently. The early monitoring of these entities is effective in controlling the outbreak of infectious diseases. However, unlike biological samples, environmental water samples are very complex and contain several physiologically relevant biomarkers for waterborne pathogens. This complexity in detecting the pathogens and biomarkers with accuracy demands the development of biosensors with included multiplexing [28]. In the view of low resources in privileged countries, the development of biosensor in the future must be with the spirit of reusing recyclable materials. It would also be an environmentally sustainable approach. In certain emergency cases, such as contamination of water stream with toxins or other hazardous pollutants that need to be monitored immediately and with complete accuracy, the development of a highly sensitive sensor/signal transmission system is required to send warning information within a reasonable time frame. This arises the future development of sensors integrated with artificial intelligence to upgrade the total system to a robotic structure [105].

References

[1] N. Yao, J. Wang, Y. Zhou, Rapid determination of the chemical oxygen demand of water using a thermal biosensor, Sensors 14 (6) (June 6, 2014) 9949–9960.

[2] S. Foorginezhad, et al., Recent advances in sensing and assessment of corrosion in sewage pipelines, Process Saf. Environ. Prot. (2020).

[3] R. Sharma, N. Verma, Y. Lugani, S. Kumar, M. Asadnia, Conventional and advanced techniques of wastewater monitoring and treatment, in: Green Sustainable Process for Chemical and Environmental Engineering and Science, Elsevier, 2021, pp. 1–48.

[4] K. Mao, H. Zhang, Y. Pan, Z. Yang, Biosensors for wastewater-based epidemiology for monitoring public health, Water Res. 191 (March 1, 2021) 116787.

[5] F. Ejeian, et al., Biosensors for wastewater monitoring: a review, Biosens. Bioelectron. 118 (October 30, 2018) 66–79.

[6] M. Asadnia, L.H. Chua, X. Qin, A. Talei, Improved particle swarm optimization–based artificial neural network for rainfall-runoff modeling, J. Hydrol. Eng. 19 (7) (2014) 1320–1329.

[7] M. Mahmud, et al., Recent progress in sensing nitrate, nitrite, phosphate, and ammonium in aquatic environment, Chemosphere (2020) 127492.

[8] M. Asadnia, et al., Mercury (II) selective sensors based on AlGaN/GaN transistors, Anal. Chim. Acta 943 (2016) 1–7.

[9] M. Asadnia, et al., Ca^{2+} detection utilising AlGaN/GaN transistors with ion-selective polymer membranes, Anal. Chim. Acta 987 (2017) 105–110.

[10] T.M. Sanders, et al., Description of ionophore-doped membranes with a blocked interface, Sens. Actuators B Chem. 250 (2017) 499–508.

[11] K. Mao, et al., Nanomaterial-based aptamer sensors for arsenic detection, Biosens. Bioelectron. 148 (2020) 111785.

[12] Y. Chen, J. Liu, Z. Yang, J.S. Wilkinson, X. Zhou, Optical biosensors based on refractometric sensing schemes: a review, Biosens. Bioelectron. 144 (2019) 111693.

[13] P. Srivastava, R. Abbassi, A.K. Yadav, V. Garaniya, F. Khan, Microbial fuel cell—integrated wastewater treatment systems, in: Integrated Microbial Fuel Cells for Wastewater Treatment, Elsevier, 2020, pp. 29—46.

[14] R. Abbassi, A.K. Yadav, Introduction to microbial fuel cells: challenges and opportunities, in: Integrated Microbial Fuel Cells for Wastewater Treatment, Elsevier, 2020, pp. 3—27.

[15] M.H. Do, et al., Microbial fuel cell-based biosensor for online monitoring wastewater quality: a critical review, Sci. Total Environ. 712 (April 10, 2020) 135612.

[16] L. Peixoto, et al., In situ microbial fuel cell-based biosensor for organic carbon, Bioelectrochemistry 81 (2) (2011) 99—103.

[17] C. Corbella, M. Hartl, M. Fernandez-gatell, J. Puigagut, MFC-based biosensor for domestic wastewater COD assessment in constructed wetlands, Sci. Total Environ. 660 (2019) 218—226.

[18] M. Do, et al., Challenges in the application of microbial fuel cells to wastewater treatment and energy production: a mini review, Sci. Total Environ. 639 (2018) 910—920.

[19] B. Bucur, C. Purcarea, S. Andreescu, A. Vasilescu, Addressing the selectivity of enzyme biosensors: solutions and perspectives, Sensors 21 (9) (April 26, 2021).

[20] Z. Changani, A. Razmjou, A. Taheri-Kafrani, M.J.A.M.I. Asadnia, Domino P-μMB: a new approach for the sequential immobilization of enzymes using polydopamine/polyethyleneimine chemistry and microfabrication, Adv. Mat. 7 (13) (2020) 1901864.

[21] A.R.M. Izzah Binti Mohammad, K. Liang, M. Asadnia, V. Chen, MOF-based enzymatic microfluidic biosensor via surface patterning and biomineralization, ACS Appl. Mater. Interfaces 11 (2) (2019) 1807—1820.

[22] Z. Mehrabi, A. Taheri-Kafrani, M. Asadnia, A.J.S. Razmjou, P. Technology, Bienzymatic modification of polymeric membranes to mitigate biofouling, Sep. Purif. Technol. 237 (2020) 116464.

[23] A.-E. Becquerel, Recherches sur les effets de la radiation chimique de la lumiere solaire au moyen des courants electriques, C. R. Acad. Sci. 9 (145) (1839) 1.

[24] M. Becquerel, Mémoire sur les effets électriques produits sous l'influence des rayons solaires, C. R. Acad. Hebd. Seances Acad. Sci. 9 (1839) 561—567.

[25] H. Han, et al., Three dimensional-TiO_2 nanotube array photoanode architectures assembled on a thin hollow nanofibrous backbone and their performance in quantum dot-sensitized solar cells, Chem. Commun. 49 (27) (2013) 2810—2812.

[26] A. Devadoss, P. Sudhagar, C. Terashima, K. Nakata, A. Fujishima, Photoelectrochemical biosensors: new insights into promising photoelectrodes and signal amplification strategies, J. Photochem. Photobiol. C Photochem. Rev. 24 (2015) 43—63.

[27] L. Zhang, H.H. Mohamed, R. Dillert, D. Bahnemann, Kinetics and mechanisms of charge transfer processes in photocatalytic systems: a review, J. Photochem. Photobiol. C Photochem. Rev. 13 (4) (2012) 263—276.

[28] J. Rainbow, et al., Integrated electrochemical biosensors for detection of waterborne pathogens in low-resource settings, Biosensors 10 (4) (2020) 36.

[29] C.I. Justino, A.C. Duarte, T.A. Rocha-Santos, Recent progress in biosensors for environmental monitoring: a review, Sensors 17 (12) (2017) 2918.

[30] L. Xu, Y. Zhao, C. Fan, Z. Fan, F. Zhao, First study to explore the feasibility of applying microbial fuel cells into constructed wetlands for COD monitoring, Bioresour. Technol. 243 (2017) 846—854.

[31] S. Sevda, et al., Biosensing capabilities of bioelectrochemical systems towards sustainable water streams: technological implications and future prospects, J. Biosci. Bioeng. 129 (6) (June 2020) 647—656.

[32] A. Adekunle, V. Raghavan, B. Tartakovsky, A comparison of microbial fuel cell and microbial electrolysis cell biosensors for real-time environmental monitoring, Bioelectrochemistry 126 (2019) 105–112.

[33] J. Chouler, M. Di Lorenzo, Water quality monitoring in developing countries; can microbial fuel cells be the answer? Biosensors 5 (3) (2015) 450–470.

[34] P. Srivastava, A.K. Yadav, V. Garaniya, R. Abbassi, Constructed wetland coupled microbial fuel cell technology: development and potential applications, in: Microbial Electrochemical Technology, Elsevier, 2019, pp. 1021–1036.

[35] S. Foorginezhad, M. Mohseni-Dargah, Z. Falahati, R. Abbassi, A. Razmjou, M. Asadnia, Sensing advancement towards safety assessment of hydrogen fuel cell vehicles, J. Power Sources 489 (2021) 229450.

[36] P. Srivastava, R. Abbassi, A.K. Yadav, V. Garaniya, M.A.F. Jahromi, A review on the contribution of an electron in electroactive wetlands: electricity generation and enhanced wastewater treatment, Chemosphere (2020) 126926.

[37] P. Srivastava, A. Belford, R. Abbassi, M. Asadnia, V. Garaniya, A.K. Yadav, Low-power energy harvester from constructed wetland-microbial fuel cells for initiating a self-sustainable treatment process, Sustain. Energy Technol. Assess. 46 (2021) 101282.

[38] J. Chouler, M.D. Monti, W.J. Morgan, P.J. Cameron, M. Di Lorenzo, A photosynthetic toxicity biosensor for water, Electrochim. Acta 309 (2019) 392–401.

[39] M. Grattieri, K. Hasan, S.D. Minteer, Bioelectrochemical systems as a multipurpose biosensing tool: present perspective and future outlook, ChemElectroChem 4 (4) (2017) 834–842.

[40] N.E. Stein, H.V. Hamelers, C.N. Buisman, Stabilizing the baseline current of a microbial fuel cell-based biosensor through overpotential control under non-toxic conditions, Bioelectrochemistry 78 (1) (2010) 87–91.

[41] M. Di Lorenzo, T.P. Curtis, I.M. Head, K. Scott, A single-chamber microbial fuel cell as a biosensor for wastewaters, Water Res. 43 (13) (July 2009) 3145–3154.

[42] H. Sun, Y. Zhang, S. Wu, R. Dong, I. Angelidaki, Innovative operation of microbial fuel cell-based biosensor for selective monitoring of acetate during anaerobic digestion, Sci. Total Environ. 655 (2019) 1439–1447.

[43] A. Kaur, et al., Microbial fuel cell type biosensor for specific volatile fatty acids using acclimated bacterial communities, Biosens. Bioelectron. 47 (2013) 50–55.

[44] A. Adekunle, V. Raghavan, B. Tartakovsky, On-line monitoring of heavy metals-related toxicity with a microbial fuel cell biosensor, Biosens. Bioelectron. 132 (2019) 382–390.

[45] A. Khan, et al., A novel biosensor for zinc detection based on microbial fuel cell system, Biosens. Bioelectron. 147 (2020) 111763.

[46] O. Modin, B.-M. Wilén, A novel bioelectrochemical BOD sensor operating with voltage input, Water Res. 46 (18) (2012) 6113–6120.

[47] S.-B. Quek, L. Cheng, R. Cord-Ruwisch, Detection of low concentration of assimilable organic carbon in seawater prior to reverse osmosis membrane using microbial electrolysis cell biosensor, Desalin. Water Treat. 55 (11) (2015) 2885–2890.

[48] I.S. Chang, et al., Continuous determination of biochemical oxygen demand using microbial fuel cell type biosensor, Biosens. Bioelectron. 19 (6) (2004) 607–613.

[49] S.R. Bazaz, A.A. Mehrizi, S. Ghorbani, S. Vasilescu, M. Asadnia, M.E. Warkiani, A hybrid micromixer with planar mixing units, RSC Adv. 8 (58) (2018) 33103–33120.

[50] A.G.P. Kottapalli, M. Asadnia, J. Miao, M. Triantafyllou, Soft polymer membrane microsensor arrays inspired by the mechanosensory lateral line on the blind cavefish, J. Intell. Mater. Syst. Struct. 26 (1) (2015) 38–46.

[51] M. Rafeie, M. Welleweerd, A. Hassanzadeh-Barforoushi, M. Asadnia, W. Olthuis, M. Ebrahimi Warkiani, An easily fabricated three-dimensional threaded lemniscate-shaped micromixer for a wide range of flow rates, Biomicrofluidics 11 (1) (2017) 014108.

[52] M.A.P. Mahmud, et al., Recent progress in sensing nitrate, nitrite, phosphate, and ammonium in aquatic environment, Chemosphere 259 (2020).

[53] H. Karimi-Maleh, F. Karimi, M. Alizadeh, A.L. Sanati, Electrochemical sensors, a bright future in the fabrication of portable kits in analytical systems, Chem. Rec. 20 (7) (2020) 682–692.

[54] H. Karimi-Maleh, K. Cellat, K. Arıkan, A. Savk, F. Karimi, F. Sen, Palladium–Nickel nanoparticles decorated on Functionalized-MWCNT for high precision non-enzymatic glucose sensing, Mater. Chem. Phys. 250 (2020) 123042.

[55] S. Gupta, P. Srivastava, A.K. Yadav, Simultaneous removal of organic matters and nutrients from high-strength wastewater in constructed wetlands followed by entrapped algal systems, Environ. Sci. Pollut. Res. 27 (1) (2020) 1112–1117, https://doi.org/10.1007/s11356-019-06896-z.

[56] Y. Gao, F. Yin, W. Ma, S. Wang, Y. Liu, H. Liu, Rapid detection of biodegradable organic matter in polluted water with microbial fuel cell sensor: method of partial coulombic yield, Bioelectrochemistry 133 (June 2020) 107488.

[57] B.H. Kim, I.S. Chang, G.C. Gil, H.S. Park, H.J. Kim, Novel BOD (biological oxygen demand) sensor using mediator-less microbial fuel cell, Biotechnol. Lett. 25 (7) (2003) 541–545.

[58] M. Kim, et al., Practical field application of a novel BOD monitoring system, J. Environ. Monit. 5 (4) (2003) 640–643.

[59] T. Yamashita, et al., A novel open-type biosensor for the in-situ monitoring of biochemical oxygen demand in an aerobic environment, Sci. Rep. 6 (1) (2016) 1–9.

[60] S. Kharkwal, Y.C. Tan, M. Lu, H.Y. Ng, Development and long-term stability of a novel microbial fuel cell BOD sensor with MnO_2 catalyst, Int. J. Mol. Sci. 18 (2) (2017) 276.

[61] D. Wang, et al., Open external circuit for microbial fuel cell sensor to monitor the nitrate in aquatic environment, Biosens. Bioelectron. 111 (2018) 97–101.

[62] M. Raud, E. Lember, E. Jõgi, T. Kikas, Nitrosomonas sp. based biosensor for ammonium nitrogen measurement in wastewater, Biotechnol. Bioproc. Eng. 18 (5) (2013) 1016–1021.

[63] R.C. Kwan, P.Y. Hon, R. Renneberg, Amperometric determination of ammonium with bienzyme/poly (carbamoyl) sulfonate hydrogel-based biosensor, Sensor. Actuator. B Chem. 107 (2) (2005) 616–622.

[64] M.H. Do, et al., Performance of a dual-chamber microbial fuel cell as biosensor for on-line measuring ammonium nitrogen in synthetic municipal wastewater, Sci. Total Environ. 795 (2021) 148755.

[65] P. Srivastava, A.K. Yadav, V. Garaniya, T. Lewis, R. Abbassi, S.J. Khan, Electrode dependent anaerobic ammonium oxidation in microbial fuel cell integrated hybrid constructed wetlands: a new process, Sci. Total Environ. 698 (2020) 134248.

[66] L.H. Larsen, L.R. Damgaard, T. Kjær, T. Stenstrøm, A. Lynggaard-Jensen, N.P. Revsbech, Fast responding biosensor for on-line determination of nitrate/nitrite in activated sludge, Water Res. 34 (9) (2000) 2463–2468.

[67] M. Grattieri, K. Hasan, S.D. Minteer, Bioelectrochemical systems as a multipurpose biosensing tool: present perspective and future outlook, ChemElectroChem 4 (4) (2016) 834–842.

[68] J. Liu, Z. Cao, Y. Lu, Functional nucleic acid sensors, Chem. Rev. 109 (5) (2009) 1948–1998.

[69] W. Zhou, R. Saran, J. Liu, Metal sensing by DNA, Chem. Rev. 117 (12) (2017) 8272–8325.

[70] Z. Xu, et al., Flat microliter membrane-based microbial fuel cell as "on-line sticker sensor" for self-supported in situ monitoring of wastewater shocks, Bioresour. Technol. 197 (2015) 244–251.

[71] Y. Shen, M. Wang, I.S. Chang, H.Y. Ng, Effect of shear rate on the response of microbial fuel cell toxicity sensor to Cu (II), Bioresour. Technol. 136 (2013) 707–710.

[72] L. Yang, S. Zhang, X. Liu, Y. Tang, Y. Zhou, D.K. Wong, Detection signal amplification strategies at nanomaterial-based photoelectrochemical biosensors, J. Mater. Chem. B 8 (35) (2020) 7880–7893.

[73] Q. Zhou, D. Tang, Recent advances in photoelectrochemical biosensors for analysis of mycotoxins in food, Trac. Trends Anal. Chem. 124 (2020) 115814.

[74] Z. Qiu, D. Tang, Nanostructure-based photoelectrochemical sensing platforms for biomedical applications, J. Mater. Chem. B 8 (13) (2020) 2541–2561.

[75] H. Si, et al., A real-time on-line photoelectrochemical sensor toward chemical oxygen demand determination based on field-effect transistor using an extended gate with 3D TiO2 nanotube arrays, Sensor. Actuator. B Chem. 289 (2019) 106–113.

[76] Z. Dai, N. Hao, M. Xiong, X. Han, Y. Zuo, K. Wang, Portable photoelectrochromic visualization sensor for detection of Chemical Oxygen Demand, Anal. Chem. 92 (19) (2020) 13604–13609.

[77] D.N. Roxby, et al., Microalgae living sensor for metal ion detection with nanocavity-enhanced photoelectrochemistry, Biosens. Bioelectron. 165 (2020) 112420.

[78] Y. Zhang, et al., Sensitive chemiluminescence immunoassay for *E. coli* O157: H7 detection with signal dual-amplification using glucose oxidase and laccase, Anal. Chem. 86 (2) (2014) 1115–1122.

[79] D.-M. Han, Z.-Y. Ma, W.-W. Zhao, J.-J. Xu, H.-Y. Chen, Ultrasensitive photoelectrochemical sensing of Pb^{2+} based on allosteric transition of G-Quadruplex DNAzyme, Electrochem. Commun. 35 (2013) 38–41.

[80] Q. Liu, et al., A visible light photoelectrochemical biosensor coupling enzyme-inhibition for organophosphates monitoring based on a dual-functional Cd 0.5 Zn 0.5 S-reduced graphene oxide nanocomposite, Analyst 139 (5) (2014) 1121–1126.

[81] J. Tian, Y. Li, J. Dong, M. Huang, J. Lu, Photoelectrochemical TiO_2 nanotube arrays biosensor for asulam determination based on in-situ generation of quantum dots, Biosens. Bioelectron. 110 (2018) 1–7.

[82] W.-W. Zhao, J.-J. Xu, H.-Y. Chen, Photoelectrochemical enzymatic biosensors, Biosens. Bioelectron. 92 (2017) 294–304.

[83] K. Mao, K. Zhang, W. Du, W. Ali, X. Feng, H. Zhang, The potential of wastewater-based epidemiology as surveillance and early warning of infectious disease outbreaks, Curr. Opin. Environ. Sci. Health 17 (2020) 1–7.

[84] A. Abbaspour, F. Norouz-Sarvestani, A. Noori, N. Soltani, Aptamer-conjugated silver nanoparticles for electrochemical dual-aptamer-based sandwich detection of staphylococcus aureus, Biosens. Bioelectron. 68 (2015) 149–155.

[85] C. Ajonina, C. Buzie, J. Möller, R. Otterpohl, The detection of Entamoeba histolytica and Toxoplasma gondii in wastewater, J. Toxicol. Environ. Health Part A 81 (1–3) (2018) 1–5.

[86] N. Yildirim, F. Long, A.Z. Gu, Aptamer based E-coli detection in waste waters by portable optical biosensor system, in: 2014 40th Annual Northeast Bioengineering Conference (NEBEC), IEEE, 2014, pp. 1–3.

[87] W.-h. Wu, et al., Aptasensors for rapid detection of *Escherichia coli* O157: H7 and *Salmonella typhimurium*, Nanoscale Res. Lett. 7 (1) (2012) 1–7.

[88] X. Yang, L. Feng, X. Qin, Preparation of the Cf-GQDs-Escherichia coli O157: H7 bioprobe and its application in optical imaging and sensing of *Escherichia coli* O157: H7, Food Anal. Methods 11 (8) (2018) 2280–2286.

[89] A. Güner, E. Çevik, M. Senel, L. Alpsoy, An electrochemical immunosensor for sensitive detection of *Escherichia coli* O157: H7 by using chitosan, MWCNT, polypyrrole with gold nanoparticles hybrid sensing platform, Food Chem. 229 (2017) 358–365.

[90] A.M. Foudeh, H. Trigui, N. Mendis, S.P. Faucher, T. Veres, M. Tabrizian, Rapid and specific SPRi detection of *L. pneumophila* in complex environmental water samples, Anal. Bioanal. Chem. 407 (18) (2015) 5541–5545.

[91] M.G. Manera, et al., SPR based immunosensor for detection of *Legionella pneumophila* in water samples, Opt Commun. 294 (2013) 420–426.

[92] A. Meneghello, A. Sonato, G. Ruffato, G. Zacco, F. Romanato, A novel high sensitive surface plasmon resonance *Legionella pneumophila* sensing platform, Sensor. Actuator. B Chem. 250 (2017) 351–355.

[93] L. Xu, D. Li, S. Ramadan, Y. Li, N. Klein, Facile biosensors for rapid detection of COVID-19, Biosens. Bioelectron. (2020) 112673.

[94] X. Zhu, et al., Multiplex reverse transcription loop-mediated isothermal amplification combined with nanoparticle-based lateral flow biosensor for the diagnosis of COVID-19, Biosens. Bioelectron. 166 (2020) 112437.

[95] G. Qiu, Z. Gai, Y. Tao, J. Schmitt, G.A. Kullak-Ublick, J. Wang, Dual-functional plasmonic photothermal biosensors for highly accurate severe acute respiratory syndrome coronavirus 2 detection, ACS Nano 14 (5) (2020) 5268–5277.

[96] J. Jiao, C. Duan, L. Xue, Y. Liu, W. Sun, Y. Xiang, DNA nanoscaffold-based SARS-CoV-2 detection for COVID-19 diagnosis, Biosens. Bioelectron. 167 (2020) 112479.

[97] X. Zhang, et al., Electrical probing of COVID-19 spike protein receptor binding domain via a graphene field-effect transistor, Physics (2020) arXiv preprint arXiv:2003.12529.

[98] Z.S. Miripour, et al., Real-time diagnosis of reactive oxygen species (ROS) in fresh sputum by electrochemical tracing; correlation between COVID-19 and viral-induced ROS in lung/respiratory epithelium during this pandemic, Biosens. Bioelectron. 165 (2020) 112435.

[99] G. Ponti, M. Maccaferri, C. Ruini, A. Tomasi, T. Ozben, Biomarkers associated with COVID-19 disease progression, Crit. Rev. Clin. Lab Sci. 57 (6) (2020) 389–399.

[100] S.M. Russell, A. Alba-Patiño, E. Barón, M. Borges, M. Gonzalez-Freire, R. De La Rica, Biosensors for managing the COVID-19 cytokine storm: challenges ahead, ACS Sensors 5 (6) (2020) 1506–1513.

[101] M. Abdollahzadeh, et al., Low humid transport of anions in layered double hydroxides membranes using polydopamine coating, J. Membr. Sci. 624 (2021) 118974.

[102] H. Ahmadi, et al., Incorporation of natural lithium-ion trappers into graphene oxide nanosheets, Adv. Mater. (2020) 2000665.

[103] A. Razmjou, M. Asadnia, E. Hosseini, A.H. Korayem, V.J. N.c. Chen, Design principles of ion selective nanostructured membranes for the extraction of lithium ions, Nat. Commun. 10 (1) (2019) 1–15.

[104] A. Razmjou, et al., Lithium ion-selective membrane with 2D subnanometer channels, Water Res. 159 (2019) 313–323.

[105] N.A. Abdullah, S. Ramli, N.H. Mamat, S. Khan, C. Gomes, Chemical and biosensor technologies for wastewater quality management, Int. J. Adv. Res. Publ 1 (6) (2017) 1–10.

Chapter 12

Machine learning in surface plasmon resonance for environmental monitoring

Masoud Mohseni-Dargah[1,2], Zahra Falahati[3], Bahareh Dabirmanesh[1], Parisa Nasrollahi[4], Khosro Khajeh[1]

[1]*Department of Biochemistry, Faculty of Biological Sciences, Tarbiat Modares University, Tehran, Iran;* [2]*School of Engineering, Macquarie University, Sydney, NSW, Australia;* [3]*Department of Biological Sciences, Institute for Advanced Studies in Basic Sciences (IASBS), Zanjan, Iran;* [4]*Department of Nanobiotechnology, Faculty of Biological Sciences, Tarbiat Modares University, Tehran, Iran*

1. Introduction

Biosensors are known as analytical instruments with biological sensing compartments and used as diagnostic/detection devices. They typically consist of three main parts: a bioreceptor (recognizing the analyte, leading to a signal), a transducer, a signal processing system (containing electronics converting analog to digital), and a display. They possess several positives such as high levels of speed, low amounts of cost, nonablative features, and on-site diagnosis [1–4]. The sensors have been broadly utilized in food safety, basic biological research [5–8], drug screening [9], disease and biomedical diagnosis [10], environmental monitoring [3,11–17], and other applications [18–22].

Surface plasmon resonance (SPR) is an optical biosensor to study molecular interactions, namely antibody–antigen binding, ligand–analyte interaction, and receptor characterization. SPR measures any alterations in the refractive index happening in a metal evanescent field after the interaction between ligands (immobilized on the surface) and analytes (the molecule of interest to be detected). This technique has great features, for instance, high sensitivity, label-free and real-time sensing, convenient sample preparation, and cost-effectiveness, compared with other detection methods, and indicates several applications including biomolecular interaction analysis, medical diagnosis, food safety, and environmental monitoring [5,9,12,16,17,23–25]. SPR shows a high level of potency in monitoring environmental pollutants in real time with a small volume of samples at low concentrations (pM to μM) [26].

Artificial Intelligence and Data Science in Environmental Sensing
https://doi.org/10.1016/B978-0-323-90508-4.00012-5

Artificial intelligence (AI), human intelligence simulation by machines, has emerged to help biosensors improve their accuracy and fill the existing gap between data acquisition and analysis, by approaches like classification algorithms and pattern analysis [16,17,24,27,28]. After the integration of AI into biosensors, a new concept named "AI biosensor" was introduced [27]. Furthermore, to analyze the data from AI-based biosensors, machine learning (ML) is used, which can make conventional biosensors intelligent and assist to overcome the existing challenges of biosensors [1]. There are several ML-based methods and algorithms to learn from data toward finding beneficial information or predictive models of an event [27]. Principal component analysis (PCA), autoencoder (AE), ridge regression, k-nearest neighbor (KNN), k-means clustering, t-distributed stochastic neighbor embedding (t-SNE), clustering using representatives (CURE), non-negative matrix factorization (NMF), generative adversarial network (GAN), and various neural networks (NNs) are the ML algorithms used in SPR-based research.

As environmental pollutants are dangerous for human health, and SPR has shown great promise in environmental monitoring, obtained data should be well manipulated and the response reliability is to be guaranteed. To reach this aim, AI, ML, and modeling can assist and confirm the test quality. In this chapter, at first, among various types of biosensors, the principle and use of SPR regarding environmental sensing will be discussed. Then, various ML methods in SPR will be briefly explained. Finally, the pieces of research related to different ML algorithms used in SPR are aimed to be reviewed.

2. Surface plasmon resonance

A surface plasmon is an electron oscillation along the interface of metal with its dielectric medium. When an incident light passes through the prism, covered with a thin metal layer on top (usually Au or Ag), at angles higher than the critical angle, it undergoes total internal reflection (TIR). Under TIR conditions, at a specific angle, the photon penetrates the metal and transfers its energy to the oscillating electrons that induce SPR. This will occur when the incident wave vector (K_i) equals the surface plasmon wave vector (K_p) as shown in Fig. 12.1. The light alone cannot directly excite the plasmon and it gets its extra momentum from the prism to satisfy the resonance matching condition.

When the photon energy is transferred into the oscillated electrons, it will generate resonance and a sharp dip in the reflected light intensity. The angle at which the maximum decrease in the intensity (dip) is observed is called a resonance angle or SPR angle (Fig. 12.1A). Meanwhile, on the opposite side, an evanescent wave will be formed that decays exponentially with distance from the surface. The evanescent wave will detect any changes in the

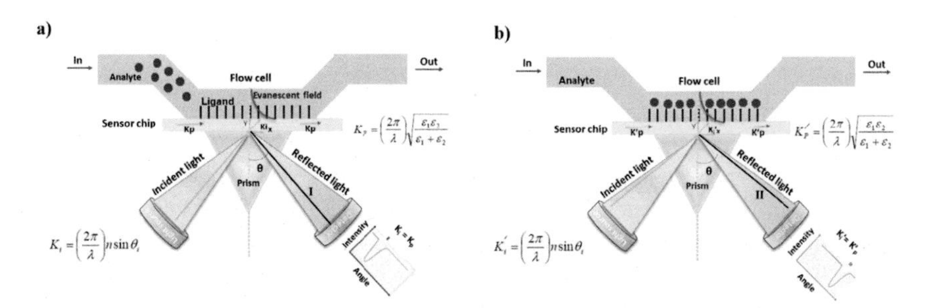

FIGURE 12.1 Principle of surface plasmon resonance and changes in the resonance angle during binding. K_i, wave vector of the incident light that is parallel to the prism; K_p, surface plasmon wave vector; θ, an incident angle; λ, wavelength of the incident light; ε_1, dielectric permittivity constants of the metal film; ε_2, dielectric medium.

refractive index close to the metal (within 200 nm). Therefore, any changes like mass increase, as a result of binding, will cause an alteration in the refractive index on the other side that affects electron oscillation. This change in the velocity of the plasmon alters the incident light vector needed for inducing SPR and leads to a shift in the resonance angle (Fig. 12.1B). Monitoring a change in the SPR dip angle over time creates a sensorgram, a graph of response against time, which visualizes the different steps of binding (Fig. 12.2).

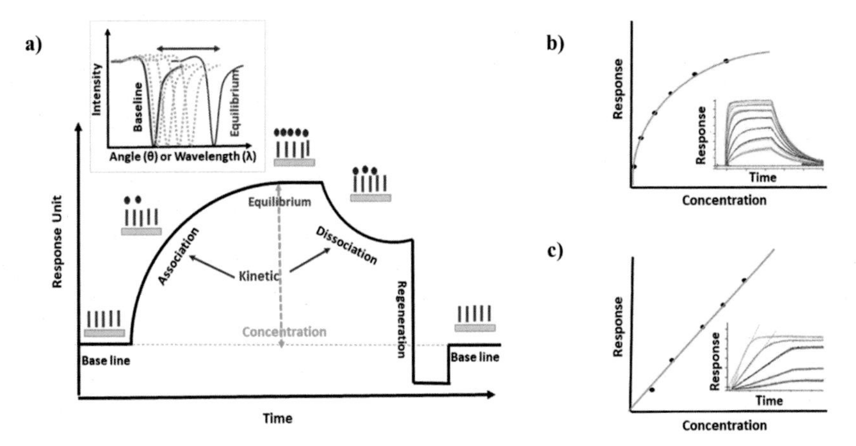

FIGURE 12.2 Schematic of a surface plasmon resonance (SPR) sensorgram. (A) An ideal sensorgram consists of four steps including association, steady state or equilibrium, dissociation, and regeneration. The inset displays the SPR angle shift (dip) caused by refractive index variation, resulting from biomolecular interactions. The plot for determining (B) kinetic parameters and (C) analyte concentrations, following the injection of different concentrations of analytes (insets B and C).

2.1 Sensorgram

A simple SPR analysis requires the immobilization of a ligand onto the sensor chip and the addition of an analyte over the ligand. Upon injecting the analyte to its target ligand on the surface chip, the association phase of a sensorgram will appear. When the injection time is long enough, a steady state is reached, where binding and dissociating molecules are in equilibrium. As soon as the injection is ended, the dissociation phase starts until reaching the baseline (Fig. 12.2A). However, some assays need an additional regeneration step to completely dissociate the analyte and reach the baseline. From the shape of the binding curve, kinetic parameters like association and dissociation rate constants (k_{on}, k_{off}) can be fitted and calculated. The equilibrium dissociation constant (K_D) can be computed from the kinetic rate constant ratio. To determine a kinetic constant, a sensorgram from multiple concentrations of the analyte must be fitted to a kinetic model using a mathematic algorithm (Fig. 12.2B). The SPR sensorgram can also evaluate analyte concentrations. To calculate the concentration, the initial binding rate of biomolecule interaction under limited mass transfer is used. Thus, any changes in the binding are proportionally related to the analyte concentration (Fig. 12.2C).

2.2 Other types of SPR platforms

Various efforts have been made so far to improve the sensitivity and limit of detection. For example, applying novel materials and detection schemes is promising for reducing noise and inducing responses. Therefore, the type of matrices on the sensor surface and the coupling strategies for attaching the ligand on the sensor surface are important factors that could influence the outcome [29].

SPR biosensing based on nanoparticles (NPs) has gained popularity due to their localized surface plasmon resonance (LSPR). LSPR is produced by metal nanoparticles, usually gold or silver, and shows a strong absorption band in the UV-Vis region. LSPR is more sensitive to binding and less sensitive to bulk refractive index changes, due to its smaller decay length (5–10 nm). LSPR sensors can be categorized into (1) colloidal nanoparticles in solution, (2) flat substrate, and (3) optical fibers coated with nanoparticles. Since no prism is required to couple the light, LSPR instruments are more affordable, much less complex, smaller, and easier to use and maintain. Additionally, they enable on-site measurements [30–35].

Besides the mentioned angular sensing, there are other reported sensing modes including intensity, wavelength, and phase interrogations. In the intensity interrogation SPR, the intensity of the reflected light will be measured in a particular incident angle or wavelength using a CCD camera. This technique is suitable for obtaining an image, and it is known as SPR imaging (SPRi). SPRi follows the same general principles of SPR.

Finally, it is noteworthy to state that SPR apparatuses have emerged as potential candidates for the rapid detection of chemical and biological components in various fields such as environmental monitoring. Due to its attractive features and capabilities, it may have crucial roles in promoting and protecting public health and ecosystems. In the following, the topics related to environmental hazards and sensing by SPR will be discussed.

3. Environmental hazard monitoring by SPR

Today, the issue of identifying and monitoring environmental pollution is a very important global challenge for environmental safety and human well-being. It is worth noting that the identification of these contaminants should be essentially at the level of environmental safety assessment (mg/l or even lower) [36]. In the following, some of the significant organic and inorganic contaminants are discussed.

3.1 Detection of pesticides

Pesticide contamination in the ecosystem is one of the most consequential worries for human and public health in the world. The first SPR biosensor in association with the detection of pesticides was developed in the early 1990s in which the atrazine detection was done, by the use of the inhibition assay, monoclonal antibody, and the atrazine derivative linked to the dextran matrix on the SPR chip. Subsequently, the atrazine assessment was improved through a sandwich assay and secondary antibody to the level of 1–0.05 ng/mL with a test time of 15 min [37].

Different studies individually reported portable SPR sensors for the detection of carbaryl chlorpyrifos, atrazine, and dichlorodiphenyltrichloroethane (DDT) in water samples, applying an inhibition format and a sensor surface coated with the alkanethiolate self-assembled monolayer (SAM) [38–41]. Due to the high diversity of pesticides, multianalyte analysis can be an ideal method. As an instance, a multianalytical analysis of chlorpyrifos, DDT, and carbaryl was presented. In this way, derivatives of all these pesticides listed were bounded to carboxylic terminal groups on a SAM of alkanethiolates. Then, a sample was combined with antibodies related to one target and introduced into the flow cell of the sensor. After generating a sensor response, a sample mixed with antibodies related to another analyte was introduced without prior regeneration of the sensor surface [42].

Recently, molecularly imprinted polymer (MIP) and their combination with SPR sensors have become popular for monitoring pesticides. Utilizing MIP and integrated magnetic NPs in manufacturing SPR biosensors led to the construction of dense and accessible recognition sites for highly sensitive identification of chlorpyrifos [43]. In 2019, Çakır and Baysal [44] reported the feasibility of dimethoate and carbofuran identification by an SPR sensor accompanied with MIP nanofilms based on reflectivity changes. This designed SPR platform was much better in terms of sensitivity, selectivity, and LOD compared to the LC-MS/MS method. Another report was provided for the online monitoring of organophosphorus pesticides in water or soil samples, using the functional plasmonic fiber and metalorganic framework (MOF-5) layer based on the plasmon absorption band shift [45]. Some SPR biosensors used for the monitoring of pesticides in the environment are shown in Table 12.1.

3.2 Detection of phenolic compounds

Phenolic compounds (bisphenol A, nonylphenol, 2,4-dichlorophenol, and phenol) are predominantly distributed in the environment as wastewater of various industries. Phenolic compounds are listed as dangerous substances to living organisms including humans. SPR-based sensors are a valuable tool for phenol detection because of their convenient features.

In one study for detecting bisphenol A, Soh et al. [50] designed an SPR sensor based on an inhibition assay and a monoclonal antibody. The same research group introduced another sensor for the very sensitive identification of 2,4-dichlorophenol as the major dioxin precursor. Detection was performed according to the competition between the analyte existing in a sample and the added BSA−2,4-dichlorophenol complex [51]. Choi et al. [52] presented a direct detection format for the detection of phenol employing the *E. coli* O157:H7 strain as a biorecognition element with the detection limit of 5 ppm. The use of enzymes like laccase, tyrosinase, and horseradish peroxidase as a recognition element has also been reported for the detection of phenol [12,53].

3.3 Detection of heavy metal ions

Heavy metals are toxic and destructive compounds for organisms including humans and are responsible for several lethal diseases. Currently, rapid, high precision, and real-time measurements of lethal metal ions are one of the immense needs to encounter environmental hazards.

Among metallic nanoparticles, gold nanoparticles (GNPs) have specific optical properties due to SPR and LSPR phenomena. Alterations in the size, shape, or geometry of GNPs result in changes in the confinement of local electrons, which lead to alterations in the maximum absorption of SPR and LSPR and the color of the solution. The interaction of GNPs with specific metal ions induces the aggregation of GNPs and leads to a change in color and a shift in the plasmon absorption peak [54].

Given that heavy metals are small analytes and their detection is difficult by portable SPR, the introduction of other technologies, for instance, LSPR, is

TABLE 12.1 Some surface plasmon resonance biosensors for the environmental monitoring of pesticides.

Analyte	Probe anchored	Detection matrix	LOD	Detection format	Ref.
Atrazine	Alkanethiolate SAM	Water	20 pg/mL	Inhibition	[38]
Atrazine	Streptavidin–biotin chemistry	Water	1 pg/mL	Inhibition	[46]
Chlorpyrifos	Alkanethiolate SAM	Water	50 pg/mL	Inhibition	[47]
Carbaryl	Alkanethiolate SAM	Water	1 ng/mL	Inhibition	[47]
Dichlorodiphenyltrichloroethane (DDT)	Alkanethiolate SAM	Water	15 pg/mL	Inhibition	[40]
2,4-Dichlorophenoxyacetic acid (2,4-D)	Immobilization of 2,4-D–BSA conjugate through physical adsorption	Buffer	0.5 ng/mL	Inhibition	[48]
2,4-Dichlorophenoxyacetic acid (2,4-D)	Alkanethiolate SAM	Buffer	10 ppt	Inhibition	[49]
Dimethoate and carbofuran	Nanofilm and MIP	Water	8.37 ng/L 7.11 ng/L	Reflectivity change	[44]
Paraoxon and fenitrothion	Gold layer and MOF-5	Water, soil	10^{-12} M	Wavelength change	[45]

MIP, molecularly imprinted polymer; *MOF*, metalorganic framework; *SAM*, self-assembled monolayer.

quite helpful in increasing sensitivity. The mercury analysis of tap water using a portable sensor could be fulfilled by gold nanorods immobilized on functionalized glass substrates. Au—mercury interactions form amalgam at the longitudinal ends of the mercury and result in structural alterations in gold nanorods and the aspect ratio, leading to a shift in the LSPR peak maximum in the absorption spectra [55].

Using dyes has also gained scholars' attention in SPR design for metal ion detection. For instance, Ock et al. [56] designed an SPR sensor for the direct detection of Cu^{2+} ions. The sensor contained a thin polymer layer containing squarylium dye (SQ), in which the binding of Cu^{2+} ions led to an alteration in refractive index absorption properties.

3.4 Detection of pathogen microorganisms

The early and rapid detection of pathogenic microorganisms in the environment and food to prevent infection, disease, and biological warfare has always been a challenging issue. For instance, bacterial pathogens including *Legionella pneumophila*, *Salmonella typhimurium*, *Yersinia enterocolitica*, and *Escherichia coli* O157:H7 cause various illnesses related to potable water systems [57].

Traditional analytical approaches for microorganisms are based on the colony-forming unit (CFU), polymerase chain reaction (PCR), and enzyme-linked immunosorbent assay (ELISA). On the contrary to the SPR sensor, these methods may require some enrichment steps or fluorescent labeling and a longer time to reach the desired conclusion. Many studies have evaluated the use of SPR-based immunosensors coupled with a specific antigen—antibody reaction as a promising tool for bacteria detection [58].

In addition to antibodies, there are several other biorecognition elements for use in SPR. Singh et al. [59] showed SPR-based detection of *Salmonella* using DNA self-assembly with very high sensitivity. Over time, other technologies including MIP, SPRi, microfluidics, grating-coupled (GC)-SPR, and fiber optic surface plasmon resonance (FOSPR) entered the field of SPR sensor design to monitor pathogen microorganisms in the environment. Meneghello et al. [60] represented a novel approach for the diagnosis of *L. pneumophila*, based on GC-SPR technology, to develop a bench type, portable, and real-time SPR setup with enhanced detection sensitivity (1000-fold) compared with the standard fluorescence method. A study also introduced a FOSPR sensor for the detection of *E. coli* O157:H7 in water based on antimicrobial peptides (AMP) (magainin I, as a recognition element, and silver nanoparticles-reduced graphene oxide (AgNPs-rGO) nanocomposites as signal amplification agents) [61]. An overview of SPR sensors developed to identify pathogen microorganisms have been collected in Table 12.2.

TABLE 12.2 Overview of SPR sensors developed to identify pathogen microorganisms.

Bacteria type	Detection format	Immobilization method	Linker molecule	LOD (CFU/mL)	Ref
Escherichia coli O157:H7	Direct	Physical adsorption	NeutrAvidin	10^2–10^3	[62]
E. coli O157:H7	Direct Sandwich Direct	Covalent binding (SAM)	– – Protein G	10^6 10^3 10^4	[63]
Salmonella enteritis *E. coli*	Direct	Covalent binding (SAM)	Protein A	25 23	[64]
E. coli	Direct	Physical adsorption	Streptavidin	90	[65]
Salmonella typhimurium	Direct	Physical adsorption	NeutrAvidin	10^6	[66]
E. coli O157:H7	Direct	Covalent binding (SAM)	Streptavidin	1.4×10^4	[67]
Staphylococcus aureus	Sandwich	Covalent binding (SAM)	–	10^5	[68]
Listeria monocytogenes	Inhibition	Covalent binding (SAM)	–	10^5	[69]
Salmonella	Direct	Covalent binding (SAM)	–	2 fM	[59]
Legionella pneumophila	–	Physical adsorption	Streptavidin	10^4	[70]
L. pneumophila	Direct	Covalent binding (SAM)	Protein A	10^3	[71]
L. pneumophila	Direct	Grating-coupled	–	10	[60]
E. coli O157:H7	Direct	Covalent binding (AgNPs-rGO)	–	5×10^2	[61]

AgNPs-rGO, silver nanoparticles–reduced graphene oxide; *GC-SPR*, grating-coupled surface plasmon resonance; *SAM*, self-assembled monolayer.

4. Machine learning algorithms in SPR

According to applications of sensors, they are usually categorized into different groups including physical [72,73], chemical [74,75], and biosensors [3,76–78]. Biosensors have been broadly employed in food safety, healthcare systems, disease diagnosis, biosafety, and environmental sensing [27].

In the field of AI, computers are used for modeling intelligent behavior with human intervention to a minimal extent [79]. AI-based biosensors contain three major elements including information collection, signal conversion, and AI-based data processing, which are involved in receiving data, producing electrical output signals, and data processing, respectively [27]. AI data processing covers the learning process from data, which involves an appropriate analysis of data, making conclusions from data, and the determination of misinterpretation in data analysis [80]. To reach the aforementioned purposes, ML should come to assist by reducing the amount of data, maintaining data consistency, and evaluating reliability and accuracy [27].

ML, as a computer program able to gain knowledge from raw data, can bring benefits to biosensors. It can process big data obtained from complicated matrices/samples and contribute to receiving proper results from noisy sensing data with low resolution. Furthermore, ML visualizes data to decipher hidden associations between sensing signals and sample-related parameters and considers interrelations between bioevents and signals [1]. Regarding raw data analysis, ML utilizes various approaches including categorization, anomaly detection, noise reduction, object identification, and pattern recognition. In brief, signals are sorted by target analyte-based algorithms. As contamination and interferences affect biosensors, the received signal should be checked to see if it looks right. In this case, ML comes to assist and make corrections to variations of sensor performances, as well. Considering noise in biosensing, ML models can be trained to differentiate signals from noise. Finally, latent objects and patterns can be detected by ML to conveniently and efficiently interpret sensing data [1]. Moreover, ML can help biosensors through reading out in direct, automatic, accurate, and rapid manners, and desirable biosensors would be achieved using ML [1].

ML contains several methods and algorithms to learn from data and then obtain beneficial information or provide predictive models for an event [27]. Supervised learning and unsupervised learning algorithms are known as two significant paradigms in ML. While in the former one, labeled training data are employed for estimating or mapping the input data to the output that is desired, in the latter, there is no labeled data with specific outputs. Several types of emerging and advanced ML are being noticed for their (possible) applications in SPR, which will be briefly described (Table 12.3).

TABLE 12.3 A list of the machine learning algorithms used in surface plasmon resonance.

Algorithms	Advantages	Disadvantages	References
κNN	Understandable and easily implemented; no training required; nonparametric algorithm; operates properly with small features	Not suitable for high-dimensional data; the k value needs to be established	[1]
K-means clustering	Easy implementation and interpretation; convergence is guaranteed	Initial points affect results; the k value needs to be established; vulnerable to outliers; large datasets lead to high computational costs	[1]
PCA	Low noise sensitivity; declined needs for capacity and memory; decreased complexity in image sorting; smaller database representation	Difficult to analyze the covariance matrix accurately; simplest invariance could be detected only if training data explicitly provide the information	[81]
Ridge regression	Strong mathematical basis; easily interpreted; handles correlated predictive attributes better compared with ordinary least squares	The number of instances must be higher compared with that of attributes; outlier sensitive; normalized data required; in the presence of nonlinear relation between the target and predictive attributes, information is used poorly	[82]
t-SNE	Preserves local neighborhoods; keeps local and global data structures; works well with linear and nonlinear data; calculates the similarity of points in high-dimensional and low-dimensional space	Computationally expensive; not always preserves global structure; the clusters of various classes are crowded and cannot be distinguished (crowding problem)	[83–86]
AE	Labeled data not required for training; many variations lead to robust representation such as denoising, contractive, K-sparse, and separable deep autoencoders	Not broadly utilized in the real world; requires pretraining; the vanishing gradient issue confronts the trained model	[87]

Continued

TABLE 12.3 A list of the machine learning algorithms used in surface plasmon resonance.—cont'd

Algorithms	Advantages	Disadvantages	References
CURE	Suitable for clustering large databases; robust to outliers; identification of clusters with nonspherical shapes and broad variations in size	Difficult to deal with databases with specific shapes	[88,89]
NMF	Optimal for learning the parts of objects; learns a parts-based representation	Unable to consider the data space geometrical structure	[90]
GANNs	Appropriate for constrained and unconstrained optimization; emphasis on effective generalization capability	No emphasis on effective feature selection; no clear description about the architecture	[91,92]
FANNs	No need for a user-determined problem-solving algorithm; holds an inherent generalization ability	Long-term modeling; needs the preprocessing of overfitting raw data	[1,93]
CNN	Automatic learning and feature extraction; reduces memory costs and parameters; suitable for high-dimensional data; compatible with transfer learning	Requires big datasets; very small datasets lead to overfitting; final models are difficult to be understood and interpreted; variable weights plus individual effects; not reproducible; expensive computation; very deep networks lead to gradient vanishing/exploding	[1]
DNN	Appropriate for massive learning data; favorable to complex nonlinear relation modeling	Requires a long time for learning; overfits to pretraining data; more hidden layers lead to more operational costs	[94]
GAN	No need for Monte Carlo approximation for training; faster sample generation	Unstable GAN training; learning for discrete data (e.g., text) generation is difficult	[87]

AE, Autoencoder; *CNN*, Convolutional Neural Network; *CURE*, Clustering Using REpresentatives; *DNN*, Deep Neural Network; *FANNs*, Feedforward Artificial Neural Networks; *GAN*, Generative adversarial network; *GANNs*, Genetic-Algorithm-Based Artificial Neural Networks; *κNN*, K-Nearest Neighbor; *NMF*, Non-Negative Matrix Factorization; *PCA*, Principal Component Analysis; *t-SNE*, T-Distributed Stochastic Neighbourh Embedding.

4.1 Supervised machine learning

4.1.1 K-Nearest neighbor

This algorithm is often used for classification purposes, and different approaches have been suggested related to KNN [95]. Selecting the neighbor number (K) is according to datasets. In this regard, smaller figures for K make boundaries distinguishable between classes; however, noise effects on classification increase [1]. In terms of corresponding applications, the algorithm underwent optimization and was broadly employed for breast cancer diagnosis [96,97] and anomaly [98].

4.1.2 Ridge regression

Ridge regression is a beneficial tool to improve the prediction process in regression situations containing correlated predictors at high levels [99]. It addresses the collinearity problem without variable removal from the independent variable original set [100]. It has been shown that this method has potentials in various fields, especially in chemometrics [99].

4.1.3 Feedforward artificial neural networks

Artificial neural networks (ANNs) consist of input, hidden, and output layers with connected neurons (nodes) to simulate the human brain. The existing nodes process and transmit input signals to the next nodes [1]. ANNs are nonlinear models to classify complicated relationships. To accurately and precisely measure catechol and hydroquinone, simultaneously, in real water samples, ANN was used as a nonlinear model calibration [101]. Zhang and Tao [102] developed degradable epidermal electronics for monitoring physiological parameters using an ANN algorithm. It is noteworthy that deep learning, using deep ANN [1], has recently appeared as a novel research direction in ML [27].

Moreover, a class of ANN has also gained researchers' attention by the name of multilayer perceptron (MLP), which contains at least three layers, including input, hidden, and output layers [103]. It is capable of distinguishing nonlinearly separable data, and training is done by the backpropagation learning algorithm [103].

4.1.4 Convolutional neural network

Convolutional neural network is a type of deep learning, suitable for image processing namely computed tomography images, magnetic resonance images, and X-ray images. It comprises convolutional, pooling, and fully connected layers. In the convolutional layer, there are some filters (kernels) sliding across preprocessed signals, and the feature map is obtained following convolution operation. The second layer reduces the dimension of the conventional layer output through pooling operation, leading to the prevention of overfitting and reducing computational intensity. The last layer deals with activation functions to make outputs nonlinear [1].

4.1.5 Deep neural network

A multilayer neural network is composed of three connected layers. The first layer is called the input layer receiving input data. In the second layer, there are several hidden layers saving association weights. Finally, there is an output layer illustrating output data [94]. Based on the complexity of a given problem, the hidden layer will be deepened. In general, the multilayer network containing several hidden layers is known as a deep neural network (DNN) [94]. This type of neural network algorithm simulates the brain hierarchical structure, in which neurons serve as data processors interconnected for information exchange [104].

4.1.6 Genetic algorithm–based artificial neural networks

The combination of genetic algorithm (GA) and neural networks has been utilized for prediction and classification aims. Usually, ANN and GA are used as a prime modeling tool and an optimizer of ANN weights or topology. For example, GA was employed for identifying the best combination of ANN topology for the classification of DNA sequences [105]. In another study, GA optimized the weight and structure of ANN for the classification of a multiclass microarray [106]. GANNs are appropriate for restricted and unrestricted optimization, and the process of deduction comes from biological evolution and natural selection [91].

4.1.7 Generative adversarial network

Generally, two models (a generative model G and a distribution model D) make a deep learning method termed GAN. The G model covers data distribution while the other one assesses the probability of a sample coming from training data. GAN is capable of mimicking various data distribution, indicating a great potential and use in various fields such as speech, image, music, and prose [87].

4.2 Unsupervised machine learning

4.2.1 K-means clustering

This technique is a popular data clustering method in which datasets with "n" data points are sorted into "k" groups or clusters. The algorithm consists of four steps: the placement of k points in the space indicated by clustered objects; the assignment of each object to the group containing the closest centroid; the recalculation of k centroid positions; and repeating the second and third steps until the point that centroids have no movement.

4.2.2 Principal component analysis

As an unsupervised algorithm, PCA is used to achieve dimensionality reduction by replacing a group of variables with principal components. PCA has been used in several studies. For example, Kuhner et al. [107] extracted vibrational data using PCA from fructose and glucose biosensors. Moreover, the algorithm was recently used to cluster body fluid samples in the technique SPRi [108].

4.2.3 T-distributed stochastic neighbor embedding

This cutting-edged method is used for dimensionality reduction in massive datasets [109]. The t-SNE produces a 2D or 3D visualization for sample interrelations, leading to the identification of close similarities between samples through the relative site of mapped points. The method puts dissimilar samples at far distances and keeps similar ones proximate in the mapping.

4.2.4 Autoencoder

This type of ANN learns features from unlabeled data and gives efficient representations in lower-dimensional space compared with original data. AE is composed of an encoder, mapping inputs to lower-dimensional data, and a decoder, reconstructing inputs from lower-dimensional data. AE has shown great potentials, for instance, in anomaly detection, image denoising, and biosensor design [110].

4.2.5 Clustering using representatives

CURE represents clusters through fixed numbers of points well-scattered. Then, the points will experience shrinkage toward the centers of their clusters by a determined factor. The CURE method is capable of reducing the dataset size and computational costs. This clustering algorithm is robust to outliers, and it can identify clusters with variable sizes and nonspherical forms. CURE can handle massive databases with no damage to clustering quality [88].

4.2.6 Non-negative matrix factorization

NMF is beneficial for clustering and classification and has gained much attention due to its desired performance and theoretical interpretation. NMF is an approach to receive data representation by the use of nonnegativity constraints. This leads to a part-based representation since they permit the additive combination of original data, instead of subtractive ones.

5. Applications of ML in SPR

Different approaches including mathematical models, ML methods, and signal descriptors should all be used in order to effectively analyze the obtained parameters of a sensorgram. Mathematical models, such as the Langmuir model, provide quantitative measurements of kinetics and affinities. The particle swarm optimization (PSO) technique through nonlinear differential equations and finite element simulations can all be employed to simulate interactions like sensorgram association/dissociation phases. However, the application of these approaches is reliant on expensive, time-consuming, and sophisticated experiments. Furthermore, such methods are ineffective for classifying sensorgrams, detecting the presence of compounds, or even assessing the quality of SPR responses [111].

For complicated matrices or samples, ML can successfully analyze large sensor data. Another advantage of using this method in biosensors is the ability to generate appropriate analytical findings from noisy data with a low resolution that might be substantially overlapped [1]. Smart SPR sensors can be designed to make choices automatically depending on specific and common properties of an experimental procedure as described by a sensorgram. To reach this aim, it should be noticed that a smart module, performing learning duties, must be integrated with the sensor without altering the primary design or setup of the sensor [111]. In the following, some recent applications of AI and applied ML-based approaches in SPR-based sensors will be described (summarized in Table 12.4).

ML algorithms were recently utilized by Gomes, Souza et al. in order to define, categorize, arrange, recognize, verify quality, and run an automated test on sensorgrams that show diagnosis procedures, resulting in the production of an intelligent SPR sensor based on sensorgram data [111]. The classification and recognition of the sensorgrams were performed using the new temporal sensorgram descriptor in conjunction with the KNN algorithm. The ML approach employed hypothesis testing through confidence intervals to detect substances, assess the quality, and conduct automatic diagnostics. Accordingly, the smart sensor can testify the quality of its response by segmenting the permanent and transitional regimes of the sensorgram effectively using linear regression. The smart sensor can also detect substances and predict the behavioral irregularities in the sensor [111]. Using the suggested ML approaches, it is practical to analyze the quality of SPR sensors and verify the conditions that they can operate in with a determined level of confidence. Finally, all of the techniques provided will be beneficial to ensure that the responses obtained from smart SPR are reliable and may lead to obtaining the related final audits of regulatory organizations.

More previously, Ong et al. [121] demonstrated an optical sensing system that is based on a simple, practical, and cost-friendly SPR sensor arrangement (so-called BU-SPR) [122] that incorporates a field programmable gate array. In their research, data categorization was accomplished by combining the CURE technique with the KNN technique. By decreasing the training set, the CURE data clustering technique dramatically decreased the execution time of the KNN algorithm. The suggested system was designed to be versatile, modular, and easy to operate, with little required maintenance [121].

Numerical methods are typically used to calculate SPRs, which demand a lot of computing power and/or time-consuming processes [120,123,124]. To overcome this challenge, recent research applied ML approaches to predict the

TABLE 12.4 Different machine learning (ML)-based techniques applied for developing smart SPR.

ML model	Applications	SPR type	References
NMF	Decoupling complicated LSPR signals	LSPR	[112]
DNN	Predicting the optical characteristics of plasmonic nanostructures; creating an exact connection between plasmonic geometric characteristics and resonance spectra	LSPR	[113]
DNN	Evaluating far-field and near-field optical properties of Au nanospheres	LSPR	[114]
μ-GA	Improving the detection sensitivity of gold nanostructures	LSPR	[115]
ANN+GA	Evaluating the link between synthesis and material characteristics	SPR	[116]
GANNs	Evaluating the correlation between synthesis properties and the wavelength of SGNPS grown via the seed-mediated growth method	SPR	[91]
GAN+MLP	Estimating the confinement loss behavior of the PCF	PCF-based SPR	[117]
NNSF	Automatically recognizing and filtering noise sources	SPR	[118]
MLPNN	Optimizing the characteristics of a plasmonic sensor	LSPR	[119]
t-SNE+AE+MLP+k-means	Anticipating the reflectance features of metamaterial-based SPR biosensors	SPR	[110]
KNN+ridge regression+MLP	Predicting the SPR wavelength of GCNCs with a defined size and deep concavity	SPR	[120]
KNN	Detecting substances, assessing the quality, conducting automatic diagnostics, and self-testing	SPR	[111]
CURE+KNN	Categorizing data	SPR	[121]

μ-GA, micro–genetic algorithm; *AE*, autoencoder; *BSA*, bovine serum albumin; *CURE*, clustering using representatives; *DNN*, deep neural network; *GAN*, generative adversarial networks; *GANNs*, genetic algorithm–based artificial neural networks; *GCNCs*, gold concave nanocubes; *KNN*: K-nearest neighbor; *LSPR*, localized surface plasmon resonance; *MLP*, multilayer perceptron; *NMF*, non-negative matrix factorization; *NNSF*, neural network SPR filter; *PCF*, photonic crystal fiber; *SGNPS*, sea urchin–like gold nanoparticles; *SPR*, surface plasmon resonance; *t-SNE*, t-stochastic neighbor embedding.

SPR of perfect and concave gold nanocubes (GCNCs). They established a computational technique combining discrete dipole approximation (DDA) and ML-based approaches to anticipate the SPR wavelength of GCNCs with a defined size and deep concavity, allowing them to forecast the position of the dipole SPR of GCNCs. The SPR wavelength dataset was obtained using DDA, in which two parameters including the edge length and the depth radius of the concavity were calculated. Moreover, three alternative ML algorithms including the KNN model, the ridge regression, and the MLPNN model were utilized with the identical set of variables for training, testing, and cross-validation. Finally, based on their findings, the authors come to the conclusion that MLPNN was the best method with significant accuracy compared to the other applied ML methods for obtaining accurate results, which prevent overfitting and enable generalization [120].

Moon et al. [110] used ML techniques to investigate the performance of plasmonic biosensor devices which utilize metamaterials, which have been noticed due to their special electromagnetic features like a negative index of refraction and artificial magnetism [125]. They revealed that the sensing signal of SPR-based biosensors has been amplified using metamaterials with negative permeability and permittivity. In order to guarantee that the resonance is effective for SPR biosensors, metamaterials with different reflectance properties must be developed. Moreover, to anticipate the reflectance features of metamaterial-based SPR biosensors, AE and MLP have been used. Following the dimensional reduction with t-SNE and AE, the metamaterials were clustered using k-means. The clustering of metamaterials can considerably speed up researchers' ability to create optimal sensing systems without performing demanding experiments. Finally, according to their findings, ML was proven to enhance detection sensitivity by more than 13-fold higher than traditional SPR biosensing with suitable resonance properties, using metamaterial [110].

ML techniques have been employed to build and forecast a broad range of optical devices, including plasmonic sensors [113], waveguides [126,127], plasmonic colors [128,129], metasurfaces [130,131], and photonic crystals [132,133]. Irrespective of the complication of the physical procedure, ANN models, like MLPNN, can effectively determine the association between input and output, and reliable findings have been obtained in different fields of nanophotonics so far [134]. When it comes to optimizing the characteristics of a plasmonic sensor, even by utilizing ML-based approaches, it remains challenging [119]. To address this problem, PSO has been shown to be practical for parameter optimization by using global search, which brings substantial benefits, particularly, when dealing with multiparameter optimization issues [135]. It has been demonstrated that using the PSO method in conjunction with the applied MLPNN offers enormous promise for the creation of nanophotonic devices like LSPR with high performance [119].

ANN-based methods have been also applied for data filtering. In this regard, reducing noise in SPR sensor signals by the use of a smart data filter system based on ANNs is another application of AI in SPR, which has been investigated recently [118]. In this study, a neural network SPR filter (NNSF) based on ANNs was applied in order to automatically recognize and filter noise sources that commonly exist in SPR sensor responses. The smart filter was able to reduce intrinsic distortions of SPR curves, using experimental curves as input information that resulted in curves with improved characteristics. The distortion coefficient matrix (DCM) measures the distortion of experimental curves compared to theoretical databases. Besides, DCM allows choosing the NNSF structure (a dynamic committee or a simple ANN). Based on their results, it has been shown that in both angular and spectral forms, this method was efficient in draining noise from all elements incorporated in the SPR sensor instrument [118]. NNSF brings various merits for intelligent SPR sensor designs due to some reasons. For example, it does not need additional electronic components [136] for signal filtering, and physical structure alterations are not required [137], leading to lowering instrumentation and maintenance expenses [118].

Furthermore, ANNs were utilized in a study to estimate the confinement loss behavior of the multichannel photonic crystal fiber (PCF)-based SPR sensors. Various propagation features of a PCF design were computed using a feedforward MLP class of ANNs. Moreover, for training an ANN model and augmenting the real dataset, GANs were applied. The suggested network could achieve both reasonable accuracy and higher speed [117].

Regarding the fact that morphologies and sizes of nanoparticles adjust the modes and field intensity of SPR in the visible and near-infrared regions, these NPs can be applied in plasmonics, catalytic reactions, detection, and clinical approaches [138]. Nowadays, thermotherapy by GNPs for cancer treatment has attracted researchers' attention [139]. In this regard, sea urchin–like gold nanoparticles (SGNPs), which are considered attractive candidates for the cancer thermotherapy approach, can be produced on the surface of Au NPs using the Au seed–mediated growth method [140,141]. The wavelength of surface plasmon of these NPs can be identified by their generated shapes [141]. Because of the complexity of SGNP growth using common seed-mediated methods [140,142] and the need for an accurate estimation of the wavelength parameter of their surface plasmon, GANNs have been recently utilized in order to evaluate the correlation between synthesis characteristics and the wavelength of SGNPs grown via the seed-mediated growth approach [91]. The GANN technique assisted the growth method with facilitating parameter selection that finally resulted in achieving a more powerful electromagnetic field [91]. Moreover, they indicated that four parameters can impact the training model accuracy: (1) the convergence of experimentation and training information; (2) the number of neurons present in the hidden layer; (3) the medium mating ratio, which is useful for building a well-trained model in order to optimize the solution; and (4) the duration of the learning cycle [91].

ML has also been used to evaluate the link between synthesis and material characteristics. Previously, AI techniques were hybridized including ANN and GA in predictive modeling and optimization [143]. Through the research, UV/ozone processing was used to create Au NPs decorated on zinc oxide nanorods (NRs). Following this, different parameters including the absorption, the efficiency of light to plasmon, plasmon to hot electron conversions, and the quality (Q) factor of Au@ZnO nanocomposites were further investigated to better realize associated SPR effects [116]. Applying this technique enabled the researchers to properly anticipate the dataset without consuming a significant amount of time and money and efficiently evaluate the influence of SPR produced from the Au@ZnO heterostructure.

Moreover, a micro-GA was applied for optimizing an SPR biosensor, and the hot embossing nanoimprint process was the selected method for manufacturing the sensor. Their study suggested a design strategy and manufacturing methodology for improving the detection sensitivity of gold nanostructure-based SPR biosensors [115].

DNN is another noticeable ML approach. It has been shown that deep learning, created by AI, is a strong and fast approach for creating an exact connection between plasmonic geometric characteristics and resonance spectra like LSPR spectra [113].

Regarding the theoretical instruction of NP design, the far-field and near-field optical characteristics of plasmonic nanoparticles may generally be determined through numerical simulation techniques. However, due to the urgent need for computer resources, these methods undergo incompatible controversies between precision and speed [114]. As a result, ML approaches, particularly DNN, have been used to develop alignments between both the far-field spectra or near-field distribution and dimensional characteristics of different forms of plasmonic nanoparticles like nanospheres, NRs, and dimers [144–146]. A parallel study, in which the DNN method has been compared with the numerical simulation methods, like finite-difference time-domain (FDTD), revealed that the predicted spectra were aligned concerning different parameters such as intensity, wavelengths, and the location of the dipolar LSPR mode [114]. Accordingly, the obtained findings, which were consistent with prior research on ML prediction [78,146,147], indicated that the difficult and time-consuming FDTD simulations may be substituted by an ML-based approach like the DNN model to determine the far-field and near-field optical characteristics of Au nanospheres [114].

The sensitivity of the resonant frequency to modifications in geometric parameters of the nanostructure, material features, and the surrounding medium gives the LSPR technique a significant capability for molecular sensing and the development of various fields like clinical diagnosis, environmental sensing, biological agents, and other fields. The NN is considered a useful technique that helps to convert complex physical problems into mathematical correlations of networks and uncover the precise link between data pairs, acting as a simulation of physical interaction. This technique has been used recently to accurately anticipate the optical characteristics of plasmonic nanostructures, in which a simple plasmonic sensor construction made of periodic gold nanodisk arrays was investigated. As a result, the strategy allowed spectra to be obtained for any combination of geometric factors. Using this method results in a general approach for reducing the simulation costs of plasmonic sensors while also directing the design and manufacturing of associated photonic devices [113].

As mentioned earlier, ML-based approaches have applications in improving LSPR. NMF is another ML-based technique that has been applied in combination with other experimental and theoretical methods like boundary element methods in order to study LSPR in aluminum nanocrystals (icosahedral Al-core $(Al2O3)_{shell}$ NPs). By applying the NMF approach, the researchers could decouple complicated LSPR signals from icosahedron-shaped Al NPs, besides correlating these data to the geometry of the particles [112]. Using these combined methods first enabled researchers to map the coupling efficiency factor of electron beams with LSPR on the NPs, which is indicative of the energy transfer pathway from the source of excitation to the plasmonic NPs. Furthermore, this strategy led to expanding the knowledge of LSPR localization in nanocatalysts (nanoengineered ones) and a dielectric coating [112].

Image processing is another application of AI. The SPRi technique, which is commonly applied to observe biological events and estimate the binding constants of interactions [41,148], was utilized in research to analyze heat and mass transfer mechanisms [149]. It has been demonstrated that one of the major barriers to increasing the sensitivity of SPRi is a weak signal-to-noise ratio, caused by speckles and stray interference structures in the obtained images [149,150]. To tackle this challenge, an image processing algorithm was developed by Pavlov, Vedyashkina et al. [149].

6. Conclusion and future perspectives

Various attractive characteristics of SPR techniques including the possibility of monitoring reactions in real time, being label-free, tiny sample sizes, reusable sensor chips, the possibility of detecting complex samples, and reproducibility of data have made SPR an attention-grabbing technique for various purposes. Additionally, AI, specifically ML, has revolutionized many fields of science and technology by redefining the mechanism of massive data processing. ML is one of the approaches that has been applied to remedy the existing issues. However, few investigations have been focused on the use of AI and ML to develop smart SPR sensors. Thus, there is quite a gap in combining ML approaches and SPR sensing.

Among different ML-based approaches, ANNs, DNNs, and KNN are the most common models that have been applied to SPR development. This might be indicative of the strength of these techniques. Nonetheless, other methods probably need more investigations to be further applicable. Perhaps in the forthcoming future, various novel algorithms will be designed for better SPR manufacturing.

Regarding the noticeable applications of SPR sensors in environmental sensing, it is worthwhile to make efforts to use ML and develop SPR sensors in environmental monitoring with more accuracy and beneficial properties. However, it seems that smart SPR sensors are mostly studied and utilized for diagnostic, biological, and medical purposes. Although SPR has been widely employed for environmental hazard detection, to the best of our knowledge, ML-based SPR for environmental monitoring has not yet been noticed by researchers. It is predicted that more ML-integrated SPR sensors will be developed for monitoring environmental hazards in real time.

References

[1] F. Cui, Y. Yue, Y. Zhang, Z. Zhang, H.S. Zhou, Advancing biosensors with machine learning, ACS Sensors 5 (11) (2020) 3346–3364.

[2] M. Asadnia, et al., From biological cilia to artificial flow sensors: biomimetic soft polymer nanosensors with high sensing performance, Sci. Rep. 6 (1) (2016) 1–13.

[3] F. Ejeian, et al., Biosensors for wastewater monitoring: a review, Biosens. Bioelectron. 118 (2018) 66–79, https://doi.org/10.1016/j.bios.2018.07.019.

[4] M.A. Parvez Mahmud, N. Huda, S.H. Farjana, M. Asadnia, C. Lang, Recent advances in nanogenerator-driven self-powered implantable biomedical devices, Adv. Energy Mater. 8 (2) (2018) 1701210, https://doi.org/10.1002/aenm.201701210.

[5] H. Salehabadi, K. Khajeh, B. Dabirmanesh, M. Biglar, S. Mohseni, M. Amanlou, Surface plasmon resonance based biosensor for discovery of new matrix metalloproteinase-9 inhibitors, Sensor. Actuator. B Chem. 263 (2018) 143–150, https://doi.org/10.1016/j.snb.2018.02.073.

[6] S. Ebrahimi, Z.E. Nataj, S. Khodaverdian, A. Khamsavi, Y. Abdi, K. Khajeh, An ion-sensitive field-effect transistor biosensor based on SWCNT and aligned MWCNTs for detection of ABTS, IEEE Sensor. J. 20 (24) (2020) 14590–14597, https://doi.org/10.1109/JSEN.2020.3009536.

[7] H. Salehabadi, K. Khajeh, B. Dabirmanesh, M. Biglar, M. Amanlou, Evaluation of angiotensin converting enzyme inhibitors by SPR biosensor and theoretical studies, Enzym. Microb. Technol. 120 (2019) 117–123, https://doi.org/10.1016/j.enzmictec.2018.10.010.

[8] N. Shahbazi, S. Hosseinkhani, K. Khajeh, B. Ranjbar, Structural and functional study of a simple, rapid, and label-free DNAzyme-based DNA biosensor for optimization activity, Biopolymers 107 (10) (2017) e23028, https://doi.org/10.1002/bip.23028.

[9] S. Jabbari, et al., A novel enzyme based SPR-biosensor to detect bromocriptine as an ergoline derivative drug, Sensor. Actuator. B Chem. 240 (2017) 519–527, https://doi.org/10.1016/j.snb.2016.08.165.

[10] M. Shamsipur, M. Shanehasz, K. Khajeh, N. Mollania, S.H. Kazemi, A novel quantum dot–laccase hybrid nanobiosensor for low level determination of dopamine, Analyst 137 (23) (2012) 5553–5559, https://doi.org/10.1039/C2AN36035G.

[11] S. Ghasempur, S.-F. Torabi, S.-O. Ranaei-Siadat, M. Jalali-Heravi, N. Ghaemi, K. Khajeh, Optimization of peroxidase-catalyzed oxidative coupling process for phenol removal from wastewater using response surface methodology, Environ. Sci. Technol. 41 (20) (2007) 7073–7079, https://doi.org/10.1021/es070626q.

[12] S. Jabbari, B. Dabirmanesh, K. Khajeh, Specificity enhancement towards phenolic substrate by immobilization of laccase on surface plasmon resonance sensor chip, J. Mol. Catal. B Enzym. 121 (2015) 32–36, https://doi.org/10.1016/j.molcatb.2015.07.016.

[13] M. Zeinoddini, K. Khajeh, F. Behzadian, S. Hosseinkhani, A.-R. Saeedinia, H. Barjesteh, Design and characterization of an aequorin-based bacterial biosensor for detection of toluene and related compounds, Photochem. Photobiol. 86 (5) (2010) 1071–1075, https://doi.org/10.1111/j.1751-1097.2010.00775.x.

[14] S.H. Kazemi, K. Khajeh, Electrochemical studies of a novel biosensor based on the CuO nanoparticles coated with horseradish peroxidase to determine the concentration of phenolic compounds, J. Iran. Chem. Soc. 8 (1) (2011) S152–S160, https://doi.org/10.1007/BF03254292.

[15] F. Hajipour, et al., Developing a fluorescent hybrid nanobiosensor based on quantum dots and azoreductase enzyme for methyl red monitoring (in eng), Iran. Biomed. J. 25 (1) (2021) 8–20, https://doi.org/10.29252/ibj.25.1.8.

[16] M. Asadnia, L.H. Chua, X. Qin, A. Talei, Improved particle swarm optimization–based artificial neural network for rainfall-runoff modeling, J. Hydrol. Eng. 19 (7) (2014) 1320–1329.

[17] M. Asadnia, M.S. Yazdi, A. Khorasani, An improved particle swarm optimization based on neural network for surface roughness optimization in face milling of 6061-T6 Aluminum, Int. J. Appl. Eng. Res. 5 (19) (2010) 3191–3201.

[18] M. Asadnia, et al., High temperature characterization of PZT (0.52/0.48) thin-film pressure sensors, J. Micromech. Microeng. 24 (1) (2013) 015017.

[19] S.R. Bazaz, A.A. Mehrizi, S. Ghorbani, S. Vasilescu, M. Asadnia, M.E. Warkiani, A hybrid micromixer with planar mixing units, RSC Adv. 8 (58) (2018) 33103–33120.

[20] R. Hagihghi, A. Razmjou, Y. Orooji, M.E. Warkiani, M. Asadnia, A miniaturized piezoresistive flow sensor for real-time monitoring of intravenous infusion, J. Biomed. Mater. Res. B Appl. Biomater. 108 (2) (2020) 568–576.

[21] A.G.P. Kottapalli, M. Asadnia, J. Miao, M. Triantafyllou, Soft polymer membrane microsensor arrays inspired by the mechanosensory lateral line on the blind cavefish, J. Intell. Mater. Syst. Struct. 26 (1) (2015) 38–46.

[22] A. Razmjou, et al., Preparation of iridescent 2D photonic crystals by using a mussel-inspired spatial patterning of ZIF-8 with potential applications in optical switch and chemical sensor, ACS Appl. Mater. Interfaces 9 (43) (2017) 38076–38080.

[23] S. Mohseni, T.T. Moghadam, B. Dabirmanesh, S. Jabbari, K. Khajeh, Development of a label-free SPR sensor for detection of matrixmetalloproteinase-9 by antibody immobilization on carboxymethyldextran chip, Biosens. Bioelectron. 81 (2016) 510–516, https://doi.org/10.1016/j.bios.2016.03.038.

[24] M. Asadnia, A.M. Khorasani, M.E. Warkiani, An accurate PSO-GA based neural network to model growth of carbon nanotubes, J. Nanomater. 2017 (2017).

[25] S. Foorginezhad, M. Mohseni-Dargah, K. Firoozirad, V. Aryai, A. Razmjou, R. Abbassi, V. Garaniya, A. Beheshti, M. Asadnia, Recent advances in sensing and assessment of corrosion in sewage pipelines, Process Saf. Environ. Protect. 147 (2021) 192–213.

[26] W. Daniyal, Y.W. Fen, N.I.M. Fauzi, H.S. Hashim, N.S.M. Ramdzan, N.A.S. Omar, Recent advances in surface plasmon resonance optical sensors for potential application in environmental monitoring, Sensor. Mater. 32 (2020) 4191–4200.

[27] X. Jin, C. Liu, T. Xu, L. Su, X. Zhang, Artificial intelligence biosensors: challenges and prospects, Biosens. Bioelectron. (2020) 112412.

[28] M. Farahnakian, M.R. Razfar, M. Moghri, M. Asadnia, The selection of milling parameters by the PSO-based neural network modeling method, Int. J. Adv. Manuf. Technol. 57 (1) (2011) 49–60.

[29] P.B. Tiwari, et al., SPRD: a surface plasmon resonance database of common factors for better experimental planning, BMC Mol. Cell Biol. 22 (1) (2021) 1–8.

[30] M. Angelopoulou, S. Kakabakos, P. Petrou, Label-free biosensors based onto monolithically integrated onto silicon optical transducers, Chemosensors 6 (4) (2018) 52.

[31] D.M. Kim, J.S. Park, S.-W. Jung, J. Yeom, S.M. Yoo, Biosensing applications using nanostructure-based localized surface plasmon resonance sensors, Sensors 21 (9) (2021) 3191.

[32] J. Liu, M. Jalali, S. Mahshid, S. Wachsmann-Hogiu, Are plasmonic optical biosensors ready for use in point-of-need applications? Analyst 145 (2) (2020) 364–384.

[33] S. Lee, H. Song, H. Ahn, S. Kim, J.-r. Choi, K. Kim, Fiber-optic localized surface plasmon resonance sensors based on nanomaterials, Sensors 21 (3) (2021) 819.

[34] B.D. Gupta, R. Kant, Recent advances in surface plasmon resonance based fiber optic chemical and biosensors utilizing bulk and nanostructures, Opt. Laser. Technol. 101 (2018) 144–161.

[35] J. Homola, Surface Plasmon Resonance Based Sensors, Springer Science & Business Media, 2006.

[36] R.M. Balakrishnan, P. Uddandarao, K. Raval, R. Raval, A perspective of advanced biosensors for environmental monitoring, in: Tools, Techniques and Protocols for Monitoring Environmental Contaminants, Elsevier, 2019, pp. 19–51.

[37] M. Minunni, M. Mascini, Detection of pesticide in drinking water using real-time biospecific interaction analysis (BIA), Anal. Lett. 26 (7) (1993) 1441–1460.

[38] M. Farré, et al., Part per trillion determination of atrazine in natural water samples by a surface plasmon resonance immunosensor, Anal. Bioanal. Chem. 388 (1) (2007) 207–214.

[39] E. Mauriz, A. Calle, L.M. Lechuga, J. Quintana, A. Montoya, J.J. Manclus, Real-time detection of chlorpyrifos at part per trillion levels in ground, surface and drinking water samples by a portable surface plasmon resonance immunosensor, Anal. Chim. Acta 561 (1–2) (2006) 40–47.

[40] E. Mauriz, et al., Optical immunosensor for fast and sensitive detection of DDT and related compounds in river water samples, Biosens. Bioelectron. 22 (7) (2007) 1410–1418.

[41] J. Homola, Surface plasmon resonance sensors for detection of chemical and biological species, Chem. Rev. 108 (2) (2008) 462–493.

[42] E. Mauriz, A. Calle, J.J. Manclus, A. Montoya, L.M. Lechuga, Multi-analyte SPR immunoassays for environmental biosensing of pesticides, Anal. Bioanal. Chem. 387 (4) (2007) 1449–1458.

[43] G.-H. Yao, R.-P. Liang, C.-F. Huang, Y. Wang, J.-D. Qiu, Surface plasmon resonance sensor based on magnetic molecularly imprinted polymers amplification for pesticide recognition, Anal. Chem. 85 (24) (2013) 11944–11951.

[44] O. Çakır, Z. Baysal, Pesticide analysis with molecularly imprinted nanofilms using surface plasmon resonance sensor and LC-MS/MS: comparative study for environmental water samples, Sensor. Actuator. B Chem. 297 (2019) 126764.

[45] E. Miliutina, et al., Plasmon-active optical fiber functionalized by metal organic framework for pesticide detection, Talanta 208 (2020) 120480.

[46] T.-k. Lim, M. Oyama, K. Ikebukuro, I. Karube, Detection of atrazine based on the SPR determination of P450 mRNA levels in *Saccharomyces cerevisiae*, Anal. Chem. 72 (13) (2000) 2856–2860.

[47] E. Mauriz, et al., Single and multi-analyte surface plasmon resonance assays for simultaneous detection of cholinesterase inhibiting pesticides, Sensor. Actuator. B Chem. 118 (1–2) (2006) 399–407.

[48] K.V. Gobi, H. Tanaka, Y. Shoyama, N. Miura, Highly sensitive regenerable immunosensor for label-free detection of 2, 4-dichlorophenoxyacetic acid at ppb levels by using surface plasmon resonance imaging, Sensor. Actuator. B Chem. 111 (2005) 562–571.

[49] S.J. Kim, K.V. Gobi, H. Tanaka, Y. Shoyama, N. Miura, Enhanced sensitivity of a surface-plasmon-resonance (SPR) sensor for 2, 4-D by controlled functionalization of self-assembled monolayer-based immunosensor chip, Chem. Lett. 35 (10) (2006) 1132–1133.

[50] N. Soh, T. Watanabe, Y. Asano, T. Imato, Indirect competitive immunoassay for bisphenol A, based on a surface plasmon resonance sensor, Sensor. Mater. 15 (2003) 423–438.

[51] N. Soh, et al., A surface plasmon resonance immunosensor for detecting a dioxin precursor using a gold binding polypeptide, Talanta 60 (4) (2003) 733–745.

[52] J.-W. Choi, K.-W. Park, D.-B. Lee, W. Lee, W.H. Lee, Cell immobilization using self-assembled synthetic oligopeptide and its application to biological toxicity detection using surface plasmon resonance, Biosens. Bioelectron. 20 (11) (2005) 2300–2305.

[53] H.S. Hashim, Y.W. Fen, N.A.S. Omar, J. Abdullah, W.M.E.M.M. Daniyal, S. Saleviter, Detection of phenol by incorporation of gold modified-enzyme based graphene oxide thin film with surface plasmon resonance technique, Opt. Express 28 (7) (2020) 9738–9752.

[54] E. Priyadarshini, N. Pradhan, Gold nanoparticles as efficient sensors in colorimetric detection of toxic metal ions: a review, Sensor. Actuator. B Chem. 238 (2017) 888–902.

[55] C. Schopf, A. Martín, D. Iacopino, Plasmonic detection of mercury via amalgam formation on surface-immobilized single Au nanorods, Sci. Technol. Adv. Mater. 18 (1) (2017) 60–67.

[56] K. Ock, G. Jang, Y. Roh, S. Kim, J. Kim, K. Koh, Optical detection of Cu2+ ion using a SQ-dye containing polymeric thin-film on Au surface, Microchem. J. 70 (3) (2001) 301–305.

[57] D. Vogrinc, M. Vodovnik, R. MarinŠEk-Logar, Microbial biosensors for environmental monitoring, Acta Agric. Slov. 106 (2) (2015) 67–75.

[58] K.R. Srivastava, S. Awasthi, P.K. Mishra, P.K. Srivastava, Biosensors/molecular tools for detection of waterborne pathogens, in: Waterborne Pathogens, 2020, pp. 237–277.

[59] A. Singh, H.N. Verma, K. Arora, Surface plasmon resonance based label-free detection of Salmonella using DNA self assembly, Appl. Biochem. Biotechnol. 175 (3) (2015) 1330–1343.

[60] A. Meneghello, A. Sonato, G. Ruffato, G. Zacco, F. Romanato, A novel high sensitive surface plasmon resonance *Legionella pneumophila* sensing platform, Sensor. Actuator. B Chem. 250 (2017) 351–355.

[61] C. Zhou, H. Zou, M. Li, C. Sun, D. Ren, Y. Li, Fiber optic surface plasmon resonance sensor for detection of *E. coli* O157: H7 based on antimicrobial peptides and AgNPs-rGO, Biosens. Bioelectron. 117 (2018) 347–353.

[62] J. Waswa, J. Irudayaraj, C. DebRoy, Direct detection of *E. coli* O157: H7 in selected food systems by a surface plasmon resonance biosensor, LWT Food Sci. Technol. 40 (2) (2007) 187–192.

[63] A. Subramanian, J. Irudayaraj, T. Ryan, A mixed self-assembled monolayer-based surface plasmon immunosensor for detection of *E. coli* O157: H7, Biosens. Bioelectron. 21 (7) (2006) 998–1006.

[64] J.W. Waswa, C. Debroy, J. Irudayaraj, Rapid detection of Salmonella enteritidis and *Escherichia coli* using surface plasmon resonance biosensor, J. Food Process. Eng. 29 (4) (2006) 373–385.

[65] F.C. Dudak, İ.H. Boyacı, Development of an immunosensor based on surface plasmon resonance for enumeration of *Escherichia coli* in water samples, Food Res. Int. 40 (7) (2007) 803–807.

[66] Y.-b. Lan, S.-z. Wang, Y.-g. Yin, W.C. Hoffmann, X.-z. Zheng, Using a surface plasmon resonance biosensor for rapid detection of *Salmonella typhimurium* in chicken carcass, JBE 5 (3) (2008) 239–246.

[67] A.D. Taylor, J. Ladd, Q. Yu, S. Chen, J. Homola, S. Jiang, Quantitative and simultaneous detection of four foodborne bacterial pathogens with a multi-channel SPR sensor, Biosens. Bioelectron. 22 (5) (2006) 752–758.

[68] A. Subramanian, J. Irudayaraj, T. Ryan, Mono and dithiol surfaces on surface plasmon resonance biosensors for detection of *Staphylococcus aureus*, Sensor. Actuator. B Chem. 114 (1) (2006) 192–198.

[69] P. Leonard, S. Hearty, J. Quinn, R. O'Kennedy, A generic approach for the detection of whole Listeria monocytogenes cells in contaminated samples using surface plasmon resonance, Biosens. Bioelectron. 19 (10) (2004) 1331–1335.

[70] A.M. Foudeh, H. Trigui, N. Mendis, S.P. Faucher, T. Veres, M. Tabrizian, Rapid and specific SPRi detection of *L. pneumophila* in complex environmental water samples, Anal. Bioanal. Chem. 407 (18) (2015) 5541–5545.

[71] M.G. Manera, et al., SPR based immunosensor for detection of *Legionella pneumophila* in water samples, Opt. Commun. 294 (2013) 420–426.

[72] A.G.P. Kottapalli, et al., Engineering biomimetic hair bundle sensors for underwater sensing applications, AIP Conf. Proc. 1965 (1) (2018) 160003, https://doi.org/10.1063/1.5038533.

[73] A.G.P. Kottapalli, et al., Polymer MEMS sensor for flow monitoring in biomedical device applications, in: 2017 IEEE 30th International Conference on Micro Electro Mechanical Systems (MEMS), 22–26 January 2017, 2017, pp. 632–635, https://doi.org/10.1109/MEMSYS.2017.7863487.

[74] S. Foorginezhad, M. Mohseni-Dargah, Z. Falahati, R. Abbassi, A. Razmjou, M. Asadnia, Sensing advancement towards safety assessment of hydrogen fuel cell vehicles, J. Power Sources 489 (2021) 229450.

[75] S. Foorginezhad, et al., Recent advances in sensing and assessment of corrosion in sewage pipelines, Process Saf. Environ. Protect. 147 (2021/03/01) 192–213, https://doi.org/10.1016/j.psep.2020.09.009.

[76] M. Nankali, Z. Einalou, M. Asadnia, A. Razmjou, High-sensitivity 3D ZIF-8/PDA photonic crystal-based biosensor for blood component recognition, ACS Appl. Bio Mater. 4 (2) (2021/02/15) 1958–1968, https://doi.org/10.1021/acsabm.0c01586.

[77] N. Verma, A.K. Singh, R. Sharma, M. Asadnia, Chapter 22 - potent aptamer-based nanosensors for early detection of lung cancer, in: C.M. Hussain, S.K. Kailasa (Eds.), Handbook of Nanomaterials for Sensing Applications, Elsevier, 2021, pp. 505–529.

[78] R. Mokhtar-Ahmadabadi, et al., Developing a circularly permuted variant of Renilla luciferase as a bioluminescent sensor for measuring Caspase-9 activity in the cell-free and cell-based systems, Biochem. Biophys. Res. Commun. 506 (4) (2018/12/02) 1032–1039, https://doi.org/10.1016/j.bbrc.2018.11.009.

[79] P. Hamet, J. Tremblay, Artificial intelligence in medicine, Metabolism 69 (2017) S36–S40, https://doi.org/10.1016/j.metabol.2017.01.011.

[80] J. Carlson, M. Fosmire, C.C. Miller, M.S. Nelson, Determining data information literacy needs: a study of students and research faculty, Portal Libr. Acad. 11 (2) (2011) 629–657.
[81] S. Karamizadeh, S.M. Abdullah, A.A. Manaf, M. Zamani, A. Hooman, An overview of principal component analysis, J. Signal Inf. Process. 4 (3B) (2013) 173.
[82] J. Moreira, A. Carvalho, T. Horvath, A General Introduction to Data Analytics, John Wiley & Sons, 2018.
[83] L. Van Der Maaten, Fast optimization for t-SNE, in: Neural Information Processing Systems (NIPS) 2010 Workshop on Challenges in Data Visualization, vol. 100, Citeseer, 2010.
[84] D. Kobak, P. Berens, The art of using t-SNE for single-cell transcriptomics, Nat. Commun. 10 (1) (2019) 1–14.
[85] J. Che, S. Zhao, Y. Li, K. Li, Bank telemarketing forecasting model based on t-SNE-SVM, J. Serv. Sci. Manag. 13 (3) (2020) 435–448.
[86] B. Melit Devassy, S. George, Dimensionality reduction and visualisation of hyperspectral ink data using t-SNE, Forensic Sci. Int. 311 (2020/06/01) 110194, https://doi.org/10.1016/j.forsciint.2020.110194.
[87] S.K. Pandey, R.R. Janghel, Correction to: recent deep learning techniques, challenges and its applications for medical healthcare system: a review, Neural Process. Lett. 53 (2021), https://doi.org/10.1007/s11063-021-10527-5.
[88] S. Guha, R. Rastogi, K. Shim, CURE: an efficient clustering algorithm for large databases, ACM Sigmod Rec. 27 (2) (1998) 73–84.
[89] Q. Yun-Tao, S. Qing-Song, W. Qi, CURE-NS: a hierarchical clustering algorithm with new shrinking scheme, in: Proceedings. International Conference on Machine Learning and Cybernetics, 4-5 Nov. 2002, vol. 2, 2002, pp. 895–899, https://doi.org/10.1109/ICMLC.2002.1174512.
[90] D. Cai, X. He, X. Wu, J. Han, Non-negative matrix factorization on manifold, in: 2008 Eighth IEEE International Conference on Data Mining, 2008, pp. 63–72.
[91] C.-C. Wu, F. Pan, Y.-H. Su, Surface plasmon resonance of gold nano-sea-urchins controlled by machine-learning-based regulation in seed-mediated growth, Adv. Photonics Res. (2021) 2100052, https://doi.org/10.1002/adpr.202100052.
[92] D.L. Tong, R. Mintram, Genetic Algorithm-Neural Network (GANN): a study of neural network activation functions and depth of genetic algorithm search applied to feature selection, Int. J. Mach. Learn. Cybern. 1 (1) (2010) 75–87, https://doi.org/10.1007/s13042-010-0004-x.
[93] P.G. Benardos, G.C. Vosniakos, Optimizing feedforward artificial neural network architecture, Eng. Appl. Artif. Intell. 20 (3) (2007) 365–382.
[94] H. Yoo, S. Han, K. Chung, A Frequency Pattern Mining Model Based on Deep Neural Network for Real-Time Classification of Heart Conditions, third ed., vol. 8, Multidisciplinary Digital Publishing Institute, 2020, p. 234.
[95] S. Bermejo, J. Cabestany, Adaptive soft k-nearest-neighbour classifiers, Pattern Recogn. 33 (12) (2000) 1999–2005.
[96] W. Cherif, Optimization of K-NN algorithm by clustering and reliability coefficients: application to breast-cancer diagnosis, Procedia Comput. Sci. 127 (2018) 293–299.
[97] M.M. Islam, H. Iqbal, M.R. Haque, M.K. Hasan, Prediction of breast cancer using support vector machine and K-Nearest neighbors, in: 2017 IEEE Region 10 Humanitarian Technology Conference (R10-HTC), 2017, pp. 226–229.
[98] M.-Y. Su, Real-time anomaly detection systems for Denial-of-Service attacks by weighted k-nearest-neighbor classifiers, Expert Syst. Appl. 38 (4) (2011) 3492–3498.

[99] R.A. Maronna, Robust ridge regression for high-dimensional data, Technometrics 53 (1) (2011) 44–53.

[100] G.C. McDonald, Ridge regression, Wiley Interdiscip. Rev. Comput. Stat. 1 (1) (2009) 93–100.

[101] S. Boroumand, M. Arab Chamjangali, G. Bagherian, An asymmetric flow injection determination of hydroquinone and catechol: an analytic hierarchy and artificial neural network approach, Measurement 139 (2019) 454–466, https://doi.org/10.1016/j.measurement.2019.03.025.

[102] Y. Zhang, T.H. Tao, Skin-friendly electronics for acquiring human physiological signatures, Adv. Mater. 31 (49) (2019) 1905767.

[103] S. Abirami, P. Chitra, Chapter Fourteen - energy-efficient edge based real-time healthcare support system, in: P. Raj, P. Evangeline (Eds.), Advances in Computers, vol. 117, Elsevier, 2020, pp. 339–368.

[104] D. Svozil, V. Kvasnicka, J. Pospichal, Introduction to multi-layer feed-forward neural networks, Chemometr. Intell. Lab. Syst. 39 (1) (1997) 43–62.

[105] R.G. Beiko, R.L. Charlebois, GANN: genetic algorithm neural networks for the detection of conserved combinations of features in DNA, BMC Bioinformatics 6 (1) (2005) 1–12.

[106] M. Karzynski, Á. Mateos, J. Herrero, J. Dopazo, Using a genetic algorithm and a perceptron for feature selection and supervised class learning in DNA microarray data, Artif. Intell. Rev. 20 (1) (2003) 39–51.

[107] L. Kühner, R. Semenyshyn, M. Hentschel, F. Neubrech, C. Tarín, H. Giessen, Vibrational sensing using infrared nanoantennas: toward the noninvasive quantitation of physiological levels of glucose and fructose, ACS Sensors 4 (8) (2019) 1973–1979.

[108] C.S. Stravers, E.L. Gool, T.G. van Leeuwen, M.C.G. Aalders, A. van Dam, Multiplex body fluid identification using surface plasmon resonance imaging with principal component analysis, Sensor. Actuator. B Chem. 283 (2019) 355–362.

[109] M.C. Cieslak, A.M. Castelfranco, V. Roncalli, P.H. Lenz, D.K. Hartline, t-Distributed Stochastic Neighbor Embedding (t-SNE): a tool for eco-physiological transcriptomic analysis, Mar.Genomics 51 (2020/06/01) 100723, https://doi.org/10.1016/j.margen.2019.100723.

[110] G. Moon, J.-r. Choi, C. Lee, Y. Oh, K.H. Kim, D. Kim, Machine learning-based design of meta-plasmonic biosensors with negative index metamaterials, Biosens. Bioelectron. 164 (2020) 112335.

[111] J.C.M. Gomes, L.C. Souza, L.C. Oliveira, SmartSPR sensor: machine learning approaches to create intelligent surface plasmon based sensors, Biosens. Bioelectron. 172 (2021) 112760.

[112] A. Bruma, C. Wang, W.-C.D. Yang, D.F. Swearer, N.J. Halas, R. Sharma, A combined experimental and theoretical approach to measure spatially resolved local surface plasmon resonances in Aluminum nanocrystals, Microsc. Microanal. 24 (S1) (2018) 1682–1683.

[113] X. Li, J. Shu, W. Gu, L. Gao, Deep neural network for plasmonic sensor modeling, Opt. Mater. Express 9 (9) (2019) 3857–3862.

[114] J. He, C. He, C. Zheng, Q. Wang, J. Ye, Plasmonic nanoparticle simulations and inverse design using machine learning, Nanoscale 11 (37) (2019) 17444–17459.

[115] P.-H. Fu, S.-C. Lo, P.-C. Tsai, K.-L. Lee, P.-K. Wei, Optimization for gold nanostructure-based surface plasmon biosensors using a microgenetic algorithm, ACS Photonics 5 (6) (2018) 2320–2327.

[116] S.-C. Yen, Y.-L. Chen, Y.-H. Su, Materials genome evolution of surface plasmon resonance characteristics of Au nanoparticles decorated ZnO nanorods, APL Mater. 8 (9) (2020) 091109.
[117] A. Zelaci, A. Yasli, C. Kalyoncu, H. Ademgil, Generative adversarial neural networks model of photonic crystal fiber based surface plasmon resonance sensor, J. Lightwave Technol. 39 (5) (2020) 1515–1522.
[118] J.C.S. Batista, M.V.S. Costa, L.C. Oliveira, Smart noise reduction in SPR sensors response using multiple-ANN design, IEEE Sensor. J. 21 (4) (2020) 4517–4524.
[119] R. Yan, et al., Design of high-performance plasmonic nanosensors by particle swarm optimization algorithm combined with machine learning, Nanotechnology 31 (37) (2020) 375202.
[120] J.A. Arzola-Flores, A.L. González, Machine learning for predicting the surface plasmon resonance of perfect and concave gold nanocubes, J. Phys. Chem. C 124 (46) (2020) 25447–25454.
[121] Y.S. Ong, I. Grout, E. Lewis, W.S. Mohammed, Utilization of data classification in the realization of a surface Plasmon resonance readout system using an FPGA controlled RGB LED light source, IEEE Sensor. J. 18 (20) (2018) 8517–8524.
[122] M. Somarapalli, K. Koul, R. Lahon, S. Boonruang, W.S. Mohammed, Demonstration of low-cost and compact SPR optical transducer through edge light coupling, Micro Nano Lett. 12 (9) (2017) 643–646.
[123] K.L. Kelly, E. Coronado, L.L. Zhao, G.C. Schatz, The Optical Properties of Metal Nanoparticles: The Influence of Size, Shape, and Dielectric Environment, ACS Publications, 2003.
[124] K.L. Kelly, A.A. Lazarides, G.C. Schatz, Computational electromagnetics of metal nanoparticles and their aggregates, Comput. Sci. Eng. 3 (4) (2001) 67–73.
[125] D.R. Smith, J.B. Pendry, M.C.K. Wiltshire, Metamaterials and negative refractive index, Science 305 (5685) (2004) 788–792.
[126] S. Chugh, A. Gulistan, S. Ghosh, B.M.A. Rahman, Machine learning approach for computing optical properties of a photonic crystal fiber, Opt. Express 27 (25) (2019) 36414–36425.
[127] J. Tak, A. Kantemur, Y. Sharma, H. Xin, A 3-D-printed W-band slotted waveguide array antenna optimized using machine learning, IEEE Antenn. Wireless Propag. Lett. 17 (11) (2018) 2008–2012.
[128] Z. Huang, X. Liu, J. Zang, The inverse design of structural color using machine learning, Nanoscale 11 (45) (2019) 21748–21758.
[129] J. Baxter, A.C. Lesina, J.-M. Guay, A. Weck, P. Berini, L. Ramunno, Plasmonic colours predicted by deep learning, Sci. Rep. 9 (1) (2019) 1–9.
[130] Z. Liu, D. Zhu, S.P. Rodrigues, K.-T. Lee, W. Cai, Generative model for the inverse design of metasurfaces, Nano Lett. 18 (10) (2018) 6570–6576.
[131] C.C. Nadell, B. Huang, J.M. Malof, W.J. Padilla, Deep learning for accelerated all-dielectric metasurface design, Opt. Express 27 (20) (2019) 27523–27535.
[132] T. Asano, S. Noda, Optimization of photonic crystal nanocavities based on deep learning, Opt. Express 26 (25) (2018) 32704–32717.
[133] A. da Silva Ferreira, G.N. Malheiros-Silveira, H.E. Hernández-Figueroa, Computing optical properties of photonic crystals by using multilayer perceptron and extreme learning machine, J. Lightwave Technol. 36 (18) (2018) 4066–4073.
[134] K. Yao, R. Unni, Y. Zheng, Intelligent nanophotonics: merging photonics and artificial intelligence at the nanoscale, Nanophotonics 8 (3) (2019) 339–366.

[135] J. Robinson, Y. Rahmat-Samii, Particle swarm optimization in electromagnetics, IEEE Trans. Antenn. Propag. 52 (2) (2004) 397–407.

[136] K. Pang, W. Dong, B. Zhang, S. Zhan, X. Wang, Sensitivity-enhanced and noise-reduced surface plasmon resonance sensing with microwell chips, Meas. Sci. Technol. 26 (8) (2015) 085104.

[137] X. Wang, et al., Shot-noise limited detection for surface plasmon sensing, Opt. Express 19 (1) (2011) 107–117.

[138] G. Yao, et al., Clicking DNA to gold nanoparticles: poly-adenine-mediated formation of monovalent DNA-gold nanoparticle conjugates with nearly quantitative yield, NPG Asia Mater. 7 (1) (2015) e159.

[139] F. Saghatchi, M. Mohseni-Dargah, S. Akbari-Birgani, S. Saghatchi, B. Kaboudin, Cancer therapy and imaging through functionalized carbon nanotubes decorated with magnetite and gold nanoparticles as a multimodal tool (in eng), Appl. Biochem. Biotechnol. 191 (3) (2020) 1280–1293, https://doi.org/10.1007/s12010-020-03280-3.

[140] J. Li, et al., Controllable synthesis of stable urchin-like gold nanoparticles using hydroquinone to tune the reactivity of gold chloride, J. Phys. Chem. C 115 (9) (2011) 3630–3637.

[141] C.-C. Chen, et al., Presence of gold nanoparticles in cells associated with the cell-killing effect of modulated electro-hyperthermia, ACS Appl. Bio Mater. 2 (8) (2019) 3573–3581.

[142] A. Pangdam, K. Wongravee, S. Nootchanat, S. Ekgasit, Urchin-like gold microstructures with tunable length of nanothorns, Mater. Des. 130 (2017) 140–148.

[143] V. Chandwani, V. Agrawal, R. Nagar, Modeling slump of ready mix concrete using genetic algorithms assisted training of Artificial Neural Networks, Expert Syst. Appl. 42 (2) (2015) 885–893.

[144] J. Peurifoy, et al., Nanophotonic particle simulation and inverse design using artificial neural networks, Sci. Adv. 4 (6) (2018) eaar4206.

[145] D. Liu, Y. Tan, E. Khoram, Z. Yu, Training deep neural networks for the inverse design of nanophotonic structures, ACS Photonics 5 (4) (2018) 1365–1369.

[146] I. Malkiel, M. Mrejen, A. Nagler, U. Arieli, L. Wolf, H. Suchowski, Plasmonic nanostructure design and characterization via deep learning, Light Sci. Appl. 7 (1) (2018) 1–8.

[147] C. Barth, C. Becker, Machine learning classification for field distributions of photonic modes, Commun. Phys. 1 (1) (2018) 1–11.

[148] C.L. Wong, M. Olivo, Surface plasmon resonance imaging sensors: a review, Plasmonics 9 (4) (2014) 809–824.

[149] I.N. Pavlov, A.V. Vedyashkina, G.M. Yanina, The Choice of an Image Processing Algorithm for Increasing Sensitivity of the Surface Plasmon Resonance Method, first ed., vol. 1421, IOP Publishing, 2019, p. 012012.

[150] J.W. Goodman, Speckle Phenomena in Optics: Theory and Applications, Roberts and Company Publishers, 2007.

Index

Note: 'Page numbers followed by "f" indicate figures and "t" indicate tables.'

F

G

H

I

N

O

P

Q

R

W